Prealgebra Solutions Manual

Richard Rusczyk
David Patrick
Art of Problem Solving

Ravi Boppana
Advantage Testing

Art of Problem Solving®

Books • Online Classes • Videos • Interactive Resources

www.aops.com

Published by: AoPS Incorporated
                10865 Rancho Bernardo Rd Ste 100
                San Diego, CA 92127-2102
                (858) 675-4555
                books@artofproblemsolving.com

ISBN: 978-1-934124-22-2

Art of Problem Solving® is a registered trademark of AoPS Incorporated.

Visit the Art of Problem Solving website at http://www.artofproblemsolving.com

 Scan this code with your mobile device to visit the Art of Problem Solving website, to view our other books, our free videos and interactive resources, our online community, and our online school.

24® is a registered trademark of Suntex International Inc.

Cover image designed by Vanessa Rusczyk using KaleidoTile software.
Cover includes the original oil painting *Bush Poppy, El Cajon Mountain*, © 2011 Vanessa Rusczyk.

Printed in the United States of America.

Fourth Printing 2015.

Contents

CONTENTS

CHAPTER 1

Properties of Arithmetic

24® Cards

First Card $(7 + 1 - 4) \times 6$.

Second Card $(5 + 1) \times (8 - 4)$ or $(5 - 1) \times 4 + 8$.

Third Card $(1 + 1) \times 23 - 22$.

Fourth Card $5 \times 7 - 10 - 1$.

Exercises for Section 1.2

1.2.1 We can group each 99 with a 101:

$$99 + 99 + 99 + 101 + 101 + 101 = (99 + 101) + (99 + 101) + (99 + 101) = 200 + 200 + 200 = \boxed{600}.$$

1.2.2 We can group the numbers into pairs without moving them:

$$1999 + 2001 + 1999 + 2001 + 1999 + 2001 + 1999 + 2001$$
$$= (1999 + 2001) + (1999 + 2001) + (1999 + 2001) + (1999 + 2001)$$
$$= 4000 + 4000 + 4000 + 4000 = \boxed{16{,}000}.$$

1.2.3 We can pair each number in the first group with a number in the second group:

$$(3 + 13 + 23 + 33 + 43) + (7 + 17 + 27 + 37 + 47)$$
$$= (3 + 47) + (13 + 37) + (23 + 27) + (33 + 17) + (43 + 7) = 50 + 50 + 50 + 50 + 50 = \boxed{250}.$$

1.2.4 Again let's match each number in the first group with a number in the second:

$$(1 + 2 + 3 + \cdots + 49 + 50) + (99 + 98 + 97 + \cdots + 51 + 50)$$
$$= (1 + 99) + (2 + 98) + (3 + 97) + \cdots + (49 + 51) + (50 + 50)$$
$$= \underbrace{100 + 100 + 100 + \cdots + 100 + 100}_{50 \text{ numbers}} = 50 \cdot 100 = \boxed{5000}.$$

Exercises for Section 1.3

1.3.1 We reorder the numbers to group pairs that are easy to multiply:
$$25 \cdot 17 \cdot 4 \cdot 20 = (25 \cdot 4)(17 \cdot 20) = 100 \cdot 340 = \boxed{34{,}000}.$$

1.3.2 We can pair the numbers without reordering them:
$$1 \cdot 100 \cdot 2 \cdot 50 \cdot 4 \cdot 25 \cdot 5 \cdot 20 = (1 \cdot 100)(2 \cdot 50)(4 \cdot 25)(5 \cdot 20) = 100 \cdot 100 \cdot 100 \cdot 100 = \boxed{100{,}000{,}000}.$$

1.3.3 $2 \cdot 2 \cdot 2 \cdot 2 \cdot 2 \cdot 5 \cdot 5 \cdot 5 \cdot 5 \cdot 5 = (2 \cdot 5)(2 \cdot 5)(2 \cdot 5)(2 \cdot 5)(2 \cdot 5) = 10 \cdot 10 \cdot 10 \cdot 10 \cdot 10 = \boxed{100{,}000}.$

1.3.4 $1 \cdot 1995 \cdot 1 = 1995 \cdot 1 = \boxed{1995}.$

1.3.5 $1 \cdot 5 \cdot 1 \cdot 5 \cdot 1 \cdot 5 = 5 \cdot 5 \cdot 5 = \boxed{125}.$

1.3.6

(a) We can factor out the 11 to get $11 \cdot 43 + 11 \cdot 57 = 11(43 + 57) = 11 \cdot 100 = \boxed{1100}.$

(b) We use the commutative property to write $22 \cdot 6$ as $6 \cdot 22$, and then factor out 6 to get
$$22 \cdot 6 + 6 \cdot 38 = 6 \cdot 22 + 6 \cdot 38 = 6(22 + 38) = 6 \cdot 60 = \boxed{360}.$$

(c) Again, we factor: $32 \cdot 16 + 16 \cdot 48 = 16 \cdot 32 + 16 \cdot 48 = 16(32 + 48) = 16 \cdot 80 = \boxed{1280}.$

1.3.7 There are many choices of a, b, and c for which the two expressions are not equal. For instance, choose $a = 2$, $b = 0$, and $c = 0$. Then the first expression is
$$a + (b \cdot c) = 2 + (0 \cdot 0) = 2 + 0 = 2.$$
The second expression is
$$(a + b) \cdot (a + c) = (2 + 0) \cdot (2 + 0) = 2 \cdot 2 = 4.$$
The two numbers are indeed different.

1.3.8 $456 + 456 + 456 + 456 + 456 + 456 + 456 + 456 + 456 + 456 = 10 \cdot 456 = \boxed{4560}.$

1.3.9 This is multiplication by zero: $1 \cdot 2 \cdot 3 \cdot 4 \cdot 5 \cdot 6 \cdot 7 \cdot 8 \cdot 9 \cdot 0 = (1 \cdot 2 \cdot 3 \cdot 4 \cdot 5 \cdot 6 \cdot 7 \cdot 8 \cdot 9) \cdot 0 = \boxed{0}.$

1.3.10 Multiplying by zero gives 0, so $10 + 110 \cdot 0 \cdot 101 + 111 = 10 + 0 + 111 = \boxed{121}.$

Exercises for Section 1.4

1.4.1 By rearranging, $-631 + (114 + 631) = (-631 + 631) + 114 = 0 + 114 = \boxed{114}.$

1.4.2 The negative integers greater than -5 are -4, -3, -2, and -1. Their sum is
$$-4 + (-3) + (-2) + (-1) = -(4 + 3 + 2 + 1) = \boxed{-10}.$$

1.4.3   Let's group numbers with their negations:

$$-10 + (-9) + (-8) + \cdots + 9 + 10 + 11 + 12 = (-10 + 10) + (-9 + 9) + \cdots + (-1 + 1) + 0 + 11 + 12$$
$$= \underbrace{0 + 0 + \cdots + 0}_{10 \text{ zeros}} + 0 + 11 + 12$$
$$= 0 + 0 + 11 + 12 = 11 + 12 = \boxed{23}.$$

1.4.4 $210 \cdot 5 + 105 \cdot (-9) = 105 \cdot 2 \cdot 5 + 105 \cdot (-9) = 105 \cdot 10 + 105 \cdot (-9) = 105(10 + (-9)) = 105 \cdot 1 = \boxed{105}$.

1.4.5   The two terms with 438 and 719 add up to zero:

$$9342 + (-438)719 + (-9340) + (-438)(-719) = 9342 + (-(438 \cdot 719)) + (-9340) + 438 \cdot 719$$
$$= (9342 + (-9340)) + (-(438 \cdot 719) + 438 \cdot 719)$$
$$= (9342 - 9340) + 0 = 9342 - 9340 = \boxed{2}.$$

Exercises for Section 1.5

1.5.1   Subtracting a number from itself gives zero, so $85(33 \cdot 22) - 33(22 \cdot 85) = 22 \cdot 33 \cdot 85 - 22 \cdot 33 \cdot 85 = \boxed{0}$.

1.5.2   Multiplying by zero gives zero, so $(1992 + 1992)(1992 - 1992) = (1992 + 1992)(0) = \boxed{0}$.

1.5.3   To find out by how many degrees the high temperature exceeded the low temperature, we compute the high minus the low:

$$18^\circ - (-5^\circ) = 18^\circ + 5^\circ = \boxed{23^\circ}.$$

1.5.4 $3 + (-9) - (-5) = 3 + (-9) + 5 = 3 + 5 + (-9) = 8 + (-9) = 8 - 9 = -(9 - 8) = \boxed{-1}$.

1.5.5   If $y - x = 7$, then $x - y = -(y - x) = \boxed{-7}$.

1.5.6   We will place the added parts first followed by the subtracted parts:

$$100 - 2 + 101 - 4 + 102 - 6 + 103 - 8 + 104 - 10$$
$$= (100 + 101 + 102 + 103 + 104) - 2 - 4 - 6 - 8 - 10$$
$$= 510 - (2 + 4 + 6 + 8 + 10) = 510 - 30 = \boxed{480}.$$

1.5.7   Let's group each number in the first sum with a number in the second sum that has the same final two digits:

$$(1901 + 1902 + 1903 + \cdots + 1993) - (101 + 102 + 103 + \cdots + 193)$$
$$= (1901 - 101) + (1902 - 102) + (1903 - 103) + \cdots + (1993 - 193)$$
$$= \underbrace{1800 + 1800 + 1800 + \cdots + 1800}_{93 \text{ numbers}} = 93 \cdot 1800 = \boxed{167,400}.$$

1.5.8   We compute the first sum minus the second sum by conveniently pairing each number in the first sum with a number in the second sum:

$$(19 + 28 + 37 + 46 + 55 + 64 + 73 + 82 + 91) - (18 + 27 + 36 + 45 + 54 + 63 + 72 + 81 + 90)$$
$$= (19 - 18) + (28 - 27) + (37 - 36) + (46 - 45) + (55 - 54) + (64 - 63) + (73 - 72) + (82 - 81) + (91 - 90)$$
$$= 1 + 1 + 1 + 1 + 1 + 1 + 1 + 1 + 1 = \boxed{9}.$$

1.5.9 $1990 \cdot 1991 - 1989 \cdot 1990 = 1990 \cdot 1991 - 1990 \cdot 1989 = 1990(1991 - 1989) = 1990 \cdot 2 = \boxed{3980}.$

1.5.10   Let's rewrite 998 as $1000 - 2$:

$$998 \cdot 23 = (1000 - 2)23 = 1000 \cdot 23 - 2 \cdot 23 = 23{,}000 - 46 = \boxed{22{,}954}.$$

1.5.11   The sum of the first 10,000 positive even numbers is

$$2 + 4 + 6 + \cdots + 19{,}998 + 20{,}000.$$

The sum of the first 10,000 positive odd numbers is

$$1 + 3 + 5 + \cdots + 19{,}997 + 19{,}999.$$

The first sum minus the second sum is

$$(2 - 1) + (4 - 3) + (6 - 5) + \cdots + (20{,}000 - 19{,}999) = \underbrace{1 + 1 + 1 + \cdots + 1}_{10{,}000 \text{ numbers}} = \boxed{10{,}000}.$$

Exercises for Section 1.6

1.6.1   By the "reciprocal of negation" property, and because the reciprocal of 1 is 1, we have

$$\frac{1}{-1} = -\frac{1}{1} = \boxed{-1}.$$

1.6.2   The only number that is not a reciprocal is $\boxed{0}$. Zero times any number is 0, not 1, so zero can't be the reciprocal of a number.

In contrast, any number besides zero is a reciprocal. Namely, if x is a nonzero number, then x is the reciprocal of $\frac{1}{x}$.

1.6.3   Let x be the nonzero number. Twice its reciprocal is $2 \cdot \frac{1}{x}$. So the product of x and twice its reciprocal is

$$x\left(2 \cdot \frac{1}{x}\right) = 2\left(\frac{1}{x} \cdot x\right) = 2 \cdot 1 = \boxed{2}.$$

1.6.4   Let x be the positive number. When we multiply the reciprocal of x by the negation of x, we get

$$\frac{1}{x}(-x) = -\left(\frac{1}{x} \cdot x\right) = \boxed{-1}.$$

1.6.5 Let's use the distributive law:

$$(2 \cdot 3 \cdot 4)\left(\frac{1}{2} + \frac{1}{3} + \frac{1}{4}\right) = (2 \cdot 3 \cdot 4)\frac{1}{2} + (2 \cdot 3 \cdot 4)\frac{1}{3} + (2 \cdot 3 \cdot 4)\frac{1}{4}$$

$$= \left(\frac{1}{2} \cdot 2\right)(3 \cdot 4) + \left(\frac{1}{3} \cdot 3\right)(2 \cdot 4) + \left(\frac{1}{4} \cdot 4\right)(2 \cdot 3)$$

$$= 1(3 \cdot 4) + 1(2 \cdot 4) + 1(2 \cdot 3) = 3 \cdot 4 + 2 \cdot 4 + 2 \cdot 3 = 12 + 8 + 6 = \boxed{26}.$$

Exercises for Section 1.7

1.7.1 Dividing a nonzero number by itself gives 1, so

$$(2 + 4 + 6 + 8 + 10) \div (10 + 8 + 6 + 4 + 2) = (2 + 4 + 6 + 8 + 10) \div (2 + 4 + 6 + 8 + 10) = \boxed{1}.$$

1.7.2 We can cancel a common factor of 205:

$$205 \cdot 205 \div 205 = (205 \cdot 205) \div (205 \cdot 1) = 205 \div 1 = \boxed{205}.$$

1.7.3 We can cancel a factor of 100 from both numbers:

$$64{,}000 \div 800 = (640 \cdot 100) \div (8 \cdot 100) = 640 \div 8 = \boxed{80}.$$

1.7.4 The first number is 10 times the second number, so the quotient is

$$777{,}777{,}777{,}770 \div 77{,}777{,}777{,}777 = (10 \cdot 77{,}777{,}777{,}777) \div 77{,}777{,}777{,}777 = \boxed{10}.$$

1.7.5 The reciprocal of $\frac{1}{7}$ is 7, so $28 \div \frac{1}{7} = 28 \cdot 7 = \boxed{196}$.

1.7.6 The reciprocal of $\frac{1}{2}$ is 2, so $78 \div \frac{1}{2} = 78 \cdot 2 = 156$. Ten more than that is $\boxed{166}$.

1.7.7 Division is neither commutative nor associative, so we must compute the divisions from left to right:

$$\frac{1}{2} \div \frac{1}{2} \div \frac{1}{2} \div \frac{1}{2} = 1 \div \frac{1}{2} \div \frac{1}{2} = 1 \cdot 2 \div \frac{1}{2} = 2 \div \frac{1}{2} = 2 \cdot 2 = \boxed{4}.$$

1.7.8 We can cancel the common factors 31, 35, 39, and 43:

$$(27 \cdot 31 \cdot 35 \cdot 39 \cdot 43) \div (43 \cdot 39 \cdot 35 \cdot 31) = (27 \cdot 31 \cdot 35 \cdot 39 \cdot 43) \div (1 \cdot 31 \cdot 35 \cdot 39 \cdot 43) = 27 \div 1 = \boxed{27}.$$

1.7.9 $(50 \cdot 60 \cdot 70 \cdot 80) \div (5 \cdot 6 \cdot 7 \cdot 8) = (50 \div 5)(60 \div 6)(70 \div 7)(80 \div 8) = 10 \cdot 10 \cdot 10 \cdot 10 = \boxed{10{,}000}.$

1.7.10

$$(77{,}777{,}777{,}777 + 77{,}077) \div 7 = 77{,}777{,}777{,}777 \div 7 + 77{,}077 \div 7$$

$$= 11{,}111{,}111{,}111 + 11{,}011$$

$$= \boxed{11{,}111{,}122{,}122}.$$

1.7.11 We will show that the first sum is twice the second sum:

$$(124 + 104 + 84 + 64 + 44 + 24) \div (62 + 52 + 42 + 32 + 22 + 12)$$
$$= (2 \cdot 62 + 2 \cdot 52 + 2 \cdot 42 + 2 \cdot 32 + 2 \cdot 22 + 2 \cdot 12) \div (62 + 52 + 42 + 32 + 22 + 12)$$
$$= 2(62 + 52 + 42 + 32 + 22 + 12) \div (62 + 52 + 42 + 32 + 22 + 12) = \boxed{2}.$$

Review Problems

1.42 We pair the smallest number with the largest, the second smallest with the second largest, and so on:

$$90 + 91 + 92 + 93 + 94 + 95 + 96 + 97 + 98 + 99 = (90 + 99) + (91 + 98) + (92 + 97) + (93 + 96) + (94 + 95)$$
$$= 189 + 189 + 189 + 189 + 189 = 5 \cdot 189 = \boxed{945}.$$

We also might view each number after the first in the sum as 90 plus a 1-digit number, so

$$90 + 91 + 92 + 93 + 94 + 95 + 96 + 97 + 98 + 99 = 90 + (90 + 1) + (90 + 2) + (90 + 3) + \cdots + (90 + 9)$$
$$= (90 + 90 + 90 + 90 + 90 + 90 + 90 + 90 + 90 + 90)$$
$$+ (1 + 2 + 3 + 4 + 5 + 6 + 7 + 8 + 9)$$
$$= 10 \cdot 90 + 45 = 900 + 45 = \boxed{945}.$$

1.43 This can be done in your head if you mentally rearrange the numbers using the commutative and associative properties of multiplication:

$$25 \cdot (12 \cdot 8) = 12(8 \cdot 25) = 12 \cdot 200 = \boxed{2400}.$$

1.44 We pair the smallest number with the largest, the second smallest with the second largest, and so on:

$$3(101 + 103 + 105 + 107 + 109 + 111 + 113 + 115 + 117 + 119)$$
$$= 3((101 + 119) + (103 + 117) + (105 + 115) + (107 + 113) + (109 + 111))$$
$$= 3(220 + 220 + 220 + 220 + 220) = 3 \cdot 5 \cdot 220 = 3 \cdot 1100 = \boxed{3300}.$$

1.45 $((1 \cdot 2) + (3 \cdot 4) - (5 \cdot 6) + (7 \cdot 8)) \cdot (9 \cdot 0) = ((1 \cdot 2) + (3 \cdot 4) - (5 \cdot 6) + (7 \cdot 8)) \cdot 0 = \boxed{0}.$

1.46 The first expression is

$$42 + 7 - 6 \cdot 6 + 3 \cdot (-1) \cdot 0 = 42 + 7 - 36 + 0 = 42 + 7 - 36 = 49 - 36 = 13.$$

Anything times zero is zero, so the second expression is

$$(42 + 7 - 6 \cdot 6 + 3 \cdot (-1)) \cdot 0 = 0.$$

The difference between the two expressions is $13 - 0 = \boxed{13}$.

1.47 $(185 + 378 + 579) - (85 + 178 + 279) = (185 - 85) + (378 - 178) + (579 - 279) = 100 + 200 + 300 = \boxed{600}$.

1.48

$$11 + (-15) + 11 - (-15) + 11 - 15 - (11 + 15) = 11 - 15 + 11 + 15 + 11 - 15 - 11 - 15$$
$$= 11 + 11 + 11 - 11 - 15 + 15 - 15 - 15$$
$$= 11 + 11 - 15 - 15$$
$$= 22 - 15 - 15 = 7 - 15 = \boxed{-8}.$$

1.49 Subtraction is addition of a negation, so

$$6\big((25 - 98) - (19 - 98)\big) = 6\big((25 - 98) + (-(19 - 98))\big)$$
$$= 6\big((25 - 98) + (98 - 19)\big) = 6\big((25 - 19) + (98 - 98)\big) = 6(6 + 0) = 6 \cdot 6 = \boxed{36}.$$

1.50 Let's pair up the first and second numbers, the third and fourth numbers, and so on:

$$1 - 3 + 5 - 7 + 9 - 11 + 13 - 15 + 17 - 19 + 21 - 23 + 25$$
$$= (1 - 3) + (5 - 7) + (9 - 11) + (13 - 15) + (17 - 19) + (21 - 23) + 25$$
$$= (-2) + (-2) + (-2) + (-2) + (-2) + (-2) + 25$$
$$= 6(-2) + 25 = -12 + 25 = \boxed{13}.$$

1.51 We can factor: $693 \cdot 1587 - 692 \cdot 1587 = (693 - 692)1587 = 1 \cdot 1587 = \boxed{1587}$.

1.52 Let's use our rules for multiplying and subtracting negations:

$$(-20)\big((-3)(-15) - (-6)(3)\big) = (-20)\big(3 \cdot 15 - (-(6 \cdot 3))\big)$$
$$= (-20)(3 \cdot 15 + 6 \cdot 3) = (-20)(45 + 18) = (-20) \cdot 63 = -(20 \cdot 63) = \boxed{-1260}.$$

1.53 Let's rewrite 298 as $299 - 1$, and then factor out 299:

$$4(299) + 3(299) + 2(299) + 298 = 4(299) + 3(299) + 2(299) + 299 - 1$$
$$= 4(299) + 3(299) + 2(299) + 1(299) - 1$$
$$= (4 + 3 + 2 + 1)299 - 1 = 10 \cdot 299 - 1 = 2990 - 1 = \boxed{2989}.$$

1.54 This is probably easiest if we convert the operations involving reciprocals to operations involving integers using the definition of division:

$$40 \cdot \frac{1}{8} + 40 \div \frac{1}{8} + 40 \cdot \frac{1}{5} + 40 \div \frac{1}{5} = 40 \div 8 + 40 \cdot 8 + 40 \div 5 + 40 \cdot 5$$
$$= 5 + 320 + 8 + 200 = \boxed{533}.$$

1.55 $(6 \div (-3))(4 - 12) = (-(6 \div 3))(4 - 12) = (-2)(-8) = 2 \cdot 8 = \boxed{16}$.

1.56 $(-13)+(-13)\div(-13)\cdot(-13)-(-13) = (-13)+1\cdot(-13)-(-13) = (-13)+(-13)-(-13) = (-13)+0 = \boxed{-13}$.

1.57 Rewriting 123,123 as $123,000 + 123$ allows us to factor out 123:

$$123,123 = 123,000 + 123 = 123 \cdot 1000 + 123 \cdot 1 = 123(1000 + 1) = 123 \cdot 1001.$$

So, we have $123,123 \div 1001 = (123 \cdot 1001) \div (1 \cdot 1001) = 123 \div 1 = \boxed{123}$.

1.58 The reciprocal of a product is the product of reciprocals, so the reciprocal of $2 \cdot 3 \cdot \dfrac{1}{4} \cdot \dfrac{1}{9}$ is $\dfrac{1}{2} \cdot \dfrac{1}{3} \cdot 4 \cdot 9$. This simplifies as

$$\frac{1}{2} \cdot \frac{1}{3} \cdot 4 \cdot 9 = \left(4 \cdot \frac{1}{2}\right) \cdot \left(9 \cdot \frac{1}{3}\right) = (4 \div 2) \cdot (9 \div 3) = 2 \cdot 3 = \boxed{6}.$$

1.59 The reciprocal of $\frac{1}{6}$ is 6, so $\frac{1}{2} \div \frac{1}{6} = \frac{1}{2} \cdot 6 = 6 \cdot \frac{1}{2} = 6 \div 2 = \boxed{3}$.

1.60 $(3 \cdot 4) \div \left(\frac{1}{5} \cdot \frac{1}{6}\right) = 12 \div \frac{1}{30} = 12 \cdot 30 = \boxed{360}$.

1.61 Sean's sum is

$$2 + 4 + 6 + \cdots + 500 = 2(1) + 2(2) + 2(3) + \cdots + 2(250) = 2(1 + 2 + 3 + \cdots + 250).$$

Julie's sum is

$$1 + 2 + 3 + \cdots + 250.$$

So Sean's sum is twice Julie's sum. Therefore, Sean's sum divided by Julie's sum is $\boxed{2}$.

1.62 Division is not associative, so Gary is incorrect. We must compute the divisions from left to right. The correct answer is $200 \div 10 \div 2 = 20 \div 2 = \boxed{10}$.

Challenge Problems

1.63 We know that we can sum an even number of consecutive integers by grouping the integers into pairs. So, we'll first add the first 60 positive integers, and then add 61 to the resulting sum. Grouping the first 60 integers into 30 pairs gives

$$1 + 2 + \cdots + 59 + 60 = (1 + 60) + (2 + 59) + \cdots + (30 + 31)$$
$$= \underbrace{61 + 61 + \cdots + 61}_{30 \text{ numbers}} = 30 \cdot 61 = 1830.$$

So, the sum of the first 61 positive integers is $1830 + 61 = \boxed{1891}$.

1.64 Let's group the 20 numbers into 10 pairs:

$$5 + 10 + 15 + \cdots + 95 + 100 = (5 + 100) + (10 + 95) + (15 + 90) + \cdots + (50 + 55)$$
$$= \underbrace{105 + 105 + 105 + \cdots + 105}_{10 \text{ numbers}}$$
$$= 10 \cdot 105 = \boxed{1050}.$$

1.65   As much as we can, let's pair off numbers with their negations:

$$-100 + (-99) + (-98) + \cdots + 97 + 98$$
$$= -100 + (-99) + (-98 + 98) + (-97 + 97) + \cdots + (-1 + 1) + 0$$
$$= -100 + (-99) + 0 + 0 + \cdots + 0 + 0$$
$$= -100 - 99 = \boxed{-199}.$$

1.66   After factoring out the negation, we will group the 20 numbers into 10 pairs:

$$-39 + (-37) + (-35) + \cdots + (-1) = -(39 + 37 + 35 + \cdots + 1)$$
$$= -((39 + 1) + (37 + 3) + (35 + 5) + \cdots + (21 + 19))$$
$$= -(\underbrace{40 + 40 + 40 + \cdots + 40}_{\text{10 numbers}}) = -(10 \cdot 40) = \boxed{-400}.$$

1.67   We will group the 50 numbers into 25 pairs:

$$100 - 98 + 96 - 94 + 92 - 90 + \cdots + 8 - 6 + 4 - 2$$
$$= (100 - 98) + (96 - 94) + (92 - 90) + \cdots + (8 - 6) + (4 - 2)$$
$$= \underbrace{2 + 2 + 2 + \cdots + 2 + 2}_{\text{25 numbers}} = 25 \cdot 2 = \boxed{50}.$$

1.68   The factors of this product are equal to 30, 27, 24, and so on until −30. So in between, one of the factors equals zero, namely the factor $(0 + 0)$. Because one of the factors equals zero, and zero times anything is zero, the entire product is $\boxed{0}$.

1.69   Let's express 2323 in terms of 23:

$$2323 = 2300 + 23 = 23 \cdot 100 + 23 \cdot 1 = 23(100 + 1) = 23 \cdot 101.$$

So 2323 is 101 times 23. The same method works for any two-digit starting number. For example, 7474 is 101 times 74. So the answer is $\boxed{101}$.

1.70   Because 6 times 37,037 is 222,222, we know that 3 times 37,037 is 111,111. So we have

$$27 \cdot 37{,}037 = 9 \cdot 3 \cdot 37{,}037 = 9 \cdot 111{,}111 = \boxed{999{,}999}.$$

1.71   This problem shows that an operation can be commutative without being associative.

(a)   Let a and b be any numbers. Using the definition of the operation @, we have

$$a \mathbin{@} b = 2a + 2b.$$

Also

$$b \mathbin{@} a = 2b + 2a = 2a + 2b.$$

The two expressions are equal. So @ is commutative.

(b) There are many examples that show that @ is not associative. For instance, choose $a = 1$, $b = 0$, and $c = 0$. On one hand,

$$(a @ b) @ c = (1 @ 0) @ 0 = (2 \cdot 1 + 2 \cdot 0) @ 0 = 2 @ 0 = 2 \cdot 2 + 2 \cdot 0 = 4.$$

On the other hand,

$$a @ (b @ c) = 1 @ (0 @ 0) = 1 @ (2 \cdot 0 + 2 \cdot 0) = 1 @ 0 = 2 \cdot 1 + 2 \cdot 0 = 2.$$

The two expressions are not equal. So @ is not associative.

1.72 This problem shows that an operation can be associative without being commutative.

(a) Let a, b, and c be any numbers. Using the definition of the operation #, we have

$$(a \# b) \# c = a \# c = a.$$

Also,

$$a \# (b \# c) = a \# b = a.$$

The two expressions are equal. So # is associative.

(b) There are many examples that show that # is not commutative. For instance, choose $a = 0$ and $b = 1$. On one hand,

$$a \# b = 0 \# 1 = 0.$$

On the other hand,

$$b \# a = 1 \# 0 = 1.$$

The two expressions are not equal. So # is not commutative.

1.73 The key is to use the distributive property in two steps:

$$(a + b)(x + y) = a(x + y) + b(x + y) = (ax + ay) + (bx + by) = ax + ay + bx + by.$$

1.74 Because negation distributes over addition and subtraction, we have

$$\begin{aligned}
(a - (b - c)) - ((a - b) - c) &= (a - b + c) - (a - b - c) \\
&= (a - b + c) - a - (-b) - (-c) \\
&= (a - b + c) - a + b + c \\
&= (a - a) + (-b + b) + (c + c) = 0 + 0 + 2c = \boxed{2c}.
\end{aligned}$$

1.75 A string of nines is 1 less than a power of ten, so the product is

$$\underbrace{9999\ldots99}_{94 \text{ nines}} \times \underbrace{4444\ldots44}_{94 \text{ fours}} = (\underbrace{10000\ldots00}_{94 \text{ zeros}} - 1)\underbrace{4444\ldots44}_{94 \text{ fours}}$$

$$= \underbrace{10000\ldots00}_{94 \text{ zeros}} \times \underbrace{4444\ldots44}_{94 \text{ fours}} - 1 \times \underbrace{4444\ldots44}_{94 \text{ fours}}$$

$$= \underbrace{4444\ldots44}_{94 \text{ fours}}\underbrace{0000\ldots00}_{94 \text{ zeros}} - \underbrace{4444\ldots44}_{94 \text{ fours}}$$

$$= \underbrace{4444\ldots44}_{93 \text{ fours}}3\underbrace{5555\ldots55}_{93 \text{ fives}}6.$$

The product has 93 fours, 1 three, 93 fives, and 1 six. So the sum of the digits is

$$93 \times 4 + 1 \times 3 + 93 \times 5 + 1 \times 6 = 93(4 + 5) + (3 + 6) = 93 \times 9 + 9 = (93 + 1)9 = 94 \times 9 = \boxed{846}.$$

1.76

(a) We must show that $a \odot b = b \odot a$ for all numbers a and b. We compute:

$$a \odot b = a + ab + b$$

and

$$b \odot a = b + ba + a.$$

But

$$a + ab + b = b + ab + a = b + ba + a$$

by the commutativity of addition and multiplication. So $a \odot b = b \odot a$, and therefore $\odot$ is commutative.

(b) We must show that $(a \odot b) \odot c = a \odot (b \odot c)$. We compute:

$$\begin{aligned}
(a \odot b) \odot c &= (a + ab + b) \odot c \\
&= (a + ab + b) + (a + ab + b)c + c \\
&= a + ab + b + ac + abc + bc + c \\
&= a + b + c + ab + ac + bc + abc,
\end{aligned}$$

and

$$\begin{aligned}
a \odot (b \odot c) &= a \odot (b + bc + c) \\
&= a + a(b + bc + c) + (b + bc + c) \\
&= a + ab + abc + ac + b + bc + c \\
&= a + b + c + ab + ac + bc + abc.
\end{aligned}$$

These are equal, so $\odot$ is associative.

(c) We need to find a number I such that $x \odot I = x$ for all numbers x. So we compute

$$x \odot I = x + xI + I.$$

For the right-hand side above to always equal x, we need $xI + I$ to equal 0 for all x. Factoring I out of $xI + I$ gives $(x + 1)I$, which will be zero for all x if $I = \boxed{0}$.

(d) We want to find the number x so that $x \odot 1 = 0$. But

$$x \odot 1 = x + (x)(1) + 1 = 2x + 1.$$

So we would like $2x + 1 = 0$. This means that $2x = -1$, so our answer is

$$x = (-1) \div 2 = -(1 \div 2) = -\left(1 \cdot \frac{1}{2}\right) = \boxed{-\frac{1}{2}}.$$

CHAPTER 2

Exponents

24® Cards

First Card $(10 \div 2 + 7) \cdot 2$.

Second Card $(9 - 2) \cdot 2 + 10$.

Third Card $(21 + 2) \cdot 2 - 22$.

Fourth Card $2 \cdot 9 + 13 - 7$ or $2 \cdot 13 + 7 - 9$ or $(9 - 7) \cdot 13 - 2$.

Exercises for Section 2.1

2.1.1

(a) Parentheses first: $8 + 6(3 - 8)^2 = 8 + 6(-5)^2 = 8 + 6 \cdot 5^2 = 8 + 6 \cdot 25 = 8 + 150 = \boxed{158}$.

(b) Parentheses first: $5(3 + 4 \cdot 2) - 6^2 = 5(3 + 8) - 6^2 = 5 \cdot 11 - 6^2 = 5 \cdot 11 - 36 = 55 - 36 = \boxed{19}$.

(c) $92 - 45 \div (3 \cdot 5) - 5^2 = 92 - 45 \div 15 - 5^2 = 92 - 45 \div 15 - 25 = 92 - 3 - 25 = 89 - 25 = \boxed{64}$.

(d) $8\left(6^2 - 3(11)\right) \div 8 + 3 = 8(36 - 3(11)) \div 8 + 3 = 8(36 - 33) \div 8 + 3 = 8 \cdot 3 \div 8 + 3 = 24 \div 8 + 3 = 3 + 3 = \boxed{6}$.

2.1.2

(a) Parentheses first: $(7 + 5)^2 - 7^2 - 5^2 = 12^2 - 7^2 - 5^2 = 144 - 49 - 25 = \boxed{70}$.

(b) We apply the rule for the product of squares: $25^2 \cdot 16^2 = (25 \cdot 16)^2 = 400^2 = \boxed{160,000}$.

(c) We apply the rule for the quotient of squares: $480^2 \div 40^2 = (480 \div 40)^2 = 12^2 = \boxed{144}$.

(d) To square 101, we can use the fact from the text that $(a + 1)^2 = a^2 + 2a + 1$. Letting $a = 100$ gives

$$101^2 = 100^2 + 2(100) + 1 = 10,000 + 200 + 1 = \boxed{10,201}.$$

2.1.3 Replacing each x with -3 gives

$$x^2 + 2x - 6 = (-3)^2 + 2(-3) - 6 = 9 + 2(-3) - 6 = 9 + (-6) - 6 = 3 - 6 = \boxed{-3}.$$

2.1.4 Replacing x with -4 and t with 2, we get

$$x^2(x - t) = (-4)^2(-4 - 2) = (-4)^2(-6) = 16(-6) = \boxed{-96}.$$

2.1.5 The square of 2 is 4, the square of 4 is 16, and the square of 16 is 256. The square of 256 is, well, greater than 500. So the number of squarings needed is $\boxed{4}$.

2.1.6 One perfect square that is close to 5000 is $70^2 = 4900$. The next highest perfect square is 71^2. We could multiply this out, or we could use $(a + 1)^2 = a^2 + 2a + 1$, which we learned in the text. We therefore have $71^2 = 70^2 + 2 \cdot 70 + 1 = 4900 + 140 + 1 = 5041$. So, 5000 is between 70^2 and 71^2, and is closer to $71^2 = \boxed{5041}$.

2.1.7 The square of 40 is 1600. The square of 41 is $40^2 + 2 \cdot 40 + 1 = 1681$. The square of 42 is $41^2 + 2 \cdot 41 + 1 = 1681 + 82 + 1 = 1764$. The square of 43 is $42^2 + 2 \cdot 42 + 1 = 1764 + 84 + 1 = \boxed{1849}$. The square of 44 is $43^2 + 2 \cdot 43 + 1 = 1849 + 86 + 1 = 1936$, so 44^2 is greater than 1900, as are the squares of all numbers greater than 44. Therefore, 43^2 is the only perfect square between 1800 and 1900.

2.1.8 The square of 22 is 484 and the square of 23 is 529. So the positive squares less than 500 are $1^2, 2^2, 3^2, \ldots, 22^2$. (We don't count 0^2 because the problem said *positive*.) There are $\boxed{22}$ such squares.

2.1.9 The square of 31 is 961 and the square of 32 is 1024. The square of 44 is 1936 and the square of 45 is 2025. So we are interested in the squares $32^2, 33^2, \ldots, 44^2$. The number of such squares is $\boxed{13}$.

2.1.10 Each even number is 2 times an integer. So, we have

$$2^2 + 4^2 + 6^2 + \cdots + 50^2 = (2 \cdot 1)^2 + (2 \cdot 2)^2 + (2 \cdot 3)^2 + \cdots + (2 \cdot 25)^2$$
$$= 2^2 \cdot 1^2 + 2^2 \cdot 2^2 + 2^2 \cdot 3^2 + \cdots + 2^2 \cdot 25^2.$$

Since $2^2 = 4$, we have

$$2^2 \cdot 1^2 + 2^2 \cdot 2^2 + 2^2 \cdot 3^2 + \cdots + 2^2 \cdot 25^2 = 4 \cdot 1^2 + 4 \cdot 2^2 + 4 \cdot 3^2 + \cdots + 4 \cdot 25^2.$$

The squares on the right are the squares in the sum $1^2 + 2^2 + 3^2 + \cdots + 25^2$, which we are given in the problem. Factoring out 4 from the sum on the right above gives us

$$4 \cdot 1^2 + 4 \cdot 2^2 + 4 \cdot 3^2 + \cdots + 4 \cdot 25^2 = 4(1^2 + 2^2 + 3^2 + \cdots + 25^2).$$

The problem said that the sum in parentheses above is 5525. So the answer is $4 \cdot 5525$, which is $\boxed{22{,}100}$.

Exercises for Section 2.2

2.2.1 Let's compute all four powers:

$$A = 2^5 = 2 \cdot 2 \cdot 2 \cdot 2 \cdot 2 = 32, \qquad B = 3^4 = 3 \cdot 3 \cdot 3 \cdot 3 = 81, \qquad C = 4^3 = 4 \cdot 4 \cdot 4 = 64, \qquad D = 5^2 = 5 \cdot 5 = 25.$$

From smallest to largest, the powers are $\boxed{D, A, C, \text{ and } B}$.

2.2.2 The square of the cube of 2 is $(2^3)^2 = 8^2 = 64$. The cube of the square of 2 is $(2^2)^3 = 4^3 = 64$. The two numbers are equal, so their difference is $\boxed{0}$.

It's not a coincidence that $(2^3)^2$ and $(2^2)^3$ are equal. By the "power of a power" rule, both are equal to 2^6.

2.2.3 We have $3^3 + 3^3 + 3^3 = 3 \cdot 3^3 = 3^1 \cdot 3^3 = 3^{1+3} = \boxed{3^4}$.

2.2.4

(a) We have $2^4 + 2^4 + 2^4 + 2^4 = 4 \cdot 2^4 = 4 \cdot 16 = \boxed{64}$.

(b) Evaluate the powers inside the parentheses first: $(2^5 + 2^6 + 2^7) \div 2^3 = (32 + 64 + 128) \div 8 = 224 \div 8 = \boxed{28}$.

We might also have used the distributive property and the quotient of powers (same base) rule:

$$(2^5 + 2^6 + 2^7) \div 2^3 = (2^5 \div 2^3) + (2^6 \div 2^3) + (2^7 \div 2^3)$$
$$= 2^{5-3} + 2^{6-3} + 2^{7-3} = 2^2 + 2^3 + 2^4 = 4 + 8 + 16 = \boxed{28}.$$

(c) Exponentiation first, then multiplication, and finally subtraction: $3^4 - 5 \cdot 8 = 81 - 5 \cdot 8 = 81 - 40 = \boxed{41}$.

(d) Evaluate the three powers first and then subtract: $2^5 - 2^4 - 2^3 = 32 - 16 - 8 = 16 - 8 = \boxed{8}$.

(e) Because 11 is odd, we have $\left(1 - (-1)^{11}\right)^2 = (1 - (-1))^2 = (1 + 1)^2 = 2^2 = \boxed{4}$.

(f) Because -1^{2008} means $-(1^{2008})$, and because 2007 is odd, we have

$$-1^{2008} + (-1)^{2007} = -1 + (-1) = \boxed{-2}.$$

(g) $5 - 7(5^2 - 3^3)^4 = 5 - 7(25 - 27)^4 = 5 - 7(-2)^4 = 5 - 7 \cdot 2^4 = 5 - 7 \cdot 16 = 5 - 112 = \boxed{-107}$.

(h) Evaluate the four powers first: $3^5(2^3) - 2^4(3^4) = 243 \cdot 8 - 16 \cdot 81 = 1944 - 1296 = \boxed{648}$.

We also could have noted that $3^5 = 3^{1+4} = 3 \cdot 3^4$ and $2^4 = 2^{1+3} = 2 \cdot 2^3$, so

$$3^5(2^3) - 2^4(3^4) = 3 \cdot 3^4 \cdot 2^3 - 2 \cdot 2^3 \cdot 3^4 = 3(2^3 \cdot 3^4) - 2(2^3 \cdot 3^4).$$

This allows us to factor out $2^3 \cdot 3^4$:

$$3(2^3 \cdot 3^4) - 2(2^3 \cdot 3^4) = (3 - 2)(2^3 \cdot 3^4) = 1(2^3 \cdot 3^4) = 8 \cdot 81 = \boxed{648}.$$

(i) By the quotient of powers rule (same exponent), we have

$$88{,}888^4 \div 22{,}222^4 = (88{,}888 \div 22{,}222)^4 = 4^4 = \boxed{256}.$$

2.2.5 Each term in the sum equals 1, so we just have to count the number of terms to evaluate the expression. Since the exponents are the even numbers 2, 4, 6, 8, and so on up to 100, the number of terms is $100 \div 2 = 50$. So the sum is

$$1^2 + 1^4 + 1^6 + 1^8 + \cdots + 1^{100} = \underbrace{1 + 1 + 1 + 1 + \cdots + 1}_{50 \text{ terms}} = \boxed{50}.$$

2.2.6

(a) By the power of a power rule, we have $(2^3)^4 = 2^{3 \cdot 4} = \boxed{2^{12}}$.

(b) We can express repeated multiplication as a power: $4 \cdot 4 \cdot 4 \cdot 4 \cdot 4 = 4^5 = (2^2)^5 = 2^{2 \cdot 5} = \boxed{2^{10}}$. We might also think of this as, "Each 4 is a product of 2 2's, so the product of 5 4's is a product of $2 \cdot 5 = 10$ 2's."

(c) Applying the product and quotient of powers (same base) rules, we have $2^{40} \cdot 2^{13} \div 2^6 = 2^{40+13} \div 2^6 = 2^{40+13-6} = \boxed{2^{47}}$.

(d) By power of a power, $2^{10} \cdot 4^{20} \cdot 8^{30} = 2^{10} \cdot (2^2)^{20} \cdot (2^3)^{30} = 2^{10} \cdot 2^{40} \cdot 2^{90} = 2^{10+40+90} = \boxed{2^{140}}$.

(e) By power of a power, $4^4 \cdot 8^8 \cdot 16^{16} = (2^2)^4 \cdot (2^3)^8 \cdot (2^4)^{16} = 2^8 \cdot 2^{24} \cdot 2^{64} = 2^{8+24+64} = \boxed{2^{96}}$.

(f) By power of a power and quotient of powers, $4^3 \div 2^2 = (2^2)^3 \div 2^2 = 2^6 \div 2^2 = 2^{6-2} = \boxed{2^4}$.

(g) By power of a power and quotient of powers, $\frac{1}{2} \cdot 8^{50} = 8^{50} \div 2 = (2^3)^{50} \div 2^1 = 2^{150} \div 2^1 = 2^{150-1} = \boxed{2^{149}}$.

(h) Since $256 = 16^2$, we have $256 = (2^4)^2 = 2^{4 \cdot 2} = \boxed{2^8}$.

(i) Because repeated addition is multiplication, $2^{50} + 2^{50} + 2^{50} + 2^{50} = 4 \cdot 2^{50} = 2^2 \cdot 2^{50} = 2^{2+50} = \boxed{2^{52}}$.

(j) By factoring out 4^4, and then using the power of a power and product of powers rules, we have

$$4^4 + 4(4^4) + 6(4^4) + 4(4^4) + 4^4 = 1(4^4) + 4(4^4) + 6(4^4) + 4(4^4) + 1(4^4)$$
$$= (1 + 4 + 6 + 4 + 1)4^4 = 16 \cdot 4^4 = 2^4 \cdot (2^2)^4 = 2^4 \cdot 2^8 = 2^{4+8} = \boxed{2^{12}}.$$

(k) By power of a power and quotient of powers, $4^{3^3} \div (4^3)^3 = 4^{27} \div 4^9 = 4^{27-9} = 4^{18} = (2^2)^{18} = \boxed{2^{36}}$.

(l) Let's express 2^{20} in terms of 2^{19}, and then factor:

$$2^{20} - 2^{19} = 2^{1+19} - 2^{19} = 2^1 \cdot 2^{19} - 2^{19} = 2 \cdot 2^{19} - 2^{19} = 2 \cdot 2^{19} - 1 \cdot 2^{19} = (2-1)2^{19} = 1 \cdot 2^{19} = \boxed{2^{19}}.$$

2.2.7 When we expand 10^{17}, we get

$$10^{17} = \underbrace{10 \times 10 \times \cdots \times 10 \times 10}_{17 \text{ tens}} = \underbrace{100\ldots00}_{17 \text{ zeros}}.$$

So the product of 91 and 10^{17} is

$$91 \times 10^{17} = 91 \times \underbrace{100\ldots00}_{17 \text{ zeros}} = 9\underbrace{100\ldots00}_{17 \text{ zeros}}.$$

The final number has 2 nonzero digits followed by 17 zeros. So the number of digits is $2 + 17 = \boxed{19}$.

2.2.8 Let's pair each factor of 5 with a factor of 2 to form factors of 10, since it's easy to multiply by 10:

$$2^{16} \cdot 5^{13} = 2^3 \cdot 2^{13} \cdot 5^{13} = 2^3 \cdot (2 \cdot 5)^{13} = 2^3 \cdot 10^{13} = 8 \cdot 10^{13}$$

Since 10^{13} multiplied out is a 1 followed by 13 zeros, we know that $8 \cdot 10^{13}$ multiplied out is an 8 followed by 13 zeros, for a total of $\boxed{14}$ digits.

2.2.9   All three exponents are 25 times something. So we can write all three numbers as 25th powers:

$$2^{100} = 2^{4 \cdot 25} = \left(2^4\right)^{25} = 16^{25},$$

$$3^{75} = 3^{3 \cdot 25} = \left(3^3\right)^{25} = 27^{25},$$

$$5^{50} = 5^{2 \cdot 25} = \left(5^2\right)^{25} = 25^{25}.$$

Of the three bases 16, 27, and 25, the largest is 27. So the largest of the numbers on the right is 27^{25}. Therefore, $\boxed{3^{75}}$ is the largest of the numbers on the left.

2.2.10 10^{93} is a 1 followed by 93 zeros, so it has 94 digits. Subtracting 93 produces a 93-digit number in which the first 91 digits are nines and the last two digits are 07:

$$\underbrace{100\ldots00}_{93 \text{ zeros}} - 93 = \underbrace{99\ldots99}_{91 \text{ nines}}07.$$

The sum of the digits of this number is

$$\underbrace{9 + 9 + \cdots + 9 + 9}_{91 \text{ nines}} + 0 + 7 = 91 \cdot 9 + 0 + 7 = 819 + 0 + 7 = \boxed{826}.$$

2.2.11   Let's first simplify the numbers and then convert the powers of 4 to powers of 2:

$$2^{3^4} = 2^{81},$$

$$2^{4^3} = 2^{64},$$

$$3^{4^2} = 3^{16},$$

$$4^{3^2} = 4^9 = \left(2^2\right)^9 = 2^{18},$$

$$4^{2^3} = 4^8 = \left(2^2\right)^8 = 2^{16}.$$

All the numbers except the third one are powers of 2. Of the powers of 2, the largest is 2^{81}. The third number is 3^{16}, which is less than $4^{16} = (2^2)^{16} = 2^{2 \cdot 16} = 2^{32}$. So the largest number is 2^{81}, which came from $\boxed{2^{3^4}}$.

Exercises for Section 2.3

2.3.1

(a)   We evaluate the exponent first: $56 \div 4 + 3 \cdot 2^0 = 56 \div 4 + 3 \cdot 1 = 14 + 3 \cdot 1 = 14 + 3 = \boxed{17}$.

(b)   We evaluate inside the parentheses first:

$$7^4(8 - 2^3) + 11^{4(8)-32} = 7^4(8 - 8) + 11^{4(8)-32} = 7^4 \cdot 0 + 11^{4(8)-32}.$$

We don't have to compute 7^4; the product $7^4 \cdot 0$ is 0 no matter what 7^4 is. So, we have

$$7^4 \cdot 0 + 11^{4(8)-32} = 0 + 11^{4(8)-32} = 11^{4(8)-32} = 11^{32-32} = 11^0 = \boxed{1}.$$

(c) We evaluate inside the parentheses first, and then the exponents:

$$7^0 + 3^2 \cdot 4 - 2(14 - 8 \div 2) = 7^0 + 3^2 \cdot 4 - 2(14 - 4)$$
$$= 7^0 + 3^2 \cdot 4 - 2 \cdot 10 = 1 + 9 \cdot 4 - 2 \cdot 10 = 1 + 36 - 20 = 37 - 20 = \boxed{17}.$$

2.3.2 The fourth term is $3^0 + 3^1 + 3^2 + 3^3$. Its value is $3^0 + 3^1 + 3^2 + 3^3 = 1 + 3 + 9 + 27 = \boxed{40}$.

2.3.3 Let's replace each x with 2 and each y with -2:

$$x^{x+y} + y^{x-y} = 2^{2+(-2)} + (-2)^{2-(-2)} = 2^0 + (-2)^4 = 1 + 16 = \boxed{17}.$$

2.3.4 Anything raised to the 0th power is 1, so $3n^0 \cdot (7n)^0 = 3 \cdot 1 \cdot 1 = \boxed{3}$.

2.3.5 By replacing 6^0 with 1 and then factoring out x^2, we get $6^0 x^2 + 6x^2 = 1x^2 + 6x^2 = (1 + 6)x^2 = \boxed{7x^2}$.

Exercises for Section 2.4

2.4.1

(a) Because 11 is odd, we have $(-1)^{11} = -1$, so $2^{(-1)^{11}} = 2^{-1} = \boxed{\frac{1}{2}}$.

(b) Applying the product rule (same base), we have $3^7 \cdot 3^{-4} = 3^{7+(-4)} = 3^3 = \boxed{27}$.

(c) Using the quotient of powers (same base), we have $2^3 \div 2^{-4} = 2^{3-(-4)} = 2^{3+4} = 2^7 = \boxed{128}$.

(d) We have $1 \div 5^{-2} = 1 \div \frac{1}{5^2} = 1 \div \frac{1}{25}$. Since the reciprocal of $\frac{1}{25}$ is 25, we have $1 \div \frac{1}{25} = 1 \cdot 25 = \boxed{25}$.

(e) Because -5 is odd, we have $(-3)^{-5} \cdot 3^3 = -(3^{-5}) \cdot 3^3 = -(3^{-5} \cdot 3^3) = -3^{-5+3} = -3^{-2} = -\frac{1}{3^2} = \boxed{-\frac{1}{9}}$.

(f) We have $\left(\frac{1}{4}\right)^{-3} = 4^3$, so $\left(\frac{1}{4}\right)^{-3} \cdot 8^{-2} = 4^3 \cdot \frac{1}{8^2} = 64 \cdot \frac{1}{64} = \boxed{1}$.

2.4.2 Replacing each a with 2, and then using the quotient of powers rule (same base), we get

$$a \div a^{-4} = 2 \div 2^{-4} = 2^1 \div 2^{-4} = 2^{1-(-4)} = 2^5 = \boxed{32}.$$

2.4.3 1 raised to any power equals 1. Meanwhile, -1 raised to an odd power is -1 (even if the power is negative), and raised to an even power is 1. So, replacing each x with 1 and each y with -1 gives

$$15x^2 y^{-3} + 18yx^{-1} + 27xy^4 = 15 \cdot 1^2 \cdot (-1)^{-3} + 18 \cdot (-1) \cdot 1^{-1} + 27 \cdot 1 \cdot (-1)^4$$
$$= 15 \cdot 1 \cdot (-1) + 18 \cdot (-1) \cdot 1 + 27 \cdot 1 \cdot 1 = -15 + (-18) + 27 = -33 + 27 = \boxed{-6}.$$

2.4.4 Because repeated addition is multiplication, the left side simplifies to

$$3^3 + 3^3 + 3^3 = 3 \cdot 3^3 = 3^1 \cdot 3^3 = 3^{1+3} = 3^4.$$

Because $243 = 3^5$, the right side of the original equation simplifies to

$$243 \cdot 3^k = 3^5 \cdot 3^k = 3^{5+k}.$$

When we use the simplified left and right sides, the original equation becomes

$$3^4 = 3^{5+k}.$$

For that equation to be true, the exponents must be equal, which means $4 = 5 + k$. Therefore, we have $k = 4 - 5 = \boxed{-1}$.

2.4.5   First let's express 2^{12} as a power of 8. Because $2^3 = 8$, we have

$$2^{12} = 2^{3 \cdot 4} = \left(2^3\right)^4 = 8^4.$$

Next, let's express 2^{12} as a power of $\frac{1}{8}$. Because $8^{-1} = \frac{1}{8}$, we have

$$2^{12} = 8^4 = 8^{(-1)(-4)} = \left(8^{-1}\right)^{-4} = \boxed{\left(\frac{1}{8}\right)^{-4}}.$$

2.4.6   By the square of a product rule and the power of a power rule, we have

$$\left(6a^2b\right)^2 \div \left(3a^2b^3\right) = 6^2\left(a^2\right)^2 b^2 \div \left(3a^2b^3\right)$$
$$= 36a^4b^2 \div \left(3a^2b^3\right) = (36 \div 3)(a^4 \div a^2)(b^2 \div b^3) = 12a^{4-2}b^{2-3} = \boxed{12a^2b^{-1}}.$$

Review Problems

2.37

(a) We evaluate inside the parentheses first:

$$4 - 8\left((-2)^2 - 4(-3)\right) = 4 - 8\left(4 - 4(-3)\right) = 4 - 8\left(4 - (-12)\right) = 4 - 8 \cdot 16 = 4 - 128 = \boxed{-124}.$$

(b) We evaluate inside both parentheses first: $(5 - 2)^2 + (2 - 5)^3 = 3^2 + (-3)^3 = 9 - 27 = \boxed{-18}$.

(c) We evaluate inside the parentheses, then compute the powers, then multiply, then subtract:

$$5 \cdot 2^5 - (2 \cdot 3)^2 = 5 \cdot 2^5 - 6^2 = 5 \cdot 32 - 36 = 160 - 36 = \boxed{124}.$$

(d) We compute $2 \cdot 3^2$ inside the parentheses first:

$$5 + (-6)^3 \div (2 \cdot 3^2) = 5 + (-6)^3 \div (2 \cdot 9) = 5 + (-6)^3 \div 18.$$

Then, we compute $(-6)^3 = (-6)(-6)(-6) = 36(-6) = -216$, so

$$5 + (-6)^3 \div 18 = 5 + (-216) \div 18 = 5 + (-12) = \boxed{-7}.$$

2.38 The first number is $3^5 = 243$. The second number is $5^3 = 125$. So the first number exceeds the second number by $243 - 125 = \boxed{118}$.

2.39 Because -1^{2004} means $-(1^{2004})$, we have $-1^{2004} = -1$. Because 2005 is odd, we have $(-1)^{2005} = -1$. Since any power of 1 is 1, we have $1^{2006} = 1$ and $1^{2007} = 1$. So, we find that

$$-1^{2004} + (-1)^{2005} + 1^{2006} - 1^{2007} = -1 + (-1) + 1 - 1 = -2 + 1 - 1 = -1 - 1 = \boxed{-2}.$$

2.40 Because $2^6 = 64$, the equation becomes $n^2 = 64$. Because n is a positive integer, the only solution is $n = \boxed{8}$.

We also can solve this problem with the power of a power rule, by noting that $2^6 = 2^{3 \cdot 2} = (2^3)^2 = 8^2$, so $n = \boxed{8}$.

2.41 The square of 40 is 1600 and the square of 50 is 2500. Since 40^2 is too low and 50^2 is too high, the square we seek must be between 40^2 and 50^2. In the middle, the square of 45 is 2025, which is a little too high. A little lower, the square of 44 is $\boxed{1936}$.

2.42 Substituting 3 for x, we get

$$x^5 - 2x = 3^5 - 2 \cdot 3 = 243 - 2 \cdot 3 = 243 - 6 = \boxed{237}.$$

2.43 Replacing each x with -4, we get

$$-2x^3 - 3x^2 = -2(-4)^3 - 3(-4)^2 = -2(-64) - 3(16) = 128 - 48 = \boxed{80}.$$

2.44 10,000 is exactly 100^2. So the positive perfect squares less than 10,000 are $1^2, 2^2, 3^2, \ldots, 99^2$. The number of such squares is $\boxed{99}$.

2.45 The left side of the equation is

$$(1 + 2 + 3 + 4 + 5 + 6)^2 = 21^2 = 441.$$

We want the sum of the first few cubes to be 441. So let's keep adding cubes until we reach 441:

$$1^3 + 2^3 = 1 + 8 = 9,$$
$$(1^3 + 2^3) + 3^3 = 9 + 27 = 36,$$
$$(1^3 + 2^3 + 3^3) + 4^3 = 36 + 64 = 100,$$
$$(1^3 + 2^3 + 3^3 + 4^3) + 5^3 = 100 + 125 = 225,$$
$$(1^3 + 2^3 + 3^3 + 4^3 + 5^3) + 6^3 = 225 + 216 = 441.$$

Bingo! The value of n is $\boxed{6}$.

Did you notice that the answer 6 is the same as the largest number on the left side of the original equation? That's not an accident. This problem can be extended to any positive integer. That is, for any positive integer n, we have $(1 + 2 + 3 + \cdots + n)^2 = 1^3 + 2^3 + 3^3 + \cdots + n^3$. You may learn why when you study algebra later.

2.46 The cube of 10 is 1000, which is too low, and the cube of 20 is 8000, which is too high. In the middle, the cube of 15 is 3375. A little higher, the cube of 17 is 4913. So N is $\boxed{17}$.

2.47 The cube of 2 is 8 and the cube of 8 is 512. The cube of 512 is between 500^3 (which is 125,000,000) and 600^3 (which is 216,000,000). But $10^9 = 1,000,000,000$, so 512^3 is less than 10^9.

Next, we consider the cube of 512^3. To see why the cube of 512^3 is greater than 10^9, we note that 512^3 is greater than 100^3. The cube of 100^3 is $(100^3)^3 = 100^{3 \cdot 3} = 100^9$, which is clearly greater than 10^9. So, the cube of 512^3 is greater than 10^9, which means that the number of cubings needed to exceed 10^9 is $\boxed{4}$.

2.48

(a) The positive perfect cubes less than 91 are 1, 8, 27, and 64. Trying every pair of these cubes, we find that 91 is $27 + 64$. So the answer is $\boxed{27 + 64}$ or $\boxed{3^3 + 4^3}$.

(b) To start, let's express 91 as the *difference* of positive perfect cubes. The first few positive cubes are 1, 8, 27, 64, 125, and 216. Trying every pair of these cubes, we find that 91 is $216 - 125$. We can convert this difference to the sum $216 + (-125)$. So the answer is $\boxed{216 + (-125)}$ or $\boxed{6^3 + (-5)^3}$.

2.49 The first few powers of 3 are $3^0 = 1$, $3^1 = 3$, $3^2 = 9$, $3^3 = 27$, $3^4 = 81$, $3^5 = 243$, and $3^6 = 729$. So the number of them less than 333 is $\boxed{6}$.

2.50 The number 695,000 is more than 100,000 and less than 1,000,000. In other words, 695,000 is between 10^5 and 10^6. So the value of n is $\boxed{5}$.

2.51 Plugging in $x = 3$ and $y = 4$, we get

$$3x^y + 4y^x = 3 \cdot 3^4 + 4 \cdot 4^3 = 3 \cdot 81 + 4 \cdot 64 = 243 + 256 = \boxed{499}.$$

2.52

(a) By the product of powers rule (same base), we have $2^3 \cdot 4 \cdot 8 = 2^3 \cdot 2^2 \cdot 2^3 = 2^{3+2+3} = \boxed{2^8}$.

(b) By the definition of division and the quotient of powers rule (same base), we have

$$\frac{1}{2}(2^{15}) = 2^{15} \cdot \frac{1}{2} = 2^{15} \div 2 = 2^{15} \div 2^1 = 2^{15-1} = \boxed{2^{14}}.$$

(c) By the power of a power rule, we have $(2^5)^6 = 2^{5 \cdot 6} = 2^{30}$. Since $4 = 2^2$, we have $4^3 = (2^2)^3 = 2^{2 \cdot 3} = 2^6$. Finally, applying the quotient of powers (same base), we have

$$(2^5)^6 \div 4^3 = 2^{30} \div 2^6 = 2^{30-6} = \boxed{2^{24}}.$$

(d) Let's express each of the powers as something times 2^{18}:

$$2^{20} - 2^{19} - 2^{18} = 2^2 \cdot 2^{18} - 2^1 \cdot 2^{18} - 2^{18} = 4 \cdot 2^{18} - 2 \cdot 2^{18} - 1 \cdot 2^{18} = (4 - 2 - 1) \cdot 2^{18} = 1 \cdot 2^{18} = \boxed{2^{18}}.$$

2.53 Because 100 is 10^2, we have $100^3 = (10^2)^3 = 10^{2 \cdot 3} = \boxed{10^6}$.

2.54

(a) Using the product of powers (same exponent) law, we have $40^3 \cdot 5^3 = (40 \cdot 5)^3 = 200^3 = \boxed{8{,}000{,}000}$.

(b) Using the quotient of powers (same exponent) law, we have $27^5 \div 9^5 = (27 \div 9)^5 = 3^5 = \boxed{243}$.

(c) It's easy to compute powers of 10, so we group the powers of 2 and 5 to form powers of 10:

$$5^4 \cdot 3^2 \cdot 2^5 = 3^2 \cdot 2^5 \cdot 5^4 = 3^2 \cdot 2 \cdot (2^4 \cdot 5^4).$$

Applying the product of powers (same exponent) law, we have

$$3^2 \cdot 2 \cdot (2^4 \cdot 5^4) = 3^2 \cdot 2 \cdot (2 \cdot 5)^4 = 3^2 \cdot 2 \cdot 10^4 = 18 \cdot (10{,}000) = \boxed{180{,}000}.$$

(d) We could compute all the powers and add, but that's a lot of work. Instead, we note that $32 = 2^5$, so we can use the distributive property together with the quotient of powers (same base) rule:

$$
\begin{aligned}
(2^8 + 2^9 + 2^{10} + 2^{11}) \div 32 &= (2^8 + 2^9 + 2^{10} + 2^{11}) \div 2^5 \\
&= (2^8 \div 2^5) + (2^9 \div 2^5) + (2^{10} \div 2^5) + (2^{11} \div 2^5) \\
&= 2^{8-5} + 2^{9-5} + 2^{10-5} + 2^{11-5} \\
&= 2^3 + 2^4 + 2^5 + 2^6 = 8 + 16 + 32 + 64 = \boxed{120}.
\end{aligned}
$$

2.55

(a) Any number raised to the 0th power equals 1, so $2^0 + 3^0 + (-4)^0 - (2 + 3 - 4)^0 = 1 + 1 + 1 - 1 = \boxed{2}$.

(b) We evaluate first inside the parentheses and then the negative exponent:

$$(5^2 - 2^3 + 10^0 - 4^2)^{-2} = (25 - 8 + 1 - 16)^{-2} = 2^{-2} = \frac{1}{2^2} = \boxed{\frac{1}{4}}.$$

(c) We showed in the text that $\left(\frac{1}{a}\right)^{-n} = a^n$, so $\left(\frac{1}{2}\right)^{-3} = 2^3 = \boxed{8}$.

(d) Since a^{-1} is the reciprocal of a, we have $(-2)^{-1} = \frac{1}{-2} = \boxed{-\frac{1}{2}}$.

(e) Since -14 is even, we have $(-1)^{-14} = 1^{-14}$. Any power of 1 is 1, so $1^{-14} = \boxed{1}$.

(f) We showed in the text that $\frac{1}{a^{-n}} = a^n$, so $\frac{1}{4^{-3}} = 4^3 = \boxed{64}$.

2.56

(a) Since $9 = 3^2$, we have $\frac{1}{9} = \frac{1}{3^2} = \boxed{3^{-2}}$.

(b) Applying the product and quotient of powers (same base) rules, we have $3^{-4} \cdot 3^2 \div 3^3 = 3^{-4+2} \div 3^3 = 3^{-4+2-3} = \boxed{3^{-5}}$.

(c) Since $\frac{1}{a^{-n}} = a^n$, we have $\frac{1}{3^{-2}} = 3^2$, so

$$\left(\frac{1}{3^{-2}}\right)^{-3} \cdot 3^2 = (3^2)^{-3} \cdot 3^2 = 3^{2 \cdot (-3)} \cdot 3^2 = 3^{-6} \cdot 3^2 = 3^{-6+2} = \boxed{3^{-4}}.$$

(d) We have $27 = 3^3$, so $27^2 \div 3^{-3} = (3^3)^2 \div 3^{-3} = 3^{3 \cdot 2} \div 3^{-3} = 3^6 \div 3^{-3} = 3^{6-(-3)} = \boxed{3^9}$.

Challenge Problems

2.57 We could list squares until we find two that differ by 67, but our work in the text gives us a faster approach. We can get from the perfect square a^2 to the next perfect square $(a + 1)^2$ by adding a and $a + 1$ to a^2. So, the squares a^2 and $(a + 1)^2$ differ by $a + (a + 1)$. In other words, the squares of two consecutive integers differ by the sum of the integers. This means we seek a consecutive pair of integers whose sum is 67. We have $33 + 34 = 67$, so $\boxed{33}$ and 34 are the consecutive integers whose squares differ by 67.

2.58 The first few positive cubes are 1, 8, 27, 64, 125, 216, 343, 512, 729, 1000, 1331, and 1728. We can express 1729 as $1728 + 1$, which is $12^3 + 1^3$. We can also express 1729 as $1000 + 729$, which is $10^3 + 9^3$. No other pairs of positive perfect cubes add up to 1729. So A, B, C, and D are 12, 1, 10, and 9, in some order. Therefore, their sum is $\boxed{32}$.

2.59 Let's first express the number as a power of 2, and then convert it to a power of 4:

$$2^5 \cdot 8^3 \cdot 16^2 = 2^5 \cdot (2^3)^3 \cdot (2^4)^2 = 2^5 \cdot 2^9 \cdot 2^8 = 2^{5+9+8} = 2^{22} = 2^{2 \cdot 11} = (2^2)^{11} = \boxed{4^{11}}.$$

2.60 We are looking for a cube between 80,000 and 90,000. The cube of 40 is 64,000 and the cube of 50 is 125,000. In the middle, the cube of 45 is 91,125. A little lower, the cube of 44 is $\boxed{85,184}$.

2.61 Because $3^2 = 9$, we have

$$3^{16} = 3^{2 \cdot 8} = \left(3^2\right)^8 = 9^8.$$

Because $9^{-1} = \frac{1}{9}$, we have

$$3^{16} = 9^8 = 9^{(-1)(-8)} = \left(9^{-1}\right)^{-8} = \boxed{\left(\frac{1}{9}\right)^{-8}}.$$

2.62 By converting the product of squares to the square of a product, and then expressing each number as a power of 2, we get

$$2^2 \times 4^2 \times 8^2 \times 16^2 \times \cdots \times 1024^2 = (2 \times 4 \times 8 \times 16 \times \cdots \times 1024)^2$$
$$= \left(2^1 \times 2^2 \times 2^3 \times 2^4 \times \cdots \times 2^{10}\right)^2$$
$$= \left(2^{1+2+3+4+\cdots+10}\right)^2 = \left(2^{55}\right)^2 = 2^{55 \cdot 2} = \boxed{2^{110}}.$$

2.63 We convert the power of 8 to a power of 2, and then combine the 2's and 5's to form a power of 10:

$$8^{10} \cdot 5^{22} = (2^3)^{10} \cdot 5^{22} = 2^{30} \cdot 5^{22} = 2^8 \cdot 2^{22} \cdot 5^{22} = 2^8 \cdot (2 \cdot 5)^{22} = 2^8 \cdot 10^{22} = 256 \cdot \underbrace{100\ldots00}_{22 \text{ zeros}} = 256\underbrace{00\ldots00}_{22 \text{ zeros}}.$$

So the number of digits is $3 + 22 = \boxed{25}$.

2.64 The left side of the equation simplifies as follows:

$$500,000^2 \cdot 200,000^2 = (500,000 \cdot 200,000)^2$$
$$= ((5 \cdot 10^5) \cdot (2 \cdot 10^5))^2$$
$$= ((5 \cdot 2) \cdot 10^5 \cdot 10^5)^2 = (10 \cdot 10^5 \cdot 10^5)^2 = (10^1 \cdot 10^5 \cdot 10^5)^2 = (10^{11})^2 = 10^{22}.$$

So the exponent n is $\boxed{22}$.

2.65 We can expand the right side of the equation as follows:

$$6^6 = (3 \cdot 2)^6 = 3^6 \cdot 2^6.$$

The left side is $n \cdot 3^4 \cdot 2^5$, so we have

$$n \cdot 3^4 \cdot 2^5 = 3^6 \cdot 2^6.$$

To get 3^6, we multiply 3^4 by 3^2, since $3^4 \cdot 3^2 = 3^{4+2} = 3^6$. Similarly, we multiply 2^5 by 2^1 to get 2^6, since $2^5 \cdot 2^1 = 2^{5+1} = 2^6$. So, if we let $n = 3^2 \cdot 2^1$, we have

$$(3^2 \cdot 2^1) \cdot 3^4 \cdot 2^5 = 2^1 \cdot 2^5 \cdot 3^2 \cdot 3^4 = 2^{1+5} \cdot 3^{2+4} = 2^6 \cdot 3^6,$$

as desired. So, we have $n = 3^2 \cdot 2^1 = 9 \cdot 2 = \boxed{18}$.

2.66 Let's combine the 2's and 5's to form a power of 10:

$$2^{2005} \cdot 5^{2007} \cdot 3 = 2^{2005} \cdot 5^{2005} \cdot 5^2 \cdot 3$$
$$= (2 \cdot 5)^{2005} \cdot 5^2 \cdot 3$$
$$= 10^{2005} \cdot 5^2 \cdot 3 = 10^{2005} \cdot 25 \cdot 3 = 10^{2005} \cdot 75 = \underbrace{100\ldots00}_{2005 \text{ zeros}} \cdot 75 = 75\underbrace{00\ldots00}_{2005 \text{ zeros}}.$$

So the sum of the digits is $7 + 5 = \boxed{12}$.

2.67 Using the square of a product rule, let's try to express $22^2 \cdot 55^2$ as 10^2 times something:

$$22^2 \cdot 55^2 = (22 \cdot 55)^2 = 1210^2 = (10 \cdot 121)^2 = 10^2 \cdot 121^2.$$

So $10^2 \cdot N^2$ is equal to $10^2 \cdot 121^2$. That means N^2 is 121^2. Because N is positive, N must be $\boxed{121}$.

2.68

(a) We will apply the power of a product rule and then the power of a power rule:

$$(2ab^2)^3 = 2^3 a^3 (b^2)^3 = 2^3 a^3 b^6 = \boxed{8a^3 b^6}.$$

(b) This time, we will apply power of a product first, followed by product of powers (same base):

$$5a^2 b(2ab)^3 = 5a^2 b \cdot 2^3 a^3 b^3 = 5a^2 b \cdot 8a^3 b^3 = (5 \cdot 8)(a^2 a^3)(bb^3) = 40a^5(b^1 b^3) = \boxed{40a^5 b^4}.$$

2.69

(a) By grouping the a's together, the b's together, and the c's together, we get

$$a^2 b \cdot 8ab^6 c^2 = 8(a^2 \cdot a)(b \cdot b^6)c^2 = 8(a^2 \cdot a^1)(b^1 \cdot b^6)c^2 = \boxed{8a^3 b^7 c^2}.$$

(b) By applying the exponent laws, and then grouping the a's together, we get

$$(a^2)^4 (ab)^3 c^3 = a^8(a^3 b^3)c^3 = (a^8 \cdot a^3)b^3 c^3 = \boxed{a^{11} b^3 c^3}.$$

2.70 Since $3000 = 3 \cdot 1000$ and $2000 = 2 \cdot 1000$, we can express 5^{3000} and n^{2000} as numbers raised to the 1000th power. Specifically, we have

$$5^{3000} = \left(5^3\right)^{1000} = 125^{1000}.$$

Similarly, $n^{2000} = \left(n^2\right)^{1000}$. So to make n^{2000} less than 5^{3000}, we need n^2 to be less than 125. That means n is one of the integers

$$-11, -10, -9, -8, -7, -6, -5, -4, -3, -2, -1, 0, 1, 2, 3, 4, 5, 6, 7, 8, 9, 10, 11.$$

The largest of these integers is $\boxed{11}$.

2.71 The left side simplifies to
$$125 \cdot 5^5 = 5^3 \cdot 5^5 = 5^{3+5} = 5^8.$$

Because repeated addition is multiplication, the right side simplifies to

$$5^x + 5^x + 5^x + 5^x + 5^x = 5 \cdot 5^x = 5^1 \cdot 5^x = 5^{1+x}.$$

So our equation becomes $5^8 = 5^{1+x}$. The bases are the same, so we have $5^8 = 5^{1+x}$ when the exponents are equal, which is when $x = \boxed{7}$.

2.72 We can make the right side of the equation look like the left side as follows:

$$(2^3)(3^3)(4^3)(5^3) = 4^3 \cdot (2^3 \cdot 3^3 \cdot 5^3) = 4^3 \cdot (2 \cdot 3 \cdot 5)^3 = 4^3 \cdot 30^3 = \left(2^2\right)^3 \cdot 30^3 = (2^6)(30^3).$$

The left side is $(2^x)(30^3)$. So the exponent x must be $\boxed{6}$.

2.73

(a) We use the distributive property:

$$(a+b)^2 = (a+b)(a+b) = a(a+b) + b(a+b) = aa + ab + ba + bb = a^2 + ab + ab + b^2 = a^2 + 2ab + b^2.$$

(b) Let's convert the subtraction to an addition and then use part (a):

$$(a-b)^2 = (a+(-b))^2 = a^2 + 2a(-b) + (-b)^2 = a^2 + (-2ab) + b^2 = a^2 - 2ab + b^2.$$

(c) Let's use part (a) and then apply the distributive law on two different steps:

$$\begin{aligned}
(a+b)^3 &= (a+b)(a+b)^2 \\
&= (a+b)(a^2 + 2ab + b^2) \\
&= a(a^2 + 2ab + b^2) + b(a^2 + 2ab + b^2) \\
&= a \cdot a^2 + a \cdot 2ab + a \cdot b^2 + b \cdot a^2 + b \cdot 2ab + b \cdot b^2 \\
&= a^3 + 2a^2b + ab^2 + a^2b + 2ab^2 + b^3 = a^3 + 3a^2b + 3ab^2 + b^3.
\end{aligned}$$

2.74

(a) Multiplication distributes over subtraction, so applying the distributive property gives

$$(a-b)(a+b) = a(a+b) - b(a+b) = aa + ab - ba - bb = a^2 + ab - ab - b^2 = a^2 - b^2.$$

(b) Let's apply the distributive law repeatedly:

$$(a-b)(a^2 + ab + b^2) = a(a^2 + ab + b^2) - b(a^2 + ab + b^2) = a^3 + a^2b + ab^2 - a^2b - ab^2 - b^3 = a^3 - b^3.$$

(c) Again, let's apply the distributive law repeatedly:

$$(a+b)(a^2 - ab + b^2) = a(a^2 - ab + b^2) + b(a^2 - ab + b^2) = a^3 - a^2b + ab^2 + a^2b - ab^2 + b^3 = a^3 + b^3.$$

2.75 The exponents 96 and 72 are each 24 times some integer. So

$$5^{96} = 5^{4\cdot24} = \left(5^4\right)^{24} = 625^{24}.$$

Similarly, $n^{72} = (n^3)^{24}$ and $(n+1)^{72} = \left((n+1)^3\right)^{24}$. So to make 5^{96} between n^{72} and $(n+1)^{72}$, we need 625 to be between n^3 and $(n+1)^3$. The perfect cubes near 625 are $8^3 = 512$ and $9^3 = 729$. So n is $\boxed{8}$.

2.76 The one-digit positive perfect squares are $1^2 = 1$, $2^2 = 4$, and $3^2 = 9$. There are 3 such squares. So they contribute 3 digits to the sequence.

The two-digit perfect squares are $4^2 = 16$, $5^2 = 25$, $6^2 = 36$, $7^2 = 49$, $8^2 = 64$, and $9^2 = 81$. There are 6 such squares. So they contribute $6 \cdot 2 = 12$ digits to the sequence.

The three-digit perfect squares are $10^2 = 100$, $11^2 = 121$, and so on through $31^2 = 961$. There are 22 such squares. So they contribute $22 \cdot 3 = 66$ digits to the sequence.

The four-digit perfect squares through 1225 are $32^2 = 1024$, $33^2 = 1089$, $34^2 = 1156$, and $35^2 = 1225$. There are 4 such squares. So they contribute $4 \cdot 4 = 16$ digits to the sequence.

Adding up all four cases, we find that the total number of digits in the sequence is

$$3 + 12 + 66 + 16 = \boxed{97}.$$

CHAPTER 3

Number Theory

24® Cards

First Card $3 \cdot 3 \cdot 3 - 3$.

Second Card $(3 \cdot 5 - 7) \cdot 3$.

Third Card $(7 - 3) \cdot 4 + 8$.

Fourth Card $(5 \cdot 13 + 7) \div 3$.

Exercises for Section 3.1

3.1.1 We can list positive multiples of 6 by starting from 6 and counting by 6's:

$$\boxed{6, 12, 18, 24, 30, 36, 42, 48, 54, 60}.$$

3.1.2 We find the positive multiples of 7 less than 100 by starting from 7 and counting by 7's. We compute the sum of the digits of each, and find that the two with digit sum 10 are $\boxed{28 \text{ and } 91}$.

3.1.3 The desired multiples of 13 are 13, 26, 39, 52, 65, 78, and 91. These have sum

$$13 \cdot 1 + 13 \cdot 2 + 13 \cdot 3 + 13 \cdot 4 + 13 \cdot 5 + 13 \cdot 6 + 13 \cdot 7$$

$$= 13(1 + 2 + 3 + 4 + 5 + 6 + 7) = 13(28) = \boxed{364}.$$

3.1.4 The sum of the desired number and 173 must be the smallest multiple of 20 that is greater than 173, which is 180. So, the desired number is $180 - 173 = \boxed{7}$.

3.1.5 Since 1000 divided by 17 leaves remainder 14, we know that 1000 is 14 more than a multiple of 17. Therefore, $1000 - 14 = 986$ is a multiple of 17, and $986 + 17 = \boxed{1003}$ is the smallest four-digit multiple of 17. (We also might have noticed that since 1000 divided by 17 leaves a remainder of 14, we know that 1000 is 3 less than a multiple of 17.)

3.1.6   The greatest four-digit number is 9999. Dividing 9999 by 18 leaves a remainder of 9, so $9999 - 9 =$ $\boxed{9990}$ is the largest four-digit multiple of 18.

Here's a quick way to see that 9999 divided by 18 has remainder 9. The number 9999 is clearly a multiple of 9, since $9999 = 9 \cdot 1111$. But 9999 is not a multiple of 2. The multiples of 9 alternate between even numbers that are multiples of 18 (like 0, 18, 36, 54) and odd numbers that are 9 more than multiples of 18 (like 9, 27, 45). So, 9999 must be 9 more than a multiple of 18.

3.1.7   First, we find the greatest three-digit multiple of 33. Since 999 divided by 33 has remainder 9, we know that 990 is the largest three-digit multiple of 33. But the digits of 990 are not all different. So, we consider the next smaller multiple of 33, which is $990 - 33 = 957$. The digits of 957 are different, so $\boxed{957}$ is the greatest three-digit multiple of 33 that can be written with three different digits.

3.1.8   The smallest multiple of 11 between 17 and 2678 is $2 \cdot 11 = 22$. But what is the largest? Dividing 2678 by 11 gives quotient 243 and remainder 5, so the largest multiple of 11 between 17 and 2678 is $243 \cdot 11$. Therefore, the integers we must count are $2 \cdot 11, 3 \cdot 11, 4 \cdot 11, \ldots, 243 \cdot 11$. This list has all of the first 243 positive multiples of 11 except 11 itself, so there are $\boxed{242}$ integers between 17 and 2678 that are multiples of 11.

Exercises for Section 3.2

3.2.1   Numbers that end in 0 or 5 are divisible by 5. All other numbers are not divisible by 5. So, $\boxed{560{,}335 \text{ and } 60{,}231{,}060}$ are the only numbers in the list that are divisible by 5.

3.2.2   If the last two digits of a number form a number that is a multiple of 4, then the original number is a multiple of 4. Otherwise, the original number is not a multiple of 4. The only two numbers in the list for which the last two digits form a multiple of 4 are $\boxed{46{,}624 \text{ and } 60{,}231{,}060}$.

3.2.3   To test if a number is divisible by 3, we sum the digits of the number. We find:

$$
\begin{aligned}
46{,}624 : \quad & 4 + 6 + 6 + 2 + 4 = 22, \\
560{,}335 : \quad & 5 + 6 + 0 + 3 + 3 + 5 = 22, \\
60{,}231{,}060 : \quad & 6 + 0 + 2 + 3 + 1 + 0 + 6 + 0 = 18, \\
9{,}671{,}118 : \quad & 9 + 6 + 7 + 1 + 1 + 1 + 8 = 33.
\end{aligned}
$$

If the sum is a multiple of 3, then the original number is divisible by 3. Otherwise, the original number is not divisible by 3. So, the only two numbers that are divisible by 3 are $\boxed{60{,}231{,}060 \text{ and } 9{,}671{,}118}$.

3.2.4   Four of the five numbers can be written as the sum of numbers that are obviously multiples of 7:

$$
\begin{aligned}
7{,}000{,}014 &= 7{,}000{,}000 + 14, \\
14{,}035 &= 14{,}000 + 35, \\
7{,}777{,}728 &= 7{,}777{,}700 + 28, \\
49{,}763 &= 49{,}000 + 700 + 63.
\end{aligned}
$$

Any sum of multiples of 7 must also be a multiple of 7, so these four numbers are multiples of 7. The remaining number in the list is 1 less than 42,721,035. We have

$$42{,}721{,}035 = 42{,}000{,}000 + 700{,}000 + 21{,}000 + 35.$$

The numbers on the right side are clearly multiples of 7, so 42,721,035 is a multiple of 7. Since 42,721,034 is 1 less than 42,721,035, we know that $\boxed{42{,}721{,}034}$ is not divisible by 7.

3.2.5 If the last two digits of a number form a number that is a multiple of 4, then the original number is a multiple of 4. Otherwise, the original number is not a multiple of 4. The only two-digit multiples of 4 with 2 in the units place are 12, 32, 52, 72, and 92. There are 5 such multiples for each hundreds digit from 0 (the two-digit numbers) through 3. So there are $5 + 5 + 5 + 5 = \boxed{20}$ numbers less than 400 that are divisible by 4. For the record, the numbers are

$$12, 32, 52, 72, 92, 112, 132, 152, 172, 192, 212, 232, 252, 272, 292, 312, 332, 352, 372, 392.$$

3.2.6 Since 3D8 is divisible by 9, the sum of its digits must be divisible by 9. The sum of the digits of 3D8 is $3 + D + 8 = D + 11$. The only digit D for which $D + 11$ is divisible by 9 is 7, so $D = 7$ and we have $ABC - 378 = 269$. Therefore, $ABC = 378 + 269 = \boxed{647}$.

3.2.7 In order for $214{,}d07$ to be divisible by 3, the sum of its digits must be divisible by 3. This sum is $2 + 1 + 4 + d + 0 + 7 = d + 14$. The only digits d for which $d + 14$ is divisible by 3 are 1, 4, and 7. The largest of these is $\boxed{7}$.

3.2.8 We found that a number is divisible by $2 = 2^1$ if its final 1 digit is divisible by 2. We found that a number is divisible by $4 = 2^2$ if its final 2 digits form a number that is divisible by 4. We might guess that a number is divisible by $8 = 2^3$ if its final 3 digits form a number that is divisible by 8.

To prove that our guess is correct, we look back to the methods we used to prove our rules for 2 and 4. The key step in our proof for the divisibility by 2 test is the fact that 10 is divisible by 2. The key step in our proof for the divisibility by 4 test is the fact that $100 = 10^2$ is divisible by 4. So, we expect that a key step in proving the rule for divisibility by 8 is the fact that $1000 = 10^3$ is divisible by 8.

We do indeed have $1000 \div 8 = 125$, so 1000 is a multiple of 8. Therefore, any multiple of 1000 is a multiple of 8. Any number can be written as the sum of a multiple of 1000 and a nonnegative number less than 1000. For example,

$$45{,}672 = 45{,}000 + 672, \qquad 6{,}012 = 6{,}000 + 012, \qquad 4{,}562{,}904 = 4{,}562{,}000 + 904.$$

The multiple of 1000 in such a sum is a multiple of 8, so if the other number in such a sum is also a multiple of 8, then the original number is a multiple of 8. Otherwise, the original number is not a multiple of 8. So, a rule for divisibility by 8 is:

If the final 3 digits of a number form a multiple of 8, then the original number is a multiple of 8. Otherwise, the original number is not a multiple of 8.

Exercises for Section 3.3

3.3.1 We can read the primes between 80 and 90 off of the Sieve of Eratosthenes in the text. Or, we can test each odd number between 80 and 90 to see if it's prime. (The even numbers are obviously not prime.) 81 and 87 are divisible by 3, and 85 is divisible by 5. Both 83 and 89 are prime, and their sum is $\boxed{172}$.

3.3.2 All of the even numbers are multiples of 2, and 75 is a multiple of 5. That leaves 71, 73, 77, and 79. We find that $77 = 7 \cdot 11$ is not prime and is not a multiple of 2, 3, or 5. Meanwhile, 71, 73, and 79 are all prime. Therefore, $\boxed{77}$ is the only number between 70 and 80 that is not prime and is not a multiple of 2, 3, or 5.

3.3.3 1000 divided by 8 is 125. So, we seek the largest prime number that is less than 125. Even numbers greater than 2 aren't prime, so we only have to check odd numbers. We have $123 = 3 \cdot 41$, $121 = 11^2$, $119 = 7 \cdot 17$, $117 = 3 \cdot 39$, and $115 = 5 \cdot 23$, so these are all composite. Since 113 isn't divisible by 2, 3, 5, or 7, and 11^2 is greater than 113, we know that $\boxed{113}$ is the desired prime.

3.3.4 If two integers add to 40, then either both are odd or both are even. We don't have to check the even pairs, since at least one of two even numbers that add to 40 is a composite multiple of 2. So, we check all the pairs of odd numbers that add to 40:

$$40 = 1 + 39 = 3 + 37 = 5 + 35 = 7 + 33 = 9 + 31$$
$$= 11 + 29 = 13 + 27 = 15 + 25 = 17 + 23 = 19 + 21.$$

The only sums that consist of two prime numbers are $3 + 37$, $11 + 29$, and $17 + 23$. So, there are $\boxed{3}$ pairs of prime numbers that add to 40.

3.3.5

(a) The product of all prime numbers between 1 and 80 includes 2 and 5. Multiplying these gives 10, which means the product of all the primes between 1 and 80 is a multiple of 10. Therefore, the remainder is $\boxed{0}$ when this product is divided by 10.

(b) The product of all prime numbers between 1 and 80 includes exactly one 2, and has no other even numbers. So, we know that the product is even, but is not a multiple of 4 (which is 2^2). When an odd number is divided by 4, the remainder is 1 or 3, and when an even number is divided by 4, the remainder is 0 or 2. The remainder is only 0 if the number is a multiple of 4. We know that the product in the problem is even but not a multiple of 4. So, the remainder is $\boxed{2}$ when the product is divided by 4.

3.3.6 The sum of three integers is even only if the three integers are even or if two of the integers are odd and the other even. The only even prime is 2, so the only integer that is the sum of three even primes is $2 + 2 + 2 = 6$. So, we must have 2 and two odd primes that add to 20. The pairs of odd integers that add to 20 are

$$20 = 1 + 19 = 3 + 17 = 5 + 15 = 7 + 13 = 9 + 11.$$

Two of these sums consist of two prime integers ($3 + 17$ and $7 + 13$), so there are $\boxed{2}$ groups of three primes whose sum is 22.

3.3.7 The sum of two odd numbers is even, so $5P$ and $7Q$ cannot both be odd. Therefore, either $5P$ or $7Q$ is even, which means P or Q is 2.

If P is 2, then the equation becomes $10 + 7Q = 109$. This means that $7Q$ must be 99. But 99 is not a multiple of 7, so there is no integer Q for which $7Q$ is 99.

If Q is 2, then the equation becomes $5P + 14 = 109$. This means that $5P$ is 95, so P must be $95 \div 5 = \boxed{19}$

Exercises for Section 3.4

3.4.1

(a) $72 = 8 \cdot 9 = \boxed{2^3 \cdot 3^2}$.

(b) $210 = 21 \cdot 10 = (3 \cdot 7) \cdot (2 \cdot 5) = \boxed{2^1 \cdot 3^1 \cdot 5^1 \cdot 7^1}$.

(c) The sum of the digits of 243 is 9, so 243 is divisible by 9. We find $243 = 9 \cdot 27 = 3^2 \cdot (3 \cdot 9) = 3^2 \cdot (3 \cdot 3^2) = \boxed{3^5}$.

(d) Since the last digit of 539 is 9, we know that 539 isn't divisible by 2 or 5. The sum of the digits of 539 is 17, which isn't a multiple of 3, so we know that 539 is not a multiple of 3. Dividing 539 by 7 gives 77, and we have $539 = 7 \cdot 77 = 7 \cdot (7 \cdot 11) = \boxed{7^2 \cdot 11^1}$.

(e) We have $5525 = 5 \cdot 1105 = 5 \cdot (5 \cdot 221)$. The last digit of 221 tells us that 221 is not divisible by 2 or 5, and the sum of the digits of 221 tells us that 221 is not divisible by 3. Dividing 221 by 7 and by 11 both give a nonzero remainder, so 221 is not divisible by 7 or 11. Finally, we find $221 = 13 \cdot 17$, so $5525 = 5 \cdot 5 \cdot 221 = \boxed{5^2 \cdot 13^1 \cdot 17^1}$.

(f) The number formed by the last two digits of 26136 is divisible by 4, so 26136 is divisible by 4, and we have $26136 = 4 \cdot 6534 = 2^2 \cdot (2 \cdot 3267)$. The sum of the digits of 3267 is a multiple of 9, so 3267 is divisible by 9. This gives us

$$26136 = 2^2 \cdot 2 \cdot 3267 = 2^3 \cdot (9 \cdot 363) = 2^3 \cdot 3^2 \cdot (3 \cdot 121) = \boxed{2^3 \cdot 3^3 \cdot 11^2}.$$

3.4.2 We have $6886 = 2 \cdot 3443$. We find that 3443 is not divisible by 3, 5, or 7, but $3443 = 11 \cdot 313$. Dividing 313 by 11, 13, or 17 leaves a nonzero remainder, so 313 is not divisible by any of these. Since $19^2 = 361$, which is greater than 313, we don't have to test if 313 is divisible by any more primes—we know that 313 is prime. This gives us the prime factorization $6886 = 2^1 \cdot 11^1 \cdot 313^1$, so $\boxed{313}$ is the largest prime factor.

3.4.3 The prime factorization of 54,000 is

$$54000 = 54 \cdot 1000 = (6 \cdot 9) \cdot 10^3 = (2 \cdot 3 \cdot 3^2) \cdot (2 \cdot 5)^3 = 2 \cdot 3^3 \cdot 2^3 \cdot 5^3 = 2^4 \cdot 3^3 \cdot 5^3.$$

So, we have $x = 4$, $y = 3$, and $z = 3$, which means $x + y + z = \boxed{10}$.

3.4.4

(a) The four smallest primes are 2, 3, 5 and 7. Since the square is divisible by each of these primes, these four primes must appear in the prime factorization of the square. We also know that each prime in the prime factorization of a perfect square must have an even exponent. We take the smallest such powers, so

$$2^2 \cdot 3^2 \cdot 5^2 \cdot 7^2 = (2 \cdot 3 \cdot 5 \cdot 7)^2 = (210)^2 = \boxed{44100}$$

is the smallest positive perfect square that is divisible by the four smallest primes.

(b) 0 is a perfect square, and 0 is divisible by every prime. So, if we remove "positive" from the question in part (a), the answer would be $\boxed{0}$.

3.4.5 Prime factorizations are often helpful in problems about products of integers. The prime factorization of 504 is $504 = 2 \cdot 252 = 2 \cdot 2 \cdot 126 = 2^2 \cdot 2 \cdot 63 = 2^3 \cdot 9 \cdot 7 = 2^3 \cdot 3^2 \cdot 7^1$. So, we seek two numbers whose product is $2^3 \cdot 3^2 \cdot 7^1$. Together the two numbers have three 2's, two 3's, and one 7 in their prime factorizations. We know that both numbers are multiples of 6, which is $2 \cdot 3$. Therefore, each number has a 2 and a 3 in its prime factorization, which accounts for two of the 2's and both of the 3's. We also know that neither number is 6, so each number has another prime factor in addition to one 2 and one 3. So, we have

$$(2 \cdot 3 \cdot \_) \cdot (2 \cdot 3 \cdot \_) = 2^3 \cdot 3^2 \cdot 7^1.$$

Therefore, one blank must have 2 and the other must have 7, which means the larger of the two numbers is $2 \cdot 3 \cdot 7 = \boxed{42}$.

3.4.6 Again, we have a problem about a product of integers, so we start with the prime factorization of 858. We have

$$858 = 2 \cdot 429 = 2 \cdot 3 \cdot 143 = 2 \cdot 3 \cdot 11 \cdot 13.$$

We can write 858 as the product of 2 two-digit numbers by pairing each of the one-digit primes with one of the two-digit primes. There are two ways we can do this, $858 = (2 \cdot 11) \cdot (3 \cdot 13) = 22 \cdot 39$ and $858 = (2 \cdot 13) \cdot (3 \cdot 11) = 26 \cdot 33$. These give us the sums $22 + 39 = 61$ and $26 + 33 = 59$.

We can also group either 11 or 13 with both 2 and 3, producing $858 = (2 \cdot 3 \cdot 11) \cdot (13) = 66 \cdot 13$ and $(11) \cdot (2 \cdot 3 \cdot 13) = 11 \cdot 78$. These give us the sums $66 + 13 = 79$ and $11 + 78 = 89$.

The largest of our four sums is $\boxed{89}$.

3.4.7 We'll look at the prime factorizations of the first few cubes after 1^3 to see if we find anything interesting:

$$2^3, \qquad 3^3, \qquad 4^3 = (2^2)^3 = 2^{2 \cdot 3} = 2^6, \qquad 5^3, \qquad 6^3 = (2 \cdot 3)^3 = 2^3 \cdot 3^3.$$

It looks like all the exponents in the prime factorization of a perfect cube must be multiples of 3. The examples of 4^3 and 6^3 above show why this occurs. To get the prime factorization of any cube n^3, we can cube the prime factorization of n. When we cube such a prime factorization, we multiply the exponents in the prime factorization by 3. For example, the cube of $2^2 \cdot 7^1$ is

$$(2^2 \cdot 7^1)^3 = 2^{2 \cdot 3} \cdot 7^{1 \cdot 3} = 2^6 \cdot 7^3.$$

Therefore, each of the exponents in the prime factorization of a cube must be a multiple of 3. Similarly, if the exponents in the prime factorization of a number are all multiples of 3, then we can use the power of power law to see that the number is a cube. For example, we have

$$2^6 \cdot 3^6 \cdot 5^{12} = 2^{2 \cdot 3} \cdot 3^{2 \cdot 3} \cdot 5^{4 \cdot 3} = (2^2 \cdot 3^2 \cdot 5^4)^3,$$

so $2^6 \cdot 3^6 \cdot 5^{12}$ is the cube of $2^2 \cdot 3^2 \cdot 5^4$.

3.4.8

(a) Each multiple of 5 contributes a 5 to the prime factorization of 40!. There are 8 multiples of 5 from 1 to 40 ($1 \cdot 5, 2 \cdot 5, \ldots, 8 \cdot 5$). However, we have to be careful. Since $25 = 5 \cdot 5$, the 25 in 40! contributes two 5's to the prime factorization of 40!. We counted one already when counting the multiples of 5, but the second 5 in $25 = 5 \cdot 5$ gives us an extra 5. So, there are $8 + 1 = 9$ 5's in the prime factorization of 40!, which means that the desired power of 5 is $\boxed{5^9}$.

(b) An integer that has a terminal 0 is a multiple of 10, and the integer can be written as the product of 10 and another integer that has one fewer terminal 0. We can continue this process, pulling out as many 10's as the number has terminal zeros. For example,

$$56774000 = 10 \cdot 10 \cdot 10 \cdot 56774.$$

So, to count the number of terminal zeros that 40! has, we must count the greatest number of 10's we can form from the numbers in the product $1 \times 2 \times 3 \times \cdots \times 40$. We form 10's by combining factors of 2 with factors of 5, since $2 \cdot 5 = 10$. So, we focus on how many factors of 2 and factors of 5 there are in the prime factorization of 40!, to see how many 10's we can form.

There are at least 20 factors of 2 in the prime factorization of 40!, since there are 20 positive even numbers from 2 to 40. Therefore, we can pair each of the 9 factors of 5 from part (a) with a factor of 2 to make a factor of 10. So, 40! has $\boxed{9}$ terminal zeros. We can see that 40! does not have 10 or more terminal zeros by noting that a number with 10 or more terminal zeros is a multiple of $10^{10} = (2 \cdot 5)^{10} = 2^{10} \cdot 5^{10}$. Therefore, a number with at least 10 terminal zeros must have at least ten 5's in its prime factorization. But 40! only has nine 5's in its prime factorization, so it can't have 10 or more terminal zeros.

Exercises for Section 3.5

3.5.1 As in the text, we will bold and underline the powers of primes we use to determine the least common multiple in each part in which we use prime factorizations.

(a) We don't need prime factorizations for this part. The smallest two positive multiples of 21 are 21 and 42. Since 42 is also a multiple of 14 (and 21 is not), we have $\text{lcm}[14, 21] = \boxed{42}$.

(b) We have $24 = 2^3 \cdot \underline{\mathbf{3^1}}$ and $32 = \underline{\mathbf{2^5}}$, so $\text{lcm}[24, 32] = 2^5 \cdot 3^1 = \boxed{96}$.

   We also might have noted that $\text{lcm}[24, 32] = \text{lcm}[8 \cdot 3, 8 \cdot 4] = 8\,\text{lcm}[3, 4] = 8 \cdot 12 = 96$.

(c) We have $27 = \underline{\mathbf{3^3}}$ and $63 = 9 \cdot 7 = 3^2 \cdot \underline{\mathbf{7^1}}$, so $\text{lcm}[27, 63] = 3^3 \cdot 7^1 = \boxed{189}$.

   We also might have noted that $\text{lcm}[27, 63] = \text{lcm}[9 \cdot 3, 9 \cdot 7] = 9\,\text{lcm}[3, 7] = 9 \cdot 21 = 189$.

(d) We have $54 = 6 \cdot 9 = 2 \cdot 3 \cdot 3^3 = 2^1 \cdot \underline{\mathbf{3^3}}$ and $60 = 6 \cdot 10 = 2 \cdot 3 \cdot 2 \cdot 5 = \underline{\mathbf{2^2}} \cdot 3^1 \cdot \underline{\mathbf{5^1}}$, so $\text{lcm}[54, 60] = 2^2 \cdot 3^3 \cdot 5^1 = \boxed{540}$.

   We also might have noted that $\text{lcm}[54, 60] = \text{lcm}[6 \cdot 9, 6 \cdot 10] = 6\,\text{lcm}[9, 10] = 6 \cdot 90 = 540$.

(e) We have $72 = 8 \cdot 9 = \underline{\mathbf{2^3}} \cdot 3^2$ and $108 = 2 \cdot 54 = 2 \cdot 6 \cdot 9 = 2 \cdot 2 \cdot 3 \cdot 3^2 = 2^2 \cdot \underline{\mathbf{3^3}}$. So, $\text{lcm}[72, 108] = 2^3 \cdot 3^3 = 8 \cdot 27 = \boxed{216}$.

   We also might have noted that $\text{lcm}[72, 108] = \text{lcm}[36 \cdot 2, 36 \cdot 3] = 36\,\text{lcm}[2, 3] = 36 \cdot 6 = 216$.

(f) We have $5096 = 4 \cdot 1274 = 2^2 \cdot 2 \cdot 637 = 2^3 \cdot 7 \cdot 91 = 2^3 \cdot 7 \cdot 7 \cdot 13 = \underline{\mathbf{2^3}} \cdot \underline{\mathbf{7^2}} \cdot \underline{\mathbf{13^1}}$ and $117 = 9 \cdot 13 = \underline{\mathbf{3^2}} \cdot 13^1$, so $\text{lcm}[5096, 117] = 2^3 \cdot 3^2 \cdot 7^2 \cdot 13^1 = \boxed{45864}$.

3.5.2   As in the text, we will bold and underline the powers of primes we use to determine the least common multiple in each part.

(a)   We have $12 = \underline{\mathbf{2^2}} \cdot 3^1$, $18 = 2^1 \cdot \underline{\mathbf{3^2}}$, and $30 = 2^1 \cdot 3^1 \cdot \underline{\mathbf{5^1}}$, so $\text{lcm}[12, 18, 30] = 2^2 \cdot 3^2 \cdot 5^1 = \boxed{180}$.

(b)   We have $36 = 6^2 = 2^2 \cdot 3^2$, $48 = 4 \cdot 12 = 2^2 \cdot (2^2 \cdot 3) = \underline{\mathbf{2^4}} \cdot 3^1$, and $27 = \underline{\mathbf{3^3}}$, so $\text{lcm}[36, 48, 27] = 2^4 \cdot 3^3 = \boxed{432}$.

(c)   We have $\underline{24} = 2^3 \cdot 3^1$, $54 = 6 \cdot 9 = 2^1 \cdot \underline{\mathbf{3^3}}$, and $144 = 12^2 = (2^2 \cdot 3)^2 = \underline{\mathbf{2^4}} \cdot 3^2$, so $\text{lcm}[24, 54, 144] = 2^4 \cdot 3^3 = \boxed{432}$.

3.5.3   There is no greatest common multiple of a group of positive numbers! Every multiple of the least common multiple is a common multiple, so there is no limit to how large a common multiple we can find.

3.5.4   The common multiples of 8 and 12 are the multiples of $\text{lcm}[8, 12]$. Since $\text{lcm}[8, 12] = 24$, we seek the greatest multiple of 24 that is less than 200. We have $8 \cdot 24 = 192$ and $9 \cdot 24 = 216$, so the desired common multiple is $\boxed{192}$.

3.5.5   The common multiples of 2, 3, 4, 5, 6, and 7 are the multiples of $\text{lcm}[2, 3, 4, 5, 6, 7]$. Among the prime factorizations of these six numbers, the only primes are 2, 3, 5, and 7. Each is raised to at most the 1st power except for 2, which is raised to the 2nd power in 4's prime factorization. So, we have $\text{lcm}[2, 3, 4, 5, 6, 7] = 2^2 \cdot 3 \cdot 5 \cdot 7 = 420$. Therefore, we seek the smallest 4-digit multiple of 420. The three smallest positive multiples of 420 are 420, 840, and 1260, so $\boxed{1260}$ is our desired multiple.

3.5.6   Alex's numbers are the positive multiples of 6 and Matthew's are the positive multiples of 4, so the numbers they both hit are the positive common multiples of 4 and 6. The common multiples of 4 and 6 are the multiples of $\text{lcm}[4, 6]$, which is 12. So, the numbers counted by both Alex and Matthew are the positive multiples of 12 that are less than or equal to 2400. Dividing 12 into 2400 gives a quotient of 200, so the positive multiples of 12 that they both hit are $1 \cdot 12, 2 \cdot 12, 3 \cdot 12, \ldots, 200 \cdot 12$. There are $\boxed{200}$ such multiples.

3.5.7   The number of people at the party is 1 more than a multiple of each of 2, 3, 5, and 7. So, 1 less than the number of people at the party is a multiple of each of 2, 3, 5, and 7. This means that 1 less than the smallest possible number of people at the party is the least common multiple of 2, 3, 5, and 7. Since these four numbers have no prime factors in common, we have $\text{lcm}[2, 3, 5, 7] = 2 \cdot 3 \cdot 5 \cdot 7 = 210$. This gives us $210 + 1 = \boxed{211}$ as the smallest possible number of people at the party.

3.5.8   First, we consider each pair of owls separately. The 3-hour owl and the 8-hour owl hoot together every $\text{lcm}[3, 8] = 24$ hours. The 8-hour owl and the 12-hour owl hoot together every $\text{lcm}[8, 12] = 24$ hours, as well. The 3-hour owl and the 12-hour owl hoot together every 12 hours. Combining these, we see that every 12 hours, the 3-hour and the 12-hour owl hoot together, and every 24 hours, all three hoot together. At no other times does more than one owl hoot at the same time.

Since multiples of 24 are also multiples of 12, the only times that exactly two owls hoot together are at multiples of 12 hours that are not multiples of 24 hours. The multiples of 12 that are less than 80 are 12, 24, 36, 48, 60, 72. Excluding the multiples of 24 leaves 12, 36, and 60, leaving $\boxed{3}$ times at which exactly 2 of the owls hoot together.

Exercises for Section 3.6

3.6.1   We have $6 = 1 \cdot 6 = 2 \cdot 3$, so the product of the positive factors is $1 \cdot 6 \cdot 2 \cdot 3 = 6 \cdot 6 = 6^2 = \boxed{36}$.

3.6.2   We have $18 = 1 \cdot 18 = 2 \cdot 9 = 3 \cdot 6$, so the sum of the positive divisors of 18 is $1+2+3+6+9+18 = \boxed{39}$.

3.6.3   We have $32 = 1 \cdot 32 = 2 \cdot 16 = 4 \cdot 8$, so 32 has $\boxed{6}$ divisors.

3.6.4   The values of n such that $28 \div n$ is an integer are the divisors of 28. However, we must be careful; the problem didn't specify that n must be positive. So, we must count both the positive and the negative divisors. The negative divisors are simply the negations of the positive divisors, so we'll count the positive divisors and double the total. We have

$$28 = 1 \cdot 28 = 2 \cdot 14 = 4 \cdot 7,$$

so 28 has 6 positive divisors. This gives us $2 \cdot 6 = \boxed{12}$ values of n for which $28 \div n$ is an integer.

3.6.5   We have $2005 = 5 \cdot 401$. Since 401 is not divisible by any of the primes less than 20, and 23^2 is greater than 401, we know that 401 is prime. So, the only possibility for the desired sum is $5 + 401 = \boxed{406}$.

3.6.6 D is a one-digit divisor of 1673. We know that D isn't 1 because D times a three-digit number equals 1673. So, we seek a one-digit divisor of 1673 besides 1. Since 1673 is odd, we know that D can't be even. We can also use divisibility tests to quickly eliminate 3, 5, and 9 as candidates for D. That leaves 7. We have $1673 \div 7 = 239$, so the desired ABC is $\boxed{239}$.

3.6.7   Let A be the units digit in the year $199A$. In order for $199A$ to be lucky, the two-digit number $9A$ must be the product of two integers in which one integer is from 1 to 12 (the month) and the other integer is no greater than the number of days in the corresponding month.

We have $90 = 3 \cdot 30$, so 1990 is lucky. (3/30/90)

We have $91 = 7 \cdot 13$, so 1991 is lucky. (7/13/91)

We have $92 = 4 \cdot 23$, so 1992 is lucky. (4/23/92)

We have $93 = 3 \cdot 31$, so 1993 is lucky. (3/31/93)

That leaves $\boxed{1994}$ as the year that is not lucky. To see that 1994 is not lucky, note that the only pairs of positive divisors of 94 that multiply to 94 are $94 = 1 \cdot 94 = 2 \cdot 47$. Neither of these pairs leads to a valid date.

3.6.8   Perfect squares have an odd number of positive divisors, and integers that are not perfect squares have an even number of positive divisors. So, we seek the largest square that is less than 100, which is $\boxed{81}$.

3.6.9   The sum is 12,111. The largest divisor of 12,111 is 12,111 itself, but that's larger than 10,000. The largest divisor of 12,111 and the smallest positive divisor have product 12,111. This gives us the pair of divisors 12,111 and 1. Similarly, the product of the next largest divisor and the next smallest positive divisor is 12,111. So, we can find the next largest divisor by finding the next smallest divisor.

Since 12,111 is odd, 2 is not a divisor. But the sum of the digits of 12,111 is a multiple of 3, so we know that 3 is a divisor of 12,111. We have $12,111 = 3 \cdot 4,037$, so the greatest factor that is less than 10,000 is $\boxed{4,037}$.

Exercises for Section 3.7

3.7.1 As in the text, we will bold and underline the powers of primes we use to determine the greatest common divisor in each part in which we use prime factorizations.

(a) We have $32 = 2^5$ and $48 = 4 \cdot 12 = 2^2 \cdot 2^2 \cdot 3 = \mathbf{\underline{2^4}} \cdot 3^1$, so $\gcd(32, 48) = 2^4 = \boxed{16}$.

   We also might have noted that $\gcd(32, 48) = \gcd(16 \cdot 2, 16 \cdot 3) = 16 \gcd(2, 3) = 16 \cdot 1 = 16$.

(b) We have $99 = 3^2 \cdot 11^1$ and $100 = 10^2 = 2^2 \cdot 5^2$, so $\gcd(99, 100) = \boxed{1}$. We also might have noticed that it's impossible for two integers that are 1 apart to have any positive divisors in common besides 1. If we divide the larger number by any divisor of the smaller (besides 1), then the remainder will always be 1. So, the numbers cannot have any positive divisors in common besides 1.

(c) We have $315 = 3 \cdot 105 = 3 \cdot 5 \cdot 21 = 3 \cdot 5 \cdot 3 \cdot 7 = \mathbf{\underline{3^2}} \cdot 5^1 \cdot 7^1$ and $108 = 4 \cdot 27 = 2^2 \cdot 3^3$, so $\gcd(315, 108) = 3^2 = \boxed{9}$.

(d) We have $99 = 3^2 \cdot \mathbf{\underline{11^1}}$ and $726 = 6 \cdot 121 = 2^1 \cdot \mathbf{\underline{3^1}} \cdot 11^2$, so $\gcd(99, 726) = 3^1 \cdot 11^1 = \boxed{33}$.

(e) We have $365 = 5^1 \cdot 73^1$ and $1985 = 5 \cdot 397$. Since 397 is not divisible by 73, we know that 365 and 1985 don't have any prime factors in common besides 5. So, $\gcd(365, 1985) = \boxed{5}$.

(f) We have $9009 = 9 \cdot 1001$ and $14014 = 14 \cdot 1001$, so

$$\gcd(9009, 14014) = \gcd(1001 \cdot 9, 1001 \cdot 14) = 1001 \gcd(9, 14) = 1001 \cdot 1 = \boxed{1001}.$$

3.7.2 We have

$$36 = 6^2 = 2^2 \cdot 3^2, \qquad 90 = 9 \cdot 10 = 3^2 \cdot 2 \cdot 5 = \mathbf{\underline{2^1}} \cdot 3^2 \cdot 5^1, \qquad 60 = 6 \cdot 10 = 2^2 \cdot \mathbf{\underline{3^1}} \cdot 5^1,$$

so $\gcd(36, 90, 75) = 2^1 \cdot 3^1 = \boxed{6}$.

3.7.3 The positive common divisors of two numbers are the positive divisors of the greatest common divisor of the two numbers. So, the positive common divisors of a and b are the positive divisors of 8, which are $\boxed{1, 2, 4, \text{ and } 8}$.

3.7.4 Two numbers are relatively prime if their greatest common divisor is 1. Since $\gcd(5, 18) = 1$, we know that $\boxed{5 \text{ and } 18}$ are relatively prime. In each of the other three pairs, 3 is a common divisor, so the greatest common divisor is greater than 1.

3.7.5 Because $\gcd(3, 4) = 1$, a number that is a multiple of both 3 and 4 must be a multiple of 12, and any number that is a multiple of 12 must be a multiple of both 3 and 4. So, we apply our divisibility tests for 3 and for 4. The number formed by the last two digits of $661,17A$ is $7A$. The only values of A for which this number is a multiple of 4 are 2 and 6. The sum of the digits of $661,17A$ is $6 + 6 + 1 + 1 + 7 + A = 21 + A$. The only values of A for which this is a multiple of 3 are 0, 3, 6, and 9. So, the only value of A that makes $661,17A$ divisible by both 3 and 4 is $\boxed{6}$.

3.7.6 $\boxed{\text{No}}$. Consider the integers 2, 3, and 4. The numbers 2 and 3 are relatively prime, as are 3 and 4. But 2 and 4 are not relatively prime, since $\gcd(2, 4) = 2$.

3.7.7 We start by finding the positive divisors of 175. We have $175 = 1 \cdot 175 = 5 \cdot 35 = 7 \cdot 25$. So, the only divisors besides 7 that have a positive factor in common with 7 are 35 and 175. Therefore, these must be the integers adjacent to 7. So, the sum of the integers adjacent to 7 is $35 + 175 = \boxed{210}$.

Review Problems

3.47

(a) We have $693 = 3 \cdot 231 = 3 \cdot 3 \cdot 77 = \boxed{3^2 \cdot 7^1 \cdot 11^1}$. We also might have noticed that $693 = 700 - 7 = 7(100 - 1) = 7(99) = 7(3^2 \cdot 11) = 3^2 \cdot 7^1 \cdot 11^1$.

(b) We can use our divisibility tests to see quickly that 5423 is not divisible by 2, 3, or 5. Dividing 5423 by 7 leaves a remainder of 5, so 5423 is not divisible by 7. Next, we try 11 and find $5423 = 11 \cdot 493$. Now, we search for prime factors of 493. Dividing 493 by 11 leaves a remainder of 9. Dividing 493 by 13 leaves a remainder of 12. Next, we try 17, and find $493 = 17 \cdot 29$. Since 29 is prime, we have our prime factorization: $5423 = 11 \cdot 493 = \boxed{11^1 \cdot 17^1 \cdot 29^1}$.

(c) We have $35100 = 351 \cdot 100 = (9 \cdot 39) \cdot 10^2 = 3^2 \cdot 3 \cdot 13 \cdot (2 \cdot 5)^2 = \boxed{2^2 \cdot 3^3 \cdot 5^2 \cdot 13^1}$.

3.48 Since 99 is a multiple of 9, so is (99)(237). Therefore, the remainder is $\boxed{0}$ when we divide (99)(237) by 9.

3.49 Since 103 divided by 6 has remainder 1, we know that 102 is a multiple of 6. So, the next two multiples of 6 are $102 + 6 = 108$ and $108 + 6 = 114$. Their sum is $108 + 114 = \boxed{222}$.

3.50 Since 2000 divided by 73 has remainder 29, we know that 2000 is 29 more than a multiple of 73. Therefore, the largest multiple of 73 less than 2000 is $2000 - 29 = \boxed{1971}$.

We might also have started by searching for multiples of 73 near 2000. Since $73 \cdot 3 = 219$, we have $73 \cdot 30 = 2190$, so 2190 is a multiple of 73. Therefore $2190 - 73 = 2117$, $2117 - 73 = 2044$, and $2044 - 73 = 1971$ are all multiples of 73 as well, so the largest multiple of 73 less than 2000 is 1971.

3.51 A number is divisible by 9 if the sum of its digits is divisible by 9. Otherwise, the number is not divisible by 9. Summing the digits of each of the numbers in the problem gives

$$45{,}624 : \ 4 + 5 + 6 + 2 + 4 = 21,$$
$$560{,}335 : \ 5 + 6 + 0 + 3 + 3 + 5 = 22,$$
$$60{,}231{,}060 : \ 6 + 0 + 2 + 3 + 1 + 0 + 6 + 0 = 18,$$
$$9{,}671{,}011 : \ 9 + 6 + 7 + 1 + 0 + 1 + 1 = 25.$$

The only number whose digits sum to a multiple of 9 is 60,231,060. So, the only number that is divisible by 9 is $\boxed{60{,}231{,}060}$.

3.52 A number is a multiple of 4 if its final two digits form a multiple of 4. Therefore, 2544, 2564, 2572, and 2576 are multiples of 4. If the number formed by the final two digits is not a multiple of 4, then the original number is not a multiple of 4. Since 54 is not a multiple of 4, we know that $\boxed{2554}$ is not a multiple of 4.

3.53 We get a strong clue by listing the first several positive multiples of 25:

$$25, 50, 75, 100, 125, 150, 175, 200, 225, 250, 275, 300.$$

It looks like every multiple of 25 ends in 25, 50, 75, or 00. To see why, we note that we can write any number as the sum of a multiple of 100 plus a two-digit number which consists of the final two digits of the original number. For example,

$$6,781,525 = 6,781,500 + 25.$$

Since any multiple of 100 is a multiple of 25, the original number is a multiple of 25 only if the two-digit number is also a multiple of 25. The only numbers we can form with two digits that are multiples of 25 are 00, 25, 50, and 75. So, if a number ends in 00, 25, 50, or 75, then the number is divisible by 25. Otherwise, the number is not divisible by 25.

3.54 Since the number is divisible by 5 and only has 1, 3, 5, and 7 as digits, 5 must be the last digit. We know that the 7 and the 3 are not next to the 5, so the 1 must be the tens digit. Therefore, the number ends in the two digits 15. The 7 is next to the 1, so the number ends 715. Therefore, the number is $\boxed{3715}$.

3.55

(a) We have $26 = \mathbf{2^1} \cdot \mathbf{13^1}$ and $65 = \mathbf{5^1} \cdot 13^1$, so $\text{lcm}[26, 65] = 2^1 \cdot 5^1 \cdot 13^1 = \boxed{130}$.

   We also might have noted that $\text{lcm}[26, 65] = \text{lcm}[13 \cdot 2, 13 \cdot 5] = 13 \text{lcm}[2, 5] = 13 \cdot 10 = 130$.

(b) We have $96 = 32 \cdot 3 = 4 \cdot 8 \cdot 3 = 2^2 \cdot 2^3 \cdot 3 = \mathbf{2^5} \cdot 3^1$ and $72 = 8 \cdot 9 = 2^3 \cdot \mathbf{3^2}$, so $\text{lcm}[96, 72] = 2^5 \cdot 3^2 = \boxed{288}$.

   We also might have noted that $\text{lcm}[96, 72] = \text{lcm}[24 \cdot 4, 24 \cdot 3] = 24 \text{lcm}[4, 3] = 24 \cdot 12 = 288$.

(c) We have $16 = \mathbf{2^4}$, $21 = \mathbf{3^1} \cdot \mathbf{7^1}$, and $28 = 4 \cdot 7 = 2^2 \cdot 7^1$, so $\text{lcm}[16, 21, 28] = 2^4 \cdot 3^1 \cdot 7^1 = \boxed{336}$.

(d) We have $45 = 9 \cdot 5 = \mathbf{3^2} \cdot 5^1$, $60 = 6 \cdot 10 = 2 \cdot 3 \cdot 2 \cdot 5 = \mathbf{2^2} \cdot 3^1 \cdot 5^1$, and $75 = 3 \cdot 25 = 3^1 \cdot \mathbf{5^2}$, so $\text{lcm}[45, 60, 75] = 2^2 \cdot 3^2 \cdot 5^2 = (2 \cdot 3 \cdot 5)^2 = \boxed{900}$.

3.56

(a) We have $45 = 9 \cdot 5 = 3^2 \cdot \mathbf{5^1}$ and $75 = 3 \cdot 25 = \mathbf{3^1} \cdot 5^2$, so $\gcd(45, 75) = 3^1 \cdot 5^1 = \boxed{15}$.

   We also might have noted that $\gcd(45, 75) = \gcd(15 \cdot 3, 15 \cdot 5) = 15 \gcd(3, 5) = 15 \cdot 1 = 15$.

(b) We have $144 = 12^2 = (2^2 \cdot 3)^2 = 2^4 \cdot \mathbf{3^2}$ and $405 = 81 \cdot 5 = 9^2 \cdot 5 = (3^2)^2 \cdot 5 = 3^4 \cdot 5^1$, so $\gcd(144, 405) = 3^2 = \boxed{9}$.

(c) We have $238 = 2 \cdot 119 = \mathbf{2^1} \cdot 7^1 \cdot \mathbf{17^1}$ and $374 = 2 \cdot 187 = 2^1 \cdot 11^1 \cdot 17^1$, so $\gcd(238, 374) = 2^1 \cdot 17^1 = \boxed{34}$.

(d) We have

$$970 = 10 \cdot 97 = 2^1 \cdot \mathbf{5^1} \cdot 97^1, \qquad 485 = 5^1 \cdot 97^1, \qquad 1330 = 10 \cdot 133 = 2^1 \cdot 5^1 \cdot 7^1 \cdot 19^1,$$

so $\gcd(970, 485, 1330) = 5^1 = \boxed{5}$.

3.57 The multiples of 6 are the numbers that are multiples of both 2 and 3. The number $72d2$ ends in 2, so it is a multiple of 2. In order for a number to be divisible by 3, the sum of its digits must be a multiple of 3. The sum of the digits of $72d2$ is $7 + 2 + d + 2 = 11 + d$. The digits d for which $11 + d$ is a multiple of 3 are 1, 4, and 7, so the desired largest possible value of d is $\boxed{7}$.

3.58 The sum of multiples of 8 must also be a multiple of 8. We can write four of the numbers as sums of multiples of 8:

$$8,024 = 8,000 + 24,$$
$$168,640 = 160,000 + 8,000 + 640,$$
$$720,032 = 720,000 + 32,$$
$$64,856 = 64,000 + 800 + 56.$$

For the remaining number, we have

$$8{,}648{,}034 = 8{,}000{,}000 + 640{,}000 + 8{,}000 + 34.$$

The first three numbers on the right are multiples of 8, but the last number, 34, is 2 more than a multiple of 8. Therefore, we see that

$$8{,}648{,}032 = 8{,}000{,}000 + 640{,}000 + 8{,}000 + 32$$

is a multiple of 8. Since 8,648,034 is 2 more than a multiple of 8, it cannot be a multiple of 8. Therefore, the only number in the list that is not a multiple of 8 is $\boxed{8{,}648{,}034}$.

3.59 Any group of six consecutive integers includes a multiple of 2 and a multiple of 5. So, a factor of 2 and a factor of 5 appear in the prime factorization of the product, which means that the product is a multiple of $2 \cdot 5 = 10$. This means the units digit of the product must be $\boxed{0}$.

3.60 Both 11^7 and 7^5 are odd, so their sum is even. Therefore, $11^7 + 7^5$ is divisible by 2. Since 2 is the smallest prime number, $\boxed{2}$ is the smallest prime factor of $11^7 + 7^5$.

3.61 Since n is a multiple of 12, and 12 is a multiple of each of 12's divisors, we know that n must be a multiple of each of 12's divisors. The positive divisors of 12 are 1, 2, 3, 4, 6, and 12, so the other positive numbers that must be a factor of n are $\boxed{1, 2, 3, 4, \text{ and } 6}$.

3.62 Let m be the largest odd factor, and let n be the factor of 12,024 such that

$$mn = 12{,}024.$$

The prime factorizations of m and n multiply to give us the prime factorization of 12,024. We can also think of this as splitting the prime factorization of 12,024 into two pieces, m's prime factorization and n's prime factorization.

We make m as large as possible by putting all the odd primes into m's prime factorization. But m must be odd, so we must include all of the 2's in the prime factorization of n. Therefore, we can start by taking out factors of 2 from 12,024:

$$12{,}024 = 2 \cdot 6012 = 2^2 \cdot 3006 = 2^3 \cdot 1503.$$

All of the odd primes in the prime factorization of 12,024 are included in the prime factorization of 1503, so the desired largest odd factor is $\boxed{1503}$.

3.63 Any number can be written as the product of itself and 1, so any integer greater than 1 has at least 2 positive divisors, 1 and itself. A prime number cannot be written as the product of any other two positive integers, so a prime number has exactly 2 positive divisors. A composite number can be written as the product of some other pair of numbers (which may both be the same number, as in $9 = 3 \cdot 3$), so a composite number must have more than 2 positive divisors. The number 1 only has 1 divisor, itself.

Combining these observations, we see that the numbers with exactly 2 divisors are the primes. The primes less than 20 are 2, 3, 5, 7, 11, 13, 17, and 19. There are $\boxed{8}$ such numbers.

3.64 We start with the prime factorization of 392:

$$392 = 2 \cdot 196 = 2 \cdot 2 \cdot 98 = 2 \cdot 2 \cdot 2 \cdot 49 = 2^3 \cdot 7^2.$$

There's no power of 3 in this prime factorization! So, if $2^x \cdot 3^y \cdot 7^z = 392$, then y must be 0, so $xyz = \boxed{0}$.

We also could have noted that the sum of the digits of 392 is 14, which is not a multiple of 3. Therefore, 392 is not a multiple of 3, which means that the exponent of 3 in 392's prime factorization must be 0.

3.65 The numbers that are multiples of 2, 3, 6, and 8 are the multiples of lcm[2, 3, 6, 8], which is 24. Since 206,496 is a multiple of 24, the next multiple of 24 is $206{,}496 + 24 = \boxed{206{,}520}$.

3.66 A number that is 3 more than a multiple of 4 is also 1 less than a multiple of 4. Similarly, a number that is 4 more than a multiple of 5 is 1 less than a multiple of 5. So, our desired number is 1 less than a number that is a multiple of both 4 and 5. The smallest such positive number is 1 less than lcm[4, 5], which is $\text{lcm}[4, 5] - 1 = 20 - 1 = \boxed{19}$.

3.67 There are 60 seconds in a minute. The light that flashes every 1 minute 15 seconds flashes every $60 + 15 = 75$ seconds, and the other light flashes every $60 + 40 = 100$ seconds. The next time they flash together will be after lcm[75, 100] seconds. Since $75 = 3^1 \cdot 5^2$ and $100 = 2^2 \cdot 5^2$, we have $\text{lcm}[75, 100] = 2^2 \cdot 3^1 \cdot 5^2 = 300$. So, the shortest time that will elapse before the lights flash together is $\boxed{300 \text{ seconds}}$, which is the same as $\boxed{5 \text{ minutes}}$.

3.68 The greatest positive factor of any positive integer is the integer itself. The least positive factor of any integer is 1. So, the difference between the greatest positive factor of 121 and the least positive factor of 6 is $121 - 1 = \boxed{120}$.

3.69 We have a problem about the product of integers, so we start by finding the prime factorization of 6545. We have
$$6545 = 5 \cdot 1309 = 5 \cdot 7 \cdot 187 = 5 \cdot 7 \cdot 11 \cdot 17.$$

We need to split this prime factorization into two prime factorizations that each produce a two-digit integer. The product of any three of the primes 5, 7, 11, and 17 is greater than 99, so we must split the four primes into two pairs such that each pair has a two-digit product. The only one of these primes we can pair with 17 to get a two-digit product is 5, which leaves 7 paired with 11. So, our two integers must be $7 \cdot 11 = \boxed{77}$ and $5 \cdot 17 = \boxed{85}$.

3.70 In the first sequence, we start at 1 and count by 6's. In the second, we also start at 1, but we count by 8's. A number appears in both lists if its distance from 1 is a multiple of both 6 and 8. The numbers that are multiples of both 6 and 8 are the multiples of lcm[6, 8], which equals 24. So, we find the numbers that are in both lists by starting with 1 and counting by 24's. This gives us the list $1, 25, 49, 73, 97, \ldots$. The sum of the smallest three numbers in this list is $1 + 25 + 49 = \boxed{75}$.

3.71

(a) The largest divisor of a is a itself, so if a is also a divisor of b, we have $\gcd(a, b) = \boxed{a}$.

(b) The smallest positive multiple of b is b. Since a is a divisor of b, we know that b is also a multiple of a. Therefore, we have $\text{lcm}[a, b] = \boxed{b}$.

Challenge Problems

3.72 Since $3456n7$ is 5 more than a multiple of 8, we know that $3456n7 - 5 = 3456n2$ is a multiple of 8. Since $345600 = 320000 + 25600$ is a multiple of 8, the number $3456n2$ is a multiple of 8 only when the two-digit number $n2$ is a multiple of 8. The only two-digit multiples of 8 with 2 as the units digit are 32 and 72, so the possible values of n are $\boxed{3 \text{ and } 7}$. (Note that it is not generally true that a number is divisible by 8 if the number formed by its last two digits is. This fact only applies to this problem because 345600 is divisible by 8. In general, a number is divisible by 8 if the number formed by its last *three* digits is divisible by 8.)

3.73 If two positive integers are relatively prime, then the product of the integers equals their least common multiple. So, the product of the two numbers is 57. Since 57 is the product of primes $3 \cdot 19$, there are only two pairs of positive integers whose product is 57. These pairs are $1 \cdot 57$ and $3 \cdot 19$, so the two possible values of the larger number are $\boxed{19 \text{ and } 57}$.

3.74 The multiples of 12 are the numbers that are multiples of both 3 and 4. A number is a multiple of 3 if the sum of its digits is a multiple of 3. Since $0 + 1 + 2 + 3 + 4 + 5 + 6 + 7 + 8 + 9 = 45$ is a multiple of 3, any number written using each digit from 0 to 9 exactly once is a multiple of 3. So, now we only have to construct the largest such number that is divisible by 4. This means we must make the final two digits of the number form a multiple of 4. We want these two digits to be as small as possible, to save the large digits for the higher-value places in the number. For example, 9765432108 is not the smallest multiple of 4 we can form, because we'd like the 8 to be farther to the left in the number, such as in 9876532104. We can't put both 0 and 1 in the final two digits, since 10 is not a multiple of 4. But we can put 0 and 2 in the final two digits: $\boxed{9876543120}$.

3.75 We have $36 = 4 \cdot 9$, and 4 and 9 are relatively prime, so we can test for divisibility by 36 through testing for divisibility by both 4 and 9. So, we need $A55B$ to be a multiple of both 4 and 9. In order for $A55B$ to be a multiple of 4, the two-digit number $5B$ must be a multiple of 4. The only possible values of the digit B for which $5B$ is a multiple of 4 are 2 and 6. So, our number is $A552$ or $A556$.

In order for the number to be a multiple of 9, the sum of its digits must be a multiple of 9. Therefore, if our number is $A552$, then $A + 5 + 5 + 2 = A + 12$ must be a multiple of 9. The only digit for which this is true is $A = 6$, which gives us $A + B = 6 + 2 = 8$ in this case. If our number is $A556$, then $A + 5 + 5 + 6 = A + 16$ must be a multiple of 9. The only digit for which this is true is 2, and this gives us $A + B = 2 + 6 = 8$ in this case as well. So, the only possible sum of A and B is $\boxed{8}$.

3.76 We have $72 = 8 \cdot 9$, and 8 and 9 are relatively prime, so we can test for divisibility by 72 through testing for divisibility by both 8 and 9. As we saw on page 28 of this Solutions Manual, a number is a multiple of 8 if the number formed by its final three digits is a multiple of 8. So, the three-digit number $73B$ must be a multiple of 8. Since 720 is a multiple of 8, the next three multiples of 8 are $720 + 8 = 728$, $728 + 8 = 736$, and $736 + 8 = 744$. So, the only three-digit multiple of 8 of the form $73B$ is 736, which means $B = 6$.

Next, we find the value of A for which $A42736$ is a multiple of 9. The sum of the digits of $A42736$ is $A + 4 + 2 + 7 + 3 + 6 = A + 22$. The only digit A for which $A + 22$ is a multiple of 9 is $A = 5$. So, $\boxed{A = 5 \text{ and } B = 6}$.

3.77 We think about the prime factorization of 2520 and its divisors, since we can use prime factorizations to tell if one number is a divisor of another. The prime factorization of 2520 is

$$2520 = 252 \cdot 10 = 2 \cdot 126 \cdot 2 \cdot 5 = 2^2 \cdot 5 \cdot 126 = 2^2 \cdot 5 \cdot 2 \cdot 63 = 2^3 \cdot 3^2 \cdot 5^1 \cdot 7^1.$$

So, the only primes that can appear in the prime factorization of a divisor of 2520 are 2, 3, 5, and 7. Moreover, the largest power of 2 that can appear is 2^3, the largest possible power of 3 is 3^2, the largest possible power of 5 is 5^1, and the largest possible power of 7 is 7^1.

We want the divisor to be as large as possible without being divisible by 6. Including 5^1 and 7^1 in the prime factorization of the divisor doesn't affect whether or not the divisor is divisible by 6, so we include both. Since the divisor cannot be a multiple of 6, the divisor can't be a multiple of both 2 and 3. This means the divisor can't have both 2 and 3 in its prime factorization. So, when building the prime factorization of the largest factor of 2520 that is not divisible by 6, we include the larger of 2^3 and 3^2. Since $3^2 = 9$ is larger than $2^3 = 8$, we'll include 3^2 in the prime factorization of our divisor.

Therefore, the largest divisor that is not a multiple of 6 is $3^2 \cdot 5^1 \cdot 7^1 = \boxed{315}$.

3.78 $\boxed{\text{No}}$. If a number's digits sum to 18, then the number must be a multiple of 9, which is 3^2. A product of different primes can only have one factor of 3, so Jack's product can't be a multiple of 9. Therefore, the sum of the digits of Jack's product can't be 18.

3.79 We can find the least common multiple of a group of numbers by finding the prime factorizations of those numbers, and then taking the highest power of each prime that appears in at least one of these prime factorizations. So, we start with the prime factorizations of the first three numbers, and of the least common multiple we are given:

$$12 = 2^2 \cdot 3^1, \qquad 15 = 3^1 \cdot 5^1, \qquad 20 = 2^2 \cdot 5^1, \qquad 420 = 2^2 \cdot 3^1 \cdot 5^1 \cdot 7^1.$$

We know that $\text{lcm}[12, 15, 20, k] = 420 = 2^2 \cdot 3^1 \cdot 5^1 \cdot 7^1$. In forming the least common multiple, we can get the 2^2 and the 3^1 from the prime factorization of 12, and the 5^1 from the prime factorization of 15. But the 7^1 can't come from any of the 12, 15, or 20. So, it must come from k. Therefore, k must a positive multiple of 7, which means the smallest possible value of k is $\boxed{7}$.

3.80

(a) We have $18 = 2^1 \cdot 3^2$ and $24 = 2^3 \cdot 3^1$, so $\gcd(18, 24) = 2^1 \cdot 3^1 = 6$ and $\text{lcm}[18, 24] = 2^3 \cdot 3^2 = 72$. Therefore, we have

$$18 \cdot 24 = (2^1 \cdot 3^2) \cdot (2^3 \cdot 3^1) = (2^1 \cdot 2^3) \cdot (3^2 \cdot 3^1) = 2^4 \cdot 3^3,$$
$$\gcd(18, 24) \cdot \text{lcm}[18, 24] = (2^1 \cdot 3^1) \cdot (2^3 \cdot 3^2) = (2^1 \cdot 2^3) \cdot (3^1 \cdot 3^2) = 2^4 \cdot 3^3.$$

Both products equal $2^4 \cdot 3^3 = 432$.

(b) We have $35 = 5^1 \cdot 7^1$ and $42 = 2^1 \cdot 3^1 \cdot 7^1$, so $\gcd(35, 42) = 7^1 = 7$ and $\text{lcm}[35, 42] = 2^1 \cdot 3^1 \cdot 5^1 \cdot 7^1 = 210$. Therefore, we have

$$35 \cdot 42 = (5^1 \cdot 7^1) \cdot (2^1 \cdot 3^1 \cdot 7^1) = 2^1 \cdot 3^1 \cdot 5^1 \cdot (7^1 \cdot 7^1),$$
$$\gcd(35, 42) \cdot \text{lcm}[35, 42] = (7^1) \cdot (2^1 \cdot 3^1 \cdot 5^1 \cdot 7^1) = 2^1 \cdot 3^1 \cdot 5^1 \cdot (7^1 \cdot 7^1).$$

Both products equal $2^1 \cdot 3^1 \cdot 5^1 \cdot 7^2 = 1470$.

(c) We have $66 = 2^1 \cdot 3^1 \cdot 11^1$ and $84 = 2^2 \cdot 3^1 \cdot 7^1$, so $\gcd(66, 84) = 2^1 \cdot 3^1 = 6$ and $\text{lcm}[66, 84] = 2^2 \cdot 3^1 \cdot 7^1 \cdot 11^1 = 924$. Therefore, we have

$$66 \cdot 84 = (2^1 \cdot 3^1 \cdot 11^1) \cdot (2^2 \cdot 3^1 \cdot 7^1) = (2^1 \cdot 2^2) \cdot (3^1 \cdot 3^1) \cdot 7^1 \cdot 11^1,$$

$$\gcd(66, 84) \cdot \text{lcm}[66, 84] = (2^1 \cdot 3^1) \cdot (2^2 \cdot 3^1 \cdot 7^1 \cdot 11^1) = (2^1 \cdot 2^2) \cdot (3^1 \cdot 3^1) \cdot 7^1 \cdot 11^1.$$

Both products equal $2^3 \cdot 3^2 \cdot 7^1 \cdot 11^1 = 5544$.

In each part, we started with two positive integers, a and b, and found that $ab = \gcd(a, b) \cdot \text{lcm}[a, b]$. We saw that these products were the same by seeing that the prime factorizations of ab and $\gcd(a, b) \cdot \text{lcm}[a, b]$ were the same. To see why this must always happen for any a and b, let's think about a single prime factor. There are three possibilities for each prime:

Case 1: The prime does not appear in the prime factorization of a or b. If the prime is not in the prime factorization of a or b, then it will not be in the prime factorization of either $\gcd(a, b)$ or $\text{lcm}[a, b]$. So, it won't appear in either ab or $\gcd(a, b) \cdot \text{lcm}[a, b]$.

Case 2: The prime appears in the prime factorization of exactly one of a or b. Suppose the prime is p, and that the power p^n appears in the prime factorization of a, but not b. Then, p^n is in the prime factorization of ab. But what about $\gcd(a, b)$ and $\text{lcm}[a, b]$? Since p is not in the prime factorization of b, we can't have p in the prime factorization of $\gcd(a, b)$ at all. On the other hand, we must have p^n in the prime factorization of $\text{lcm}[a, b]$, since $\text{lcm}[a, b]$ must be a multiple of a. Therefore, p^n appears in the prime factorizations of both ab and $\gcd(a, b) \cdot \text{lcm}[a, b]$.

Essentially the same argument holds if p^n appears in the prime factorization of b but not a.

Case 3: The prime appears in the prime factorizations of both a and b. Suppose p^m is in the prime factorization of a, and p^n is in the prime factorization of b. Then, the prime factorization of ab includes the product of these, $p^m \cdot p^n = p^{m+n}$.

The prime factorization of $\gcd(a, b)$ includes the smaller of p^m and p^n, and the prime factorization of $\text{lcm}[a, b]$ has the larger of p^m and p^n. So, one of $\gcd(a, b)$ and $\text{lcm}[a, b]$ has p^m and the other has p^n in its prime factorization. (This is true even if $m = n$.) Therefore, the product $\gcd(a, b) \cdot \text{lcm}[a, b]$ has $p^m \cdot p^n = p^{m+n}$ in its prime factorization, just like ab does.

To see all this in action, take a look at the prime factorizations of 72 and 120:

$$72 = \underline{2^3} \cdot \mathbf{3^2}, \qquad 120 = \mathbf{2^3} \cdot \underline{\mathbf{3^1}} \cdot \mathbf{5^1}.$$

We use the underlined factors to compute $\gcd(72, 120)$ and the bold factors to compute $\text{lcm}[72, 120]$:

$$\gcd(72, 120) = \underline{2^3} \cdot \underline{3^1}, \qquad \text{lcm}[72, 120] = 2^3 \cdot 3^2 \cdot 5^1.$$

So, we see that $72 \cdot 120$ has exactly the same powers of primes as $\gcd(72, 120) \cdot \text{lcm}[72, 120]$.

Therefore, the prime factorizations of ab and $\gcd(a, b) \cdot \text{lcm}[a, b]$ are the same, which means that $ab = \gcd(a, b) \cdot \text{lcm}[a, b]$.

See if you can also use the relationships $\gcd(na, nb) = n \gcd(a, b)$ and $\text{lcm}[na, nb] = n \text{lcm}[a, b]$ to explain why $ab = \gcd(a, b) \cdot \text{lcm}[a, b]$.

3.81 We multiply a number by 10 by adding a zero to the end of the number. Similarly, we multiply by five 10's, or 10^5, by adding five zeros to the end of the number. So, to figure out how many zeros there are at the end of $80^{16} \cdot 75^8$, we write the number as the product of a power of 10 times some number that doesn't end with 0.

We get factors of 10 when we combine factors of 2 and factors of 5. So, we'll start by writing the prime factorization of $80^{16} \cdot 75^8$ to see how many factors of each we have. Since $80 = 8 \cdot 10 = 2^3 \cdot 2 \cdot 5 = 2^4 \cdot 5$ and $75 = 3 \cdot 25 = 3 \cdot 5^2$, we have

$$80^{16} \cdot 75^8 = (2^4 \cdot 5)^{16} \cdot (3 \cdot 5^2)^8$$
$$= (2^4)^{16} \cdot 5^{16} \cdot 3^8 \cdot (5^2)^8$$
$$= 2^{4 \cdot 16} \cdot 5^{16} \cdot 3^8 \cdot 5^{2 \cdot 8}$$
$$= 2^{64} \cdot 5^{16} \cdot 3^8 \cdot 5^{16}$$
$$= 2^{64} \cdot 3^8 \cdot 5^{32}.$$

We can pair each of the 32 5's with a factor of 2 to make a factor of 10:

$$2^{64} \cdot 3^8 \cdot 5^{32} = 2^{32} \cdot 2^{32} \cdot 3^8 \cdot 5^{32} = 2^{32} \cdot 3^8 \cdot (2^{32} \cdot 5^{32}) = 2^{32} \cdot 3^8 \cdot (2 \cdot 5)^{32} = 2^{32} \cdot 3^8 \cdot 10^{32}.$$

Since $2^{32} \cdot 3^8$ doesn't have a factor of 5, it doesn't end in 0. So, when we multiply this number by 10^{32}, we get a number with $\boxed{32}$ zeros at the end.

3.82 The exponents in the prime factorization of a perfect square must be even. Let's see if there's a similar fact for the exponents in the prime factorization of a perfect cube. We'll look at the prime factorizations of the first few cubes to see if we find anything interesting:

$$2^3, \qquad 3^3, \qquad 4^3 = (2^2)^3 = 2^{2 \cdot 3} = 2^6, \qquad 5^3, \qquad 6^3 = (2 \cdot 3)^3 = 2^3 \cdot 3^3.$$

It looks like all the exponents in the prime factorization of a perfect cube must be multiples of 3. The examples of 4^3 and 6^3 above show why this occurs. To get the prime factorization of any cube n^3, we can cube the prime factorization of n. When we cube such a prime factorization, we multiply the exponents in the prime factorization by 3. For example, the cube of $2^2 \cdot 7^1$ is

$$(2^2 \cdot 7^1)^3 = 2^{2 \cdot 3} \cdot 7^{1 \cdot 3} = 2^6 \cdot 7^3.$$

Therefore, each of the exponents in the prime factorization of a cube must be a multiple of 3.

Each of the exponents in the prime factorization of a square must be a multiple of 2, and each of the exponents in the prime factorization of a cube must be a multiple of 3. Therefore, if a number is a square and a cube, then each of the exponents in its prime factorization must be a multiple of both 2 and 3. This means each exponent must be a multiple of 6. As we have seen, the smallest such number (besides 1) is $2^6 = 64$. The next smallest is $3^6 = \boxed{729}$. Note that $729 = 3^6 = (3^2)^3 = 9^3$ and $729 = 3^6 = (3^3)^2 = 27^2$, so 729 is indeed both a square and a cube.

3.83 We know that we can build the prime factorizations of $\operatorname{lcm}[a, b]$ and $\gcd(a, b)$ from the prime factorizations of a and b. So, we find the prime factorizations of the given values of $\operatorname{lcm}[a, b]$ and $\gcd(a, b)$, hoping these will help us learn about the prime factorizations of a and b. We have

$$\operatorname{lcm}[a, b] = 462 = 2^1 \cdot 3^1 \cdot 7^1 \cdot 11^1, \qquad \gcd(a, b) = 33 = 3^1 \cdot 11^1.$$

Since 3^1 is in the prime factorizations of both $\operatorname{lcm}[a, b]$ and $\gcd(a, b)$, we know that 3^1 is both the largest and the smallest power of 3 that appears in the prime factorizations of a and b. Therefore, 3^1 appears in the prime factorizations of both a and b. Similarly, 11^1 is in the prime factorizations of both lcm$[a, b]$ and gcd(a, b), so 11^1 appears in the prime factorizations of both a and b.

Since 2^1 is a factor of lcm$[a, b]$ but not gcd(a, b), we know that 2^1 is in the prime factorization of one of a and b, but not the other. Similarly, 7^1 is in the prime factorization of one of a and b, but not the other.

So, the prime factorizations of a and b both include $3^1 \cdot 11^1$. The prime factorization of exactly one of a and b includes 2^1. Also, the prime factorization of exactly one of a and b includes 7^1 (possibly the same one that includes 2^1). Neither has any other prime factor besides these four, since no other prime factor appears in the prime factorization of lcm$[a, b]$. Since a is greater than b, it must be a that has 7^1 in its prime factorization. Therefore, $\boxed{7}$ is the largest prime that is a factor of a but not of b.

3.84 We have lcm$[5, 9] = 45$, so the multiples of 45 are the numbers that are multiples of both 5 and 9. Therefore, we seek the smallest positive integer that is divisible by both 5 and 9 such that the integer has only 0's and 1's as digits.

A number is divisible by 5 if it ends in 0 or 5. Our desired number can only have 0's and 1's as digits, so it must end in 0.

Since our desired number must be divisible by 9, the sum of its digits must be divisible by 9. The only nonzero digits in our desired number are 1's. So, our desired number must have 9 1's in order for the sum of its digits to be a multiple of 9.

Combining these two conditions, $\boxed{1,111,111,110}$ is the smallest positive multiple of 45 that only has 0's and 1's as digits.

3.85 We will express $10^{22} + 8$ as the sum of two numbers that are easy to divide by 9:

$$(10^{22} + 8) \div 9 = (1\underbrace{00\ldots00}_{22 \text{ zeros}} + 8) \div 9$$

$$= (\underbrace{99\ldots99}_{22 \text{ nines}} + 1 + 8) \div 9$$

$$= (\underbrace{99\ldots99}_{22 \text{ nines}} + 9) \div 9 = (\underbrace{99\ldots99}_{22 \text{ nines}} \div 9) + (9 \div 9) = \underbrace{11\ldots11}_{22 \text{ ones}} + 1 = \underbrace{11\ldots112}_{21 \text{ ones}}.$$

So the sum of the digits is $21 + 2 = \boxed{23}$.

3.86 The numbers that are divisible by both 2 and 3 are the multiples of 6. So, we want to count the positive numbers less than 1000 that are multiples of 6, but not of 5. Since 1000 divided by 6 has quotient 166 and remainder 4, the positive multiples of 6 less than 1000 are

$$6 \cdot 1, 6 \cdot 2, 6 \cdot 3, \ldots, 6 \cdot 166.$$

There are 166 such numbers. But many of them are also multiples of 5. Any time we multiply 6 by a multiple of 5, we get a number that is a multiple of both 6 and 5. We must exclude these from our count. So, we have to count the number of multiples of 5 from 1 to 166. Since $165 = 5 \cdot 33$, we see that there are 33 multiples of 5 between 1 and 166, namely $5 \cdot 1, 5 \cdot 2, 5 \cdot 3, \ldots, 5 \cdot 33$. When we multiply 6 by each of these, we get a multiple of 6 less than 1000 that is also a multiple of 5. Excluding these 33 from our previous count of 166 leaves $166 - 33 = \boxed{133}$ numbers less than 1000 that are multiples of 6 but not multiples of 5.

3.87 We'll say that a student "touches" a locker if she either opens or shuts it. So, the first student touches every locker, the 2nd student touches every 2nd locker, the 3rd student touches every 3rd locker, and so on. Since the 2nd student touches every 2nd locker starting with locker 2, the 2nd student touches every locker that is a multiple of 2. Similarly, the 3rd student touches every locker that is a multiple of 3, the 4th student touches every locker that is a multiple of 4, and so on.

Therefore, locker number n is touched by every student whose number evenly divides n. So, the number of times locker n gets touched equals the number of positive divisors of n. If a locker is touched an even number of times total, then it is closed after the whole process, since each "opening" touch pairs with a "closing" touch. Similarly, if a locker is touched an odd number of times, the last touch will leave the locker open.

Therefore, the lockers that are open at the end are the ones whose numbers have an odd number of positive divisors. The only positive numbers with an odd number of positive divisors are the positive perfect squares, so we must count the number of positive perfect squares less than 1000. Since $31^2 = 961$ is less than 1000 and $32^2 = 1024$ is greater than 1000, there are $\boxed{31}$ positive perfect squares less than 1000.

CHAPTER 4

Fractions

24® Cards

First Card $4 \cdot 4 + 4 + 4$.

Second Card $4(1 + 9) - 16$ or $4(16 - 9 - 1)$.

Third Card $(10 \cdot 10 - 4) \div 4$.

Fourth Card $16 \cdot 9 \div (10 - 4)$.

Exercises for Section 4.1

4.1.1

(a) Applying our rule for dividing into negation, we have $\frac{-6}{6} = -\frac{6}{6} = -(6 \div 6) = \boxed{-1}$.

(b) $\frac{18}{3} = 18 \div 3 = \boxed{6}$.

(c) Any nonzero number divided by itself equals 1, so $\frac{-23}{-23} = \boxed{1}$.

(d) 0 divided by any nonzero number is 0, so $\frac{0}{-5} = \boxed{0}$.

4.1.2 $\frac{16+6}{4-2} = \frac{22}{2} = 22 \div 2 = \boxed{11}$.

4.1.3 $\frac{1+5+9+13+17+21}{6} = \frac{66}{6} = 66 \div 6 = \boxed{11}$.

4.1.4 We first find the multiples of 7 that 43 is between. We have $42 = 7 \cdot 6$ and $49 = 7 \cdot 7$. So $\frac{43}{7}$ is between $\frac{42}{7} = 6$ and $\frac{49}{7} = 7$. Since 43 is closer to 42 than to 49, we know that $\frac{43}{7}$ is closer to $\boxed{6}$ than to 7.

Another way to think about this problem is to consider the location of $\frac{43}{7}$ on the number line. Suppose we break the number line between each pair of consecutive integers into 7 equal pieces. The fraction $\frac{42}{7}$, which equals 6, is at the right end of the 42nd piece to the right of 0. The fraction $\frac{43}{7}$ is at the right end of the next piece, so it is closer to $\boxed{6}$ than to any other integer.

4.1.5 Since $(-12) \div 4 = -3$, we have $xy - \frac{x}{y} = (-12)(4) - \frac{-12}{4} = -48 - (-3) = -48 + 3 = \boxed{-45}$.

4.1.6 Only B and C are to the left of 0, so these are the only two that correspond to negative fractions. Since $-\frac{8}{4} = -2$, point B corresponds to $-\frac{8}{4}$. Therefore, C corresponds to the other negative fraction, $-\frac{1}{3}$. (Notice that point C is between 0 and -1, but closer to 0 than to -1.)

Turning to our positive fractions, because 11 is greater than 7, we know that $\frac{11}{6}$ is greater than $\frac{7}{6}$. (We could also have seen this by dividing the number line between each pair of consecutive integers into six equal pieces. $\frac{11}{6}$ is at the right end of the 11th piece to the right of 0, while $\frac{7}{6}$ is at the right end of the 7th piece.) Therefore, A corresponds to $\frac{7}{6}$ and D corresponds to $\frac{11}{6}$.

4.1.7 We locate $\frac{3}{5}$ on the number line by dividing the number line between 0 and 1 into 5 equal pieces. $\frac{3}{5}$ is at the right end of the 3rd of these pieces:

If we divide each of these 5 equal pieces into 2 equal pieces, then $\frac{3}{5}$ is at the right end of the 6th of 10 equal pieces, so $\frac{3}{5} = \frac{6}{10}$:

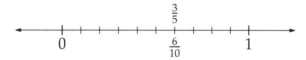

If we divide each of our original 5 equal pieces into 3 equal pieces, then $\frac{3}{5}$ is at the right end of the 9th of 15 equal pieces, so $\frac{3}{5} = \frac{9}{15}$:

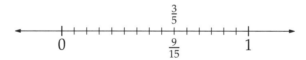

If we divide each of our original 5 equal pieces into 4 equal pieces, then $\frac{3}{5}$ is at the right end of the 12th of 20 equal pieces, so $\frac{3}{5} = \frac{12}{20}$:

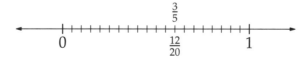

So, three fractions that equal $\frac{3}{5}$ are $\boxed{\frac{6}{10}, \frac{9}{15}, \text{ and } \frac{12}{20}}$. These are not the only possible answers. There are infinitely many other possible answers!

4.1.8 The expression $\frac{36}{n+1}$ is a positive integer if and only if $n + 1$ divides evenly into 36. Therefore, $n + 1$ must be a positive divisor of 36. The positive divisors of 36 are 1, 2, 3, 4, 6, 9, 12, 18, and 36, and subtracting 1 from each gives the corresponding values of n, which are 0, 1, 2, 3, 5, 8, 11, 17, and 35. However, the problem asks for the number of *positive* values of n, so we must exclude 0. This leaves $\boxed{8}$ values of n that satisfy the problem.

Exercises for Section 4.2

4.2.1

(a) $\frac{5}{6} \cdot \frac{11}{7} = \frac{5 \cdot 11}{6 \cdot 7} = \boxed{\frac{55}{42}}$.

(b) We have $\frac{1}{5} \cdot (-75) = \frac{1 \cdot (-75)}{5} = \frac{-75}{5} = -15$. Therefore, $\frac{1}{5} \cdot (-75) \cdot \frac{2}{3} = (-15) \cdot \frac{2}{3} = \frac{(-15) \cdot 2}{3} = \frac{-30}{3} = \boxed{-10}$.

(c) First, we notice that two of the numbers are negative and the other is positive, so the product of all three numbers is positive. This gives us $\left(-\frac{80}{7}\right)\left(\frac{14}{9}\right)\left(-\frac{63}{16}\right) = \left(\frac{80}{7}\right)\left(\frac{14}{9}\right)\left(\frac{63}{16}\right)$. Next, we notice that each numerator is a multiple of one of the denominators. So, we can rearrange a bit to find the product more easily:

$$\left(\frac{80}{7}\right)\left(\frac{14}{9}\right)\left(\frac{63}{16}\right) = \frac{80 \cdot 14 \cdot 63}{7 \cdot 9 \cdot 16} = \frac{80 \cdot 14 \cdot 63}{16 \cdot 7 \cdot 9} = \left(\frac{80}{16}\right)\left(\frac{14}{7}\right)\left(\frac{63}{9}\right) = 5 \cdot 2 \cdot 7 = \boxed{70}.$$

4.2.2 Each factor in the numerator is double one of the factors in the denominator, so rearranging gives us

$$\frac{30 \cdot 28 \cdot 26 \cdot 24}{12 \cdot 13 \cdot 14 \cdot 15} = \frac{30 \cdot 28 \cdot 26 \cdot 24}{15 \cdot 14 \cdot 13 \cdot 12} = \frac{30}{15} \cdot \frac{28}{14} \cdot \frac{26}{13} \cdot \frac{24}{12} = 2 \cdot 2 \cdot 2 \cdot 2 = \boxed{16}.$$

4.2.3 $\frac{3 \cdot 5}{9 \cdot 11} \times \frac{7 \cdot 9 \cdot 11}{3 \cdot 5 \cdot 7} = \frac{3 \cdot 5 \cdot 7 \cdot 9 \cdot 11}{9 \cdot 11 \cdot 3 \cdot 5 \cdot 7} = \frac{3 \cdot 5 \cdot 7 \cdot 9 \cdot 11}{3 \cdot 5 \cdot 7 \cdot 9 \cdot 11}$. Since the numerator and denominator of this fraction are the same, the fraction equals $\boxed{1}$.

4.2.4 $\frac{8}{9}$ of 180 is $\frac{8}{9} \cdot 180 = \frac{8 \cdot 180}{9} = 8 \cdot \frac{180}{9} = 8 \cdot 20 = 160$. Therefore, $\frac{3}{4}$ of $\frac{8}{9}$ of 180 is $\frac{3}{4}$ of 160. Computing $\frac{3}{4}$ of 160 gives $\frac{3}{4} \cdot 160 = \frac{3 \cdot 160}{4} = 3 \cdot \frac{160}{4} = 3 \cdot 40 = \boxed{120}$.

4.2.5 As in the previous problem, we can solve this problem in steps. Or, we can note that since "of" means multiply, $\frac{1}{2}$ of $\frac{2}{3}$ of $\frac{3}{4}$ of $\frac{4}{5}$ of 100 is $\frac{1}{2} \cdot \frac{2}{3} \cdot \frac{3}{4} \cdot \frac{4}{5} \cdot 100$. We can compute this quickly with some clever rearranging:

$$\begin{aligned} \frac{1}{2} \cdot \frac{2}{3} \cdot \frac{3}{4} \cdot \frac{4}{5} \cdot 100 &= \frac{1 \cdot 2 \cdot 3 \cdot 4}{2 \cdot 3 \cdot 4 \cdot 5} \cdot 100 \\ &= \frac{2 \cdot 3 \cdot 4 \cdot 1}{2 \cdot 3 \cdot 4 \cdot 5} \cdot 100 \\ &= \frac{2}{2} \cdot \frac{3}{3} \cdot \frac{4}{4} \cdot \frac{1}{5} \cdot 100 = 1 \cdot 1 \cdot 1 \cdot \frac{1}{5} \cdot 100 = \frac{100}{5} = \boxed{20}. \end{aligned}$$

4.2.6 Seeing that the denominator of each fraction, except the last one, is the numerator of another fraction, we can simplify with some clever rearranging:

$$\frac{5}{8} \cdot \frac{8}{11} \cdot \frac{11}{14} \cdot \frac{14}{17} \cdot \frac{17}{20} \cdot \frac{20}{23} = \frac{5 \cdot 8 \cdot 11 \cdot 14 \cdot 17 \cdot 20}{8 \cdot 11 \cdot 14 \cdot 17 \cdot 20 \cdot 23}$$

$$= \frac{8 \cdot 11 \cdot 14 \cdot 17 \cdot 20 \cdot 5}{8 \cdot 11 \cdot 14 \cdot 17 \cdot 20 \cdot 23}$$

$$= \frac{8}{8} \cdot \frac{11}{11} \cdot \frac{14}{14} \cdot \frac{17}{17} \cdot \frac{20}{20} \cdot \frac{5}{23}$$

$$= 1 \cdot 1 \cdot 1 \cdot 1 \cdot 1 \cdot \frac{5}{23} = \boxed{\frac{5}{23}}.$$

4.2.7

(a) Suppose we multiply $\frac{5}{6}$ by $\frac{a}{b}$. We need the final denominator to be 7, so we let b be 7. This makes our product $\frac{5}{6} \cdot \frac{a}{7} = \frac{5 \cdot a}{6 \cdot 7}$. We'd like to be able to cancel out the 6 in the denominator, so we let a be 6. This gives us $\frac{5}{6} \cdot \frac{6}{7} = \frac{5 \cdot 6}{6 \cdot 7} = \frac{6 \cdot 5}{6 \cdot 7} = \frac{6}{6} \cdot \frac{5}{7} = \frac{5}{7}$, as desired. So, the fraction we want is $\boxed{\frac{6}{7}}$. Note that this is not the only valid answer, since there are other fractions that equal $\frac{6}{7}$.

(b) From part (a), we have $\frac{5}{6} \cdot \frac{6}{7} = \frac{5}{7}$. But what do we multiply $\frac{5}{7}$ by to replace the 5 in the numerator with a 4? Looking back at how we multiplied $\frac{5}{6}$ by $\frac{6}{7}$ to cancel out the 6 in the denominator and replace it with 7, we try multiplying $\frac{5}{7}$ by $\frac{4}{5}$, to cancel the 5 and replace it with a 4:

$$\frac{5}{7} \cdot \frac{4}{5} = \frac{5 \cdot 4}{7 \cdot 5} = \frac{5 \cdot 4}{5 \cdot 7} = \frac{5}{5} \cdot \frac{4}{7} = \frac{4}{7}.$$

Success! So, multiplying $\frac{5}{6}$ by $\frac{6}{7}$ and then by $\frac{4}{5}$ gives us $\frac{4}{7}$. Instead of multiplying by $\frac{6}{7}$ and by $\frac{4}{5}$ in two steps, we can multiply by both of them at once by multiplying by $\frac{6}{7} \cdot \frac{4}{5}$, which equals $\frac{6 \cdot 4}{7 \cdot 5} = \boxed{\frac{24}{35}}$.

4.2.8 We are given $\frac{a}{b} = \frac{3}{4}$ and $\frac{b}{c} = \frac{11}{13}$ and we want $\frac{a}{c}$. We can cancel out the b's from $\frac{a}{b}$ and $\frac{b}{c}$ by multiplying them: $\frac{a}{b} \cdot \frac{b}{c} = \frac{ab}{bc} = \frac{ab}{cb} = \frac{a}{c} \cdot \frac{b}{b} = \frac{a}{c}$. Now we see how to use our given values to compute $\frac{a}{c}$:

$$\frac{a}{c} = \frac{a}{b} \cdot \frac{b}{c} = \frac{3}{4} \cdot \frac{11}{13} = \frac{3 \cdot 11}{4 \cdot 13} = \boxed{\frac{33}{52}}.$$

Exercises for Section 4.3

4.3.1

(a) $\frac{3}{5} \div 2 = \frac{3}{5} \cdot \frac{1}{2} = \boxed{\frac{3}{10}}.$

(b) $7 \div \frac{7}{8} = 7 \cdot \frac{8}{7} = \frac{7 \cdot 8}{7} = \frac{7}{7} \cdot 8 = \boxed{8}.$

(c) $\frac{14/3}{5/4} = \frac{14}{3} \div \frac{5}{4} = \frac{14}{3} \cdot \frac{4}{5} = \frac{14 \cdot 4}{3 \cdot 5} = \boxed{\frac{56}{15}}.$

(d) The quotient of two negative numbers is positive, so we have

$$\left(-\frac{5}{6}\right) \div \left(-\frac{12}{7}\right) = \frac{5}{6} \div \frac{12}{7} = \frac{5}{6} \cdot \frac{7}{12} = \frac{5 \cdot 7}{6 \cdot 12} = \boxed{\frac{35}{72}}.$$

4.3.2 When $\frac{3}{7}$ is divided by $\frac{7}{3}$, the result is $\frac{3}{7} \div \frac{7}{3} = \frac{3}{7} \cdot \frac{3}{7} = \boxed{\frac{9}{49}}$.

4.3.3 $\frac{36}{x} = 36 \div x = 36 \div \frac{3}{4} = 36 \cdot \frac{4}{3} = \frac{36}{3} \cdot 4 = 12 \cdot 4 = \boxed{48}$.

4.3.4

(a) We must determine what number is in the blank in $40 = \frac{2}{3} \cdot \underline{\quad}$, so we divide: $40 \div \frac{2}{3} = 40 \cdot \frac{3}{2} = \frac{40}{2} \cdot 3 = 20 \cdot 3 = \boxed{60}$.

(b) We must determine what number is in the blank in $\frac{9}{5} = \frac{2}{3} \cdot \underline{\quad}$, so we divide: $\frac{9}{5} \div \frac{2}{3} = \frac{9}{5} \cdot \frac{3}{2} = \boxed{\frac{27}{10}}$.

4.3.5 To get a handle on these questions, we first think about the problem without fractions. Suppose we are asked, "Dividing 6 by what number gives 3?" Then, the answer is 2, which is $6 \div 3$. Similarly, the answer to "Dividing 42 by what number is 6?" is 7, which is $42 \div 6$. So, it seems like the answer to the question,

"Dividing a by what number equals b?"

is $a \div b$. So, we expect that $a \div (a \div b)$ equals b. To see that this is correct, we write $a \div b$ as $\frac{a}{b}$, and we have

$$a \div (a \div b) = a \div \frac{a}{b} = a \cdot \frac{b}{a} = \frac{ab}{a} = \frac{a}{a} \cdot b = b.$$

So, indeed, $a \div b$ is the number that we must divide a by to get b.

(a) Using what we just learned, we must divide: $\frac{6}{7} \div \frac{3}{7} = \frac{6}{7} \cdot \frac{7}{3} = \frac{6 \cdot 7}{7 \cdot 3} = \frac{6}{3} \cdot \frac{7}{7} = \boxed{2}$. Checking our answer, we find that $\frac{6}{7} \div 2 = \frac{6}{7} \cdot \frac{1}{2} = \frac{6 \cdot 1}{7 \cdot 2} = \frac{6}{2} \cdot \frac{1}{7} = 3 \cdot \frac{1}{7} = \frac{3}{7}$, as expected.

(b) Again using what we just learned, we must divide: $\frac{6}{7} \div \frac{6}{5} = \frac{6}{7} \cdot \frac{5}{6} = \frac{6}{6} \cdot \frac{5}{7} = \boxed{\frac{5}{7}}$. Checking our answer, we find $\frac{6}{7} \div \frac{5}{7} = \frac{6}{7} \cdot \frac{7}{5} = \frac{6}{5} \cdot \frac{7}{7} = \frac{6}{5}$, as required.

(c) We must divide: $\frac{6}{7} \div \frac{2}{3} = \frac{6}{7} \cdot \frac{3}{2} = \frac{6}{2} \cdot \frac{3}{7} = 3 \cdot \frac{3}{7} = \boxed{\frac{9}{7}}$.

4.3.6 Once again, we get a handle on the problem by trying a problem without fractions. Suppose the problem were:

"Dividing 2 into what number gives a quotient of 20?"

Now, it's more clear what the answer is. We divide 2 into 40 to get a quotient of 20. Notice that $40 = 2 \cdot 20$. So, it looks like the answer to the question, "Dividing a into what number gives a quotient of b?" is simply $a \cdot b$. We can quickly check this by simplifying $(a \cdot b) \div a$:

$$(a \cdot b) \div a = (a \cdot b) \cdot \frac{1}{a} = \frac{ab}{a} = b.$$

This makes sense: if we divide the product of a and b by a, we get b. That's how division works; it undoes multiplication!

Returning to the original problem, to get a quotient of 20, we divide $\frac{3}{5}$ into $\frac{3}{5} \cdot 20 = \frac{3 \cdot 20}{5} = 3 \cdot \frac{20}{5} = 3 \cdot 4 = \boxed{12}$. Checking our answer, we have $12 \div \frac{3}{5} = 12 \cdot \frac{5}{3} = \frac{12}{3} \cdot 5 = 4 \cdot 5 = 20$, as required.

4.3.7 Suppose our original number is x. Multiplying x by $\frac{3}{4}$ gives $\left(x \cdot \frac{3}{4}\right)$. Dividing this number by $\frac{3}{5}$ gives

$$\left(x \cdot \frac{3}{4}\right) \div \frac{3}{5} = \left(x \cdot \frac{3}{4}\right) \cdot \frac{5}{3} = x \cdot \left(\frac{3}{4} \cdot \frac{5}{3}\right) = x \cdot \left(\frac{3 \cdot 5}{4 \cdot 3}\right) = x \cdot \frac{5}{4}.$$

So, the result is the same as multiplying the original number by $\boxed{\frac{5}{4}}$.

4.3.8 If we multiply the number of scoops by the size of the scoop, we get the total amount of flour:

(Number of scoops) $\cdot$ (Size of scoop) = Total amount of flour.

So, to find the number of scoops, we divide the total amount of flour by the size of each scoop. Therefore, I need $6 \div \frac{2}{3} = 6 \cdot \frac{3}{2} = \frac{18}{2} = \boxed{9}$ scoops.

Exercises for Section 4.4

4.4.1

(a) $\left(\frac{3}{5}\right)^2 = \frac{3^2}{5^2} = \boxed{\frac{9}{25}}$.

(b) Any number raised to the power 0 equals 1, so $\left(-\frac{2}{7}\right)^0 = \boxed{1}$.

(c) $\left(\frac{4}{9}\right)^{-2} = \left(\left(\frac{4}{9}\right)^{-1}\right)^2 = \left(\frac{9}{4}\right)^2 = \frac{9^2}{4^2} = \boxed{\frac{81}{16}}$.

(d) $\left(\frac{-3}{2}\right)^5 = \frac{(-3)^5}{2^5} = \frac{-243}{32} = \boxed{-\frac{243}{32}}$.

(e) Since $\left(\frac{1}{5}\right)^3 = \frac{1^3}{5^3} = \frac{1}{125}$, we have $\frac{1}{(1/5)^3} = \frac{1}{1/125} = \frac{125}{1} = \boxed{125}$.

(f) $\dfrac{(2/9)^2}{(5/3)^4} = \dfrac{2^2/9^2}{5^4/3^4} = \dfrac{4/81}{625/81} = \dfrac{4}{81} \cdot \dfrac{81}{625} = \dfrac{81}{81} \cdot \dfrac{4}{625} = \boxed{\dfrac{4}{625}}$.

4.4.2

(a) We might recognize 27 as 3^3 and 64 as 4^3, but if we don't, we can simply compute a few powers of $\frac{3}{4}$ and hope we get lucky. We have $\left(\frac{3}{4}\right)^2 = \frac{3^2}{4^2} = \frac{9}{16}$, and $\left(\frac{3}{4}\right)^3 = \frac{3^3}{4^3} = \frac{27}{64}$. So, we raise $\frac{3}{4}$ to the power $\boxed{3}$ to get $\frac{27}{64}$.

(b) We see that 9 is 3^2 and 16 is 4^2, so $\frac{16}{9} = \frac{4^2}{3^2} = \left(\frac{4}{3}\right)^2$. But we need $\frac{16}{9}$ as a power of $\frac{3}{4}$, not as a power of $\frac{4}{3}$! Since $\frac{4}{3}$ is the reciprocal of $\frac{3}{4}$, we can use negative exponents:

$$\left(\frac{4}{3}\right)^2 = \left(\left(\frac{3}{4}\right)^{-1}\right)^2 = \left(\frac{3}{4}\right)^{-2}.$$

Therefore, we raise $\frac{3}{4}$ to the power $\boxed{-2}$ to get $\frac{16}{9}$.

4.4.3 *Solution 1: Raise both fractions to the 4th power.* It's a little scary to raise 1641 to the 4th power, but maybe we'll get lucky and not have to. We have

$$\frac{(2/1641)^4}{(3/1641)^4} = \frac{2^4/1641^4}{3^4/1641^4} = \frac{2^4}{1641^4} \cdot \frac{1641^4}{3^4} = \frac{2^4}{3^4} \cdot \frac{1641^4}{1641^4}.$$

Phew! Since the numerator and denominator of $\frac{1641^4}{1641^4}$ are the same, we have $\frac{1641^4}{1641^4} = 1$. So, we have $\frac{2^4}{3^4} \cdot \frac{1641^4}{1641^4} = \frac{2^4}{3^4} = \boxed{\frac{16}{81}}$.

Solution 2: Use the fact that $\frac{a^n}{b^n} = \left(\frac{a}{b}\right)^n$. The twist here is that a and b themselves are fractions! We have

$$\frac{(2/1641)^4}{(3/1641)^4} = \left(\frac{2/1641}{3/1641}\right)^4 = \left(\frac{2}{1641} \cdot \frac{1641}{3}\right)^4.$$

Conveniently, once again the 1641's cancel out and we have $\left(\frac{2}{3}\right)^4 = \frac{2^4}{3^4} = \boxed{\frac{16}{81}}$.

4.4.4 All three fractions are the same, so we can apply the exponent laws $a^b \cdot a^c = a^{b+c}$ and $a^b/a^c = a^{b-c}$:

$$\frac{(5/3)^4(5/3)^3}{(5/3)^5} = \frac{(5/3)^{4+3}}{(5/3)^5} = \frac{(5/3)^7}{(5/3)^5} = (5/3)^{7-5} = (5/3)^2 = \boxed{\frac{25}{9}}.$$

4.4.5 One of the fractions is not the same as the other two, so it looks like we can't use the same strategy as the previous problem. However, $\frac{4}{7}$ is the reciprocal of $\frac{7}{4}$, so we can use negative exponents! Since $\frac{4}{7} = \left(\frac{7}{4}\right)^{-1}$, we have $\left(\frac{4}{7}\right)^5 = \left(\left(\frac{7}{4}\right)^{-1}\right)^5 = \left(\frac{7}{4}\right)^{-5}$. Then, we have

$$\left(\frac{7}{4}\right)^3 \left(\frac{4}{7}\right)^5 \left(\frac{7}{4}\right)^3 = \left(\frac{7}{4}\right)^3 \left(\frac{7}{4}\right)^{-5} \left(\frac{7}{4}\right)^3 = \left(\frac{7}{4}\right)^{3+(-5)} \left(\frac{7}{4}\right)^3 = \left(\frac{7}{4}\right)^{3+(-5)+3} = \left(\frac{7}{4}\right)^1 = \boxed{\frac{7}{4}}.$$

4.4.6 We have

$$\frac{a^{-n}}{b^{-n}} = a^{-n} \div b^{-n} \qquad \text{definition of fraction}$$

$$= \frac{1}{a^n} \div \frac{1}{b^n} \qquad \text{negation in exponent}$$

$$= \frac{1}{a^n} \cdot \frac{b^n}{1} \qquad \text{division by a fraction}$$

$$= \frac{b^n}{a^n}. \qquad \text{product of fractions}$$

Since $\left(\frac{a}{b}\right)^{-n}$ and $\frac{a^{-n}}{b^{-n}}$ both equal $\frac{b^n}{a^n}$, we must have $\left(\frac{a}{b}\right)^{-n} = \frac{a^{-n}}{b^{-n}}$.

Exercises for Section 4.5

4.5.1

(a) $\frac{36}{27} = \frac{4 \cdot 9}{3 \cdot 9} = \boxed{\frac{4}{3}}$.

(b) We simplify this fraction in steps:

$$\frac{256}{304} = \frac{2 \cdot 128}{2 \cdot 152} = \frac{128}{152} = \frac{2 \cdot 64}{2 \cdot 76} = \frac{64}{76} = \frac{2 \cdot 32}{2 \cdot 38} = \frac{32}{38} = \frac{2 \cdot 16}{2 \cdot 19} = \boxed{\frac{16}{19}}.$$

(c) $\frac{4800}{12000} = \frac{48 \cdot 100}{120 \cdot 100} = \frac{48}{120} = \frac{4 \cdot 12}{10 \cdot 12} = \frac{4}{10} = \boxed{\frac{2}{5}}.$

(d) We start by noticing that $\frac{1260}{1008} = \frac{2 \cdot 630}{2 \cdot 504} = \frac{630}{504} = \frac{2 \cdot 315}{2 \cdot 252} = \frac{315}{252}.$ Using our divisibility rule for 9, we see that the sum of the digits of 315 is 9, as is the sum of the digits of 252. So, 315 and 252 are both divisible by 9, and we have $\frac{315}{252} = \frac{9 \cdot 35}{9 \cdot 28} = \frac{35}{28} = \frac{5 \cdot 7}{4 \cdot 7} = \boxed{\frac{5}{4}}.$

4.5.2

(a) We simplify before multiplying. We have $\frac{24}{80} = \frac{3 \cdot 8}{10 \cdot 8} = \frac{3}{10}$ and $\frac{28}{49} = \frac{4 \cdot 7}{7 \cdot 7} = \frac{4}{7}$, so $\frac{24}{80} \cdot \frac{28}{49} = \frac{3}{10} \cdot \frac{4}{7} = \frac{12}{70} = \frac{6 \cdot 2}{35 \cdot 2} = \boxed{\frac{6}{35}}.$

(b) We have $\frac{88}{34} = \frac{44 \cdot 2}{17 \cdot 2} = \frac{44}{17}$, so

$$\frac{88}{34} \div \frac{44}{51} = \frac{44}{17} \div \frac{44}{51} = \frac{44}{17} \cdot \frac{51}{44} = \frac{44 \cdot 51}{17 \cdot 44} = \frac{44}{44} \cdot \frac{51}{17} = 1 \cdot 3 = \boxed{3}.$$

(c) The product of a negative number and a positive number is negative, so we have $\left(-\frac{84}{125}\right) \cdot \frac{100}{63} = -\left(\frac{84}{125} \cdot \frac{100}{63}\right)$. We can't simplify either of these fractions, but each numerator has divisors besides 1 in common with the other fraction's denominator. So, we write $-\left(\frac{84}{125} \cdot \frac{100}{63}\right) = -\frac{84 \cdot 100}{125 \cdot 63} = -\left(\frac{84}{63} \cdot \frac{100}{125}\right)$. We have $\frac{84}{63} = \frac{4 \cdot 21}{3 \cdot 21} = \frac{4}{3}$ and $\frac{100}{125} = \frac{4 \cdot 25}{5 \cdot 25} = \frac{4}{5}$, so $-\left(\frac{84}{63} \cdot \frac{100}{125}\right) = -\left(\frac{4}{3} \cdot \frac{4}{5}\right) = \boxed{-\frac{16}{15}}.$

(d) We start with $\frac{400}{39} \div \frac{1300}{9} = \frac{400}{39} \cdot \frac{9}{1300} = \frac{400 \cdot 9}{39 \cdot 1300} = \frac{400}{1300} \cdot \frac{9}{39}$. We have $\frac{400}{1300} = \frac{4 \cdot 100}{13 \cdot 100} = \frac{4}{13}$ and $\frac{9}{39} = \frac{3 \cdot 3}{13 \cdot 3} = \frac{3}{13}$, so $\frac{400}{1300} \cdot \frac{9}{39} = \frac{4}{13} \cdot \frac{3}{13} = \boxed{\frac{12}{169}}.$

4.5.3

(a) $\frac{4a^3b}{2ab} = \frac{4}{2} \cdot \frac{a^3}{a} \cdot \frac{b}{b} = 2 \cdot a^2 \cdot 1 = \boxed{2a^2}.$

(b) $\frac{8m^7p^{12}}{12m^5p^{15}} = \frac{8}{12} \cdot \frac{m^7}{m^5} \cdot \frac{p^{12}}{p^{15}} = \frac{2}{3} \cdot m^{7-5} \cdot p^{12-15} = \frac{2}{3}m^2p^{-3} = \frac{2}{3}m^2 \cdot \frac{1}{p^3} = \boxed{\frac{2m^2}{3p^3}}$

4.5.4 The integers 4 through 11 appear as factors in both the numerator and denominator, so they cancel out conveniently:

$$\frac{3}{4} \cdot \frac{4}{5} \cdot \frac{5}{6} \cdot \frac{6}{7} \cdot \frac{7}{8} \cdot \frac{8}{9} \cdot \frac{9}{10} \cdot \frac{10}{11} \cdot \frac{11}{12} = \frac{3 \cdot (4 \cdot 5 \cdot 6 \cdot 7 \cdot 8 \cdot 9 \cdot 10 \cdot 11)}{12 \cdot (4 \cdot 5 \cdot 6 \cdot 7 \cdot 8 \cdot 9 \cdot 10 \cdot 11)}.$$

Canceling the common factor leaves $\frac{3}{12}$, which equals $\boxed{\frac{1}{4}}.$

4.5.5 We simplify the fraction we wish to evaluate before we substitute:

$$\frac{42x^3y^6}{35x^2y^6} = \frac{42}{35} \cdot \frac{x^3}{x^2} \cdot \frac{y^6}{y^6} = \frac{6 \cdot 7}{5 \cdot 7} \cdot x^{3-2} \cdot 1 = \frac{6}{5}x.$$

When $x = \frac{5}{4}$, we have $\frac{6}{5}x = \frac{6}{5} \cdot \frac{5}{4} = \frac{30}{20} = \boxed{\frac{3}{2}}.$

Exercises for Section 4.6

4.6.1

(a) Writing the two fractions with a common denominator gives us $\frac{3}{2} = \frac{3\cdot5}{2\cdot5} = \frac{15}{10}$ and $\frac{7}{5} = \frac{7\cdot2}{5\cdot2} = \frac{14}{10}$. Since 14 is smaller than 15, we know that $\frac{14}{10}$ is smaller than $\frac{15}{10}$. This means $\boxed{\frac{7}{5}}$ is the smaller fraction.

(b) Every negative number is less than 0, and 0 is less than every positive number. So, every negative number is less than every positive number, which means that $\boxed{-\frac{3}{4}}$ is the smaller number.

(c) Since the numbers are negative, the number that is farther from 0 on the number line is the lesser number. To locate $-\frac{2}{5}$ on the number line, we go two steps of length $\frac{1}{5}$ to the left of 0. To get to $-\frac{3}{5}$, we must take one more step leftward of length $\frac{1}{5}$. So, $-\frac{3}{5}$ is to the left of $-\frac{2}{5}$ on the number line, which means that $\boxed{-\frac{3}{5}}$ is the smaller number.

4.6.2

(a) We write all three fractions with a common denominator. The least common multiple of 2, 4, and 12 is 12. We have $\frac{1}{2} = \frac{1}{2} \cdot \frac{6}{6} = \frac{6}{12}$ and $\frac{3}{4} = \frac{3}{4} \cdot \frac{3}{3} = \frac{9}{12}$, so the numbers from largest to smallest are $\boxed{\frac{3}{4}, \frac{1}{2}, \frac{5}{12}}$.

(b) The least common denominator of these three fractions is 24. We have $\frac{3}{4} \cdot \frac{6}{6} = \frac{18}{24}$, $\frac{2}{3} \cdot \frac{8}{8} = \frac{16}{24}$, and $\frac{5}{8} \cdot \frac{3}{3} = \frac{15}{24}$, so the numbers from largest to smallest are $\boxed{\frac{3}{4}, \frac{2}{3}, \frac{5}{8}}$.

(c) Since $\frac{5}{2}$ is the only positive number in the list, it is the largest number. To compare $-\frac{5}{4}$ and $-\frac{13}{3}$ to -3, we can think about what pair of consecutive integers each is between. Since 5 is between 4 and 8, we know that $\frac{5}{4}$ is between $\frac{4}{4} = 1$ and $\frac{8}{4} = 2$. So, $-\frac{5}{4}$ is between -2 and -1, which means $-\frac{5}{4}$ is greater than -3. Similarly, since 13 is between 12 and 15, we know that $\frac{13}{3}$ is between $\frac{12}{3} = 4$ and $\frac{15}{3} = 5$. This means that $-\frac{13}{3}$ is between -5 and -4, which means $-\frac{13}{3}$ is less than -3. Therefore, from largest to smallest, the numbers are $\boxed{\frac{5}{2}, -\frac{5}{4}, -3, -\frac{13}{3}}$.

4.6.3 The reciprocal of -2 is $\frac{1}{-2}$, which equals $-\frac{1}{2}$. Since -2 is less than -1 and $-\frac{1}{2}$ is greater than -1 (because $-\frac{1}{2}$ is between -1 and 0), we know that $\boxed{-2}$ is less than $-\frac{1}{2}$.

The reciprocal of -1 is $\frac{1}{-1} = -1$, which is not greater than -1. The number 0 does not have a reciprocal. The reciprocal of 1 is 1, which is not greater than 1. The reciprocal of 2 is $\frac{1}{2}$, which is less than 2. So, none of the other numbers in the list is less than its reciprocal.

4.6.4 Writing the two fractions with a common denominator gives $\frac{3}{2011} = \frac{3\cdot2012}{2011\cdot2012}$ and $\frac{3}{2012} = \frac{3\cdot2011}{2011\cdot2012}$. Since $3 \cdot 2012$ is greater than $3 \cdot 2011$, we see that $\boxed{\frac{3}{2011}}$ is greater than $\frac{3}{2012}$.

We might also have thought about the number line. To locate $\frac{3}{2011}$ on the number line, we split the number line between 0 and 1 into 2011 equal pieces, and $\frac{3}{2011}$ is at the right end of the 3rd piece. Similarly, to locate $\frac{3}{2012}$, we split the number line between 0 and 1 into 2012 equal pieces, and $\frac{3}{2012}$ is at the right end of the 3rd piece. We make smaller pieces when we make 2012 pieces than we do when we make

2011 pieces. Therefore, $\frac{3}{2012}$ is not as far from 0 as $\frac{3}{2011}$, which means $\boxed{\frac{3}{2011}}$ is greater than $\frac{3}{2012}$.

4.6.5 Since the least common multiple of 30 and 35 is 210, the least common denominator of $\frac{19}{30}$ and $\frac{22}{35}$ is 210. We have $\frac{19}{30} = \frac{19 \cdot 7}{30 \cdot 7} = \frac{133}{210}$ and $\frac{22}{35} = \frac{22 \cdot 6}{35 \cdot 6} = \frac{132}{210}$, so $\boxed{\frac{19}{30}}$ is the greater fraction.

4.6.6 We see that the numerators of the fractions are close to 5 times the denominators. So, we compare each fraction to 5. Since $\frac{505}{101} = 5$, we see that $\frac{506}{101}$ is greater than 5. Since $\frac{510}{102} = 5$, we see that $\frac{509}{102}$ is less than 5. Combining these, we know that $\boxed{\frac{506}{101}}$ is greater than $\frac{509}{102}$.

Exercises for Section 4.7

4.7.1

(a) $\frac{2}{3} + \frac{3}{4} = \frac{2 \cdot 4}{3 \cdot 4} + \frac{3 \cdot 3}{4 \cdot 3} = \frac{8}{12} + \frac{9}{12} = \boxed{\frac{17}{12}}$.

(b) $\frac{9}{8} - \frac{11}{12} = \frac{9 \cdot 3}{8 \cdot 3} - \frac{11 \cdot 2}{12 \cdot 2} = \frac{27}{24} - \frac{22}{24} = \boxed{\frac{5}{24}}$.

(c) We simplify each fraction first. We have $\frac{3100}{2700} = \frac{31 \cdot 100}{27 \cdot 100} = \frac{31}{27}$, $\frac{55}{66} = \frac{5 \cdot 11}{6 \cdot 11} = \frac{5}{6}$, and $\frac{888}{999} = \frac{8 \cdot 111}{9 \cdot 111} = \frac{8}{9}$. So, we have

$$\frac{3100}{2700} + \frac{55}{66} - \frac{888}{999} = \frac{31}{27} + \frac{5}{6} - \frac{8}{9}.$$

The least common multiple of 27, 6, and 9 is 54. Writing our three fractions with 54 as the denominator gives

$$\frac{31}{27} + \frac{5}{6} - \frac{8}{9} = \frac{31 \cdot 2}{27 \cdot 2} + \frac{5 \cdot 9}{6 \cdot 9} - \frac{8 \cdot 6}{9 \cdot 6} = \frac{62}{54} + \frac{45}{54} - \frac{48}{54} = \frac{62 + 45 - 48}{54} = \boxed{\frac{59}{54}}.$$

(d) $2^{-1} + 3^{-1} = \frac{1}{2} + \frac{1}{3} = \frac{3}{6} + \frac{2}{6} = \boxed{\frac{5}{6}}$.

(e) We start by simplifying the first two fractions. We have $\frac{24}{16} = \frac{3 \cdot 8}{2 \cdot 8} = \frac{3}{2}$ and $\frac{15}{9} = \frac{5 \cdot 3}{3 \cdot 3} = \frac{5}{3}$, so

$$\frac{24}{16} + \frac{15}{9} - \frac{7}{6} = \frac{3}{2} + \frac{5}{3} - \frac{7}{6} = \frac{9}{6} + \frac{10}{6} - \frac{7}{6} = \frac{9 + 10 - 7}{6} = \frac{12}{6} = \boxed{2}.$$

(f) Since $2 - \frac{2}{3} = \frac{6}{3} - \frac{2}{3} = \frac{4}{3}$, we have

$$\frac{8}{2 - \frac{2}{3}} = \frac{8}{\frac{4}{3}} = 8 \cdot \frac{3}{4} = \frac{24}{4} = \boxed{6}.$$

(g) $\frac{3}{4} + 6 - \frac{7}{2} = \frac{3}{4} + \frac{24}{4} - \frac{14}{4} = \frac{3 + 24 - 14}{4} = \boxed{\frac{13}{4}}$.

(h) We have $\frac{2}{3} + \frac{5}{6} = \frac{4}{6} + \frac{5}{6} = \frac{9}{6} = \frac{3}{2}$ and $\frac{3}{4} - \frac{1}{2} = \frac{3}{4} - \frac{2}{4} = \frac{1}{4}$, so

$$\frac{\frac{2}{3} + \frac{5}{6}}{\frac{3}{4} - \frac{1}{2}} = \frac{\frac{3}{2}}{\frac{1}{4}} = \frac{3}{2} \cdot \frac{4}{1} = \frac{12}{2} = \boxed{6}.$$

4.7.2 We have $\frac{1}{2} + \frac{1}{5} = \frac{5}{10} + \frac{2}{10} = \frac{7}{10}$, and the reciprocal of $\frac{7}{10}$ is $\boxed{\frac{10}{7}}$.

4.7.3 To determine how much greater one number is than another, we subtract the second number from the first. (For example, the amount by which 7 is greater than 3 is $7 - 3 = 4$.) So, the amount by which $\frac{2003}{25} + 25$ is greater than $\frac{2003+25}{25}$ is their difference. Writing $\frac{2003+25}{25}$ as the sum of two fractions gives

$$\frac{2003 + 25}{25} = \frac{2003}{25} + \frac{25}{25} = \frac{2003}{25} + 1.$$

So, now we want the difference between $\frac{2003}{25} + 25$ and $\frac{2003}{25} + 1$, which is $25 - 1 = \boxed{24}$.

4.7.4

(a) $\frac{1}{1\cdot2} + \frac{1}{2\cdot3} = \frac{1}{2} + \frac{1}{6} = \frac{3}{6} + \frac{1}{6} = \frac{4}{6} = \boxed{\frac{2}{3}}$.

(b) We use the first part to add the first two fractions:

$$\left(\frac{1}{1\cdot2} + \frac{1}{2\cdot3}\right) + \frac{1}{3\cdot4} = \frac{2}{3} + \frac{1}{3\cdot4} = \frac{2\cdot4}{3\cdot4} + \frac{1}{3\cdot4} = \frac{9}{12} = \boxed{\frac{3}{4}}.$$

(c) We use part (b) to add the first three fractions, and we have

$$\left(\frac{1}{1\cdot2} + \frac{1}{2\cdot3} + \frac{1}{3\cdot4}\right) + \frac{1}{4\cdot5} = \frac{3}{4} + \frac{1}{4\cdot5} = \frac{3\cdot5}{4\cdot5} + \frac{1}{4\cdot5} = \frac{16}{20} = \boxed{\frac{4}{5}}.$$

(d) We use part (c) to add the first four fractions, and we have

$$\left(\frac{1}{1\cdot2} + \frac{1}{2\cdot3} + \frac{1}{3\cdot4} + \frac{1}{4\cdot5}\right) + \frac{1}{5\cdot6} = \frac{4}{5} + \frac{1}{5\cdot6} = \frac{4\cdot6}{5\cdot6} + \frac{1}{5\cdot6} = \frac{25}{30} = \boxed{\frac{5}{6}}.$$

Do you notice a pattern in these answers? Will this pattern continue? If so, why?

4.7.5 *Solution 1: Compute both products.* We have $\frac{2}{3} \cdot \frac{4}{5} + \frac{2}{3} \cdot \frac{11}{10} = \frac{8}{15} + \frac{22}{30}$. Simplifying $\frac{22}{30}$ gives $\frac{22}{30} = \frac{11}{15}$, so $\frac{8}{15} + \frac{22}{30} = \frac{8}{15} + \frac{11}{15} = \boxed{\frac{19}{15}}$.

Solution 2: Factor. $\frac{2}{3}$ appears in both products, so we can factor:

$$\frac{2}{3} \cdot \frac{4}{5} + \frac{2}{3} \cdot \frac{11}{10} = \frac{2}{3}\left(\frac{4}{5} + \frac{11}{10}\right) = \frac{2}{3}\left(\frac{8}{10} + \frac{11}{10}\right) = \frac{2}{3} \cdot \frac{19}{10} = \frac{2}{10} \cdot \frac{19}{3} = \frac{1}{5} \cdot \frac{19}{3} = \boxed{\frac{19}{15}}.$$

4.7.6 The positive divisors of 12 are 1, 2, 3, 4, 6 and 12, so we must find the sum $\frac{1}{1} + \frac{1}{2} + \frac{1}{3} + \frac{1}{4} + \frac{1}{6} + \frac{1}{12}$. Since each denominator is a divisor of 12, we can use 12 as a common denominator:

$$\frac{1}{1} + \frac{1}{2} + \frac{1}{3} + \frac{1}{4} + \frac{1}{6} + \frac{1}{12} = \frac{12}{12} + \frac{6}{12} + \frac{4}{12} + \frac{3}{12} + \frac{2}{12} + \frac{1}{12} = \frac{12+6+4+3+2+1}{12} = \frac{28}{12} = \boxed{\frac{7}{3}}.$$

Notice that our answer is the sum of the divisors of 12 divided by 12. Is that a coincidence?

4.7.7

(a) To get a handle on the problem, we think about how it works if the numbers are integers. Suppose the question were "What number must we add to 3 to get 7?" Then, the answer is $7 - 3 = 4$. This makes sense; if $a + b = c$, then $a = c - b$.

Returning to the question "What number must we add to $\frac{3}{10}$ to get $\frac{7}{15}$?", we see that the answer is $\frac{7}{15} - \frac{3}{10} = \frac{14}{30} - \frac{9}{30} = \frac{5}{30} = \boxed{\frac{1}{6}}$. Checking our answer, we have $\frac{1}{6} + \frac{3}{10} = \frac{5}{30} + \frac{9}{30} = \frac{14}{30} = \frac{7}{15}$.

(b) Suppose the question were "What number must we subtract 3 from to get 1?" Then, the answer is $3 + 1 = 4$. In other words, if $a - b = c$, then $a = b + c$.

Returning to the question, "What number must we subtract $\frac{5}{9}$ from to get $\frac{1}{6}$?", our answer is $\frac{5}{9} + \frac{1}{6} = \frac{10}{18} + \frac{3}{18} = \boxed{\frac{13}{18}}$.

(c) Again, we think about the problem with integers, asking ourselves, "What number must we subtract from 5 to get 1?" Then, the answer is $5 - 1 = 4$. Similarly, suppose we have "What number must we subtract from a to get b?" Looking at our example, we expect that the answer is $a - b$. Testing this, we have

$$a - (a - b) = a - a + b = b.$$

So, $a - b$ is the number we subtract from a to get b.

Returning to the question, "What number must we subtract from $\frac{5}{6}$ to get $\frac{1}{10}$?", our answer is $\frac{5}{6} - \frac{1}{10} = \frac{50}{60} - \frac{6}{60} = \frac{44}{60} = \boxed{\frac{11}{15}}$.

4.7.8 The fraction of students that go home by bus, automobile, or bicycle is $\frac{1}{2} + \frac{1}{4} + \frac{1}{10} = \frac{10}{20} + \frac{5}{20} + \frac{2}{20} = \frac{17}{20}$. Therefore, the remaining $1 - \frac{17}{20} = \frac{20}{20} - \frac{17}{20} = \boxed{\frac{3}{20}}$ of the students walk home.

4.7.9 Let's try to find another pair of fractions with 1 as the numerator that add to $\frac{1}{2}$. We'll start by guessing that one of the fractions is $\frac{1}{3}$. The number we must add to $\frac{1}{3}$ to get $\frac{1}{2}$ is $\frac{1}{2} - \frac{1}{3}$. Computing this difference gives $\frac{1}{2} - \frac{1}{3} = \frac{3}{6} - \frac{2}{6} = \frac{1}{6}$. Success! Checking our answer we have $\frac{1}{3} + \frac{1}{6} = \frac{2}{6} + \frac{1}{6} = \frac{3}{6} = \frac{1}{2}$. There are many, many other solutions to this problem.

Exercises for Section 4.8

4.8.1

(a) $4\frac{7}{8} - 1\frac{3}{4} = (4 - 1) + \left(\frac{7}{8} - \frac{3}{4}\right) = 3 + \left(\frac{7}{8} - \frac{6}{8}\right) = 3 + \frac{1}{8} = \boxed{3\frac{1}{8}}$.

(b) $3\frac{1}{3} - 7\frac{2}{9} = 3 + \frac{1}{3} - 7 - \frac{2}{9} = (3 - 7) + \left(\frac{1}{3} - \frac{2}{9}\right) = -4 + \left(\frac{3}{9} - \frac{2}{9}\right) = -4 + \frac{1}{9}$. Here, we have to be careful. The number $-4 + \frac{1}{9}$ is $\frac{1}{9}$ to the right of -4 on the number line, so it is $\boxed{-3\frac{8}{9}}$, not $-4\frac{1}{9}$.

(c) $19\frac{3}{20} - 9\frac{13}{15} = 19 + \frac{3}{20} - 9 - \frac{13}{15} = (19 - 9) + \left(\frac{3}{20} - \frac{13}{15}\right) = 10 + \left(\frac{9}{60} - \frac{52}{60}\right) = 10 - \frac{43}{60} = 9 + 1 - \frac{43}{60} = 9 + \frac{17}{60} = \boxed{9\frac{17}{60}}$. We might also have evaluated $10 - \frac{43}{60}$ by considering the number line. Suppose we break the number line between 9 and 10 into 60 equal pieces. To get to $10 - \frac{43}{60}$, we start at 10, and then go leftward by 43 of these 60 pieces. This leaves us $60 - 43 = 17$ of these pieces to the right of 9, so we are at $9\frac{17}{60}$.

(d) We first sum $6\frac{1}{2}$ and $5\frac{1}{3}$. We have $6\frac{1}{2} + 5\frac{1}{3} = (6 + 5) + \left(\frac{1}{2} + \frac{1}{3}\right) = 11 + \left(\frac{3}{6} + \frac{2}{6}\right) = 11 + \frac{5}{6}$. So, we have

$$18 - \left(6\frac{1}{2} + 5\frac{1}{3}\right) = 18 - \left(11 + \frac{5}{6}\right) = 18 - 11 - \frac{5}{6} = 7 - \frac{5}{6} = \boxed{6\frac{1}{6}}.$$

(e) Since $5\frac{5}{12} = 5 + \frac{5}{12} = \frac{60}{12} + \frac{5}{12} = \frac{65}{12}$, we have $5\frac{5}{12} \cdot 24 = \frac{65}{12} \cdot 24 = 65 \cdot \frac{24}{12} = 65 \cdot 2 = \boxed{130}$.

We also could have used the distributive property:

$$5\frac{5}{12} \cdot 24 = \left(5 + \frac{5}{12}\right)24 = 5 \cdot 24 + \frac{5}{12} \cdot 24 = 120 + 5 \cdot \frac{24}{12} = 120 + 5 \cdot 2 = \boxed{130}.$$

(f) We subtract first: $6\frac{2}{3} - 4\frac{4}{9} = 6 + \frac{2}{3} - 4 - \frac{4}{9} = (6 - 4) + \left(\frac{2}{3} - \frac{4}{9}\right) = 2 + \left(\frac{6}{9} - \frac{4}{9}\right) = 2 + \frac{2}{9} = \frac{20}{9}$. Therefore,

$$1\frac{1}{2} \cdot \left(6\frac{2}{3} - 4\frac{4}{9}\right) = \frac{3}{2} \cdot \frac{20}{9} = \frac{3}{9} \cdot \frac{20}{2} = \frac{1}{3} \cdot 10 = \frac{10}{3} = \boxed{3\frac{1}{3}}.$$

(g) We divide first: $2\frac{1}{3} \div 3\frac{1}{2} = \frac{7}{3} \div \frac{7}{2} = \frac{7}{3} \cdot \frac{2}{7} = \frac{2}{3}$. So, $5\frac{1}{3} + 2\frac{1}{3} \div 3\frac{1}{2} = 5\frac{1}{3} + \frac{2}{3} = \boxed{6}$.

(h) The quotient of a positive number and a negative number is negative, so we have

$$3\frac{2}{3} \div \left(-6\frac{7}{8}\right) = -\left(3\frac{2}{3} \div 6\frac{7}{8}\right) = -\left(\frac{11}{3} \div \frac{55}{8}\right) = -\left(\frac{11}{3} \cdot \frac{8}{55}\right) = -\left(\frac{11}{55} \cdot \frac{8}{3}\right) = -\left(\frac{1}{5} \cdot \frac{8}{3}\right) = \boxed{-\frac{8}{15}}.$$

4.8.2 We have $2\frac{1}{2} + 3\frac{1}{3} + 4\frac{1}{4} + 5\frac{1}{5} + 6\frac{1}{6} = (2 + 3 + 4 + 5 + 6) + \left(\frac{1}{2} + \frac{1}{3} + \frac{1}{4} + \frac{1}{5} + \frac{1}{6}\right) = 20 + \left(\frac{1}{2} + \frac{1}{3} + \frac{1}{4} + \frac{1}{5} + \frac{1}{6}\right)$. We could find a common denominator of all 5 fractions, but we don't have to know exactly what the sum is. We only have to know the greatest integer less than the sum. Since $\frac{1}{3} + \frac{1}{6} = \frac{2}{6} + \frac{1}{6} = \frac{1}{2}$, we know that $\frac{1}{2} + \frac{1}{3} + \frac{1}{4} + \frac{1}{5} + \frac{1}{6} = \left(\frac{1}{2} + \frac{1}{3} + \frac{1}{6}\right) + \frac{1}{4} + \frac{1}{5} = 1 + \frac{1}{4} + \frac{1}{5}$. Since both $\frac{1}{4}$ and $\frac{1}{5}$ are less than $\frac{1}{2}$, we know that the sum $1 + \frac{1}{4} + \frac{1}{5}$ is between 1 and 2. So, the original sum is between 21 and 22. Therefore, the greatest integer less than the original sum is $\boxed{21}$.

4.8.3 We start with the easier factor, $3a - 4$. We have

$$3a - 4 = 3 \cdot 1\frac{1}{3} - 4 = 3 \cdot \frac{4}{3} - 4 = 4 - 4 = 0.$$

So, it doesn't matter what $7a^2 - 11a + 3$ equals when $a = 1\frac{1}{3}$. Since $3a - 4$ equals 0 when $a = 1\frac{1}{3}$, the product is $\boxed{0}$.

4.8.4 The amount of weight Peggy lost is the difference between her weight at the start of the season and her weight at the end of the season. This difference is

$$136\frac{3}{4} - 131\frac{7}{8} = 136 + \frac{3}{4} - 131 - \frac{7}{8} = (136 - 131) + \left(\frac{3}{4} - \frac{7}{8}\right) = 5 + \left(\frac{6}{8} - \frac{7}{8}\right) = 5 - \frac{1}{8} = \boxed{4\frac{7}{8}} \text{ pounds}.$$

4.8.5 To make one of each cake, I need $2\frac{1}{2} + 3\frac{1}{3}$ cups of flour, so for 3 of each cake, I need

$$3\left(2\frac{1}{2} + 3\frac{1}{3}\right) = 3 \cdot 2\frac{1}{2} + 3 \cdot 3\frac{1}{3} = 3\left(2 + \frac{1}{2}\right) + 3\left(3 + \frac{1}{3}\right) = 6 + \frac{3}{2} + 9 + 1 = 16 + 1\frac{1}{2} = \boxed{17\frac{1}{2} \text{ cups}}.$$

4.8.6 Each hour is 60 minutes, so the number of minutes dedicated to programs is $60 - 6\frac{1}{2} = 60 - 6 - \frac{1}{2} = 54 - \frac{1}{2} = 53\frac{1}{2}$. So, the fraction of each hour dedicated to programs is

$$\frac{53\frac{1}{2}}{60} = \frac{\frac{107}{2}}{60} = \frac{107}{2} \cdot \frac{1}{60} = \boxed{\frac{107}{120}}.$$

Review Problems

4.55

(a) $\frac{3}{14} + \frac{5}{7} - \frac{1}{21} = \frac{9}{42} + \frac{30}{42} - \frac{2}{42} = \boxed{\frac{37}{42}}$.

(b) We simplify both fractions first: $\frac{64}{96} - \frac{63}{84} = \frac{2 \cdot 32}{3 \cdot 32} - \frac{3 \cdot 21}{4 \cdot 21} = \frac{2}{3} - \frac{3}{4} = \frac{8}{12} - \frac{9}{12} = \boxed{-\frac{1}{12}}$.

(c) $\frac{36}{48} \cdot \frac{44}{66} \cdot \frac{16}{56} = \frac{3 \cdot 12}{4 \cdot 12} \cdot \frac{2 \cdot 22}{3 \cdot 22} \cdot \frac{2 \cdot 8}{7 \cdot 8} = \frac{3}{4} \cdot \frac{2}{3} \cdot \frac{2}{7} = \frac{12}{4 \cdot 3 \cdot 7} = \boxed{\frac{1}{7}}$.

(d) We pair up factors in the numerator with factors in the denominator in ways that allow us to cancel common divisors:

$$\frac{27 \cdot 14 \cdot 35}{42 \cdot 9 \cdot 28 \cdot 24} = \frac{14}{42} \cdot \frac{27}{9} \cdot \frac{35}{28} \cdot \frac{1}{24} = \frac{1}{3} \cdot 3 \cdot \frac{5}{4} \cdot \frac{1}{24} = 1 \cdot \frac{5}{4} \cdot \frac{1}{24} = \boxed{\frac{5}{96}}.$$

(e) $\left(\frac{3}{4}\right)^3 = \frac{3^3}{4^3} = \boxed{\frac{27}{64}}$.

(f) We first simplify the fraction: $\frac{18}{27} = \frac{2 \cdot 9}{3 \cdot 9} = \frac{2}{3}$. So, we have

$$\left(\frac{18}{27}\right)^{-4} = \left(\frac{2}{3}\right)^{-4} = \left(\left(\frac{2}{3}\right)^{-1}\right)^4 = \left(\frac{3}{2}\right)^4 = \frac{3^4}{2^4} = \boxed{\frac{81}{16}}.$$

(g) *Solution 1: Evaluate the expression in parentheses first.*

$$6\left(\frac{7}{12} - \frac{2}{3} + \frac{1}{4}\right) = 6\left(\frac{7}{12} - \frac{8}{12} + \frac{3}{12}\right) = 6 \cdot \frac{2}{12} = 6 \cdot \frac{1}{6} = \boxed{1}.$$

Solution 2: Apply the distributive property.

$$6\left(\frac{7}{12} - \frac{2}{3} + \frac{1}{4}\right) = 6 \cdot \frac{7}{12} - 6 \cdot \frac{2}{3} + 6 \cdot \frac{1}{4} = \frac{7}{2} - 4 + \frac{3}{2} = \frac{7}{2} + \frac{3}{2} - 4 = 5 - 4 = \boxed{1}.$$

(h) $\frac{31+71+111}{1+3+5} \cdot \frac{5+15+25}{111+71+31} = \frac{31+71+111}{111+71+31} \cdot \frac{5+15+25}{1+3+5} = 1 \cdot \frac{45}{9} = \boxed{5}$.

(i) We have $50 - (-3)^3 = 50 - (-27) = 77$ and $\left(\frac{2}{7}\right)^{-1} = \frac{7}{2}$, so $\frac{50-(-3)^3}{\left(\frac{2}{7}\right)^{-1}} = \frac{77}{\frac{7}{2}} = 77 \cdot \frac{2}{7} = \frac{77}{7} \cdot 2 = 11 \cdot 2 = \boxed{22}$.

(j) The fractions conveniently cancel out:

$$\left(5\frac{1}{3} - 2\frac{1}{4}\right) + \left(5\frac{1}{4} - 3\frac{1}{3}\right) = 5 + \frac{1}{3} - 2 - \frac{1}{4} + 5 + \frac{1}{4} - 3 - \frac{1}{3}$$

$$= 5 - 2 + 5 - 3 + \frac{1}{3} - \frac{1}{3} + \frac{1}{4} - \frac{1}{4}$$

$$= \boxed{5}.$$

(k) The square of a negative number is positive, so $\left(-1\frac{1}{4}\right)^2 = \left(1\frac{1}{4}\right)^2$. Writing the mixed number as a fraction, we find

$$\left(1\frac{1}{4}\right)^2 = \left(\frac{5}{4}\right)^2 = \frac{5^2}{4^2} = \frac{25}{16} = \boxed{1\frac{9}{16}}.$$

(l) $6\left(11\frac{2}{3} + 4\frac{1}{2}\right) = 6\left(11 + \frac{2}{3} + 4 + \frac{1}{2}\right) = 6\left(15 + \frac{2}{3} + \frac{1}{2}\right) = 6 \cdot 15 + 6 \cdot \frac{2}{3} + 6 \cdot \frac{1}{2} = 90 + 4 + 3 = \boxed{97}.$

4.56 We evaluate the numerator and denominator separately. In the numerator, we have

$$3 + x(3 + 2x) - 3^2 = 3 + (-4)(3 + 2(-4)) - 9$$
$$= 3 + (-4)(3 + (-8)) - 9$$
$$= 3 + (-4)(-5) - 9 = 3 + 20 - 9 = 14.$$

In the denominator, we have $x - 5 + x^2 = (-4) - 5 + (-4)^2 = -9 + 16 = 7$. So, the fraction equals $\frac{14}{7} = \boxed{2}$.

4.57 The denominator of each fraction is half of the integer that follows the fraction. So, the product equals

$$\frac{1}{2} \cdot 4 \cdot \frac{1}{8} \cdot 16 \cdot \frac{1}{32} \cdot 64 \cdot \frac{1}{128} \cdot 256 = \frac{4}{2} \cdot \frac{16}{8} \cdot \frac{64}{32} \cdot \frac{256}{128} = 2 \cdot 2 \cdot 2 \cdot 2 = \boxed{16}.$$

4.58 Since "of" means multiply, $\frac{1}{6}$ of $\frac{2}{7}$ of $\frac{1}{2}$ of 168 is $\frac{1}{6} \cdot \frac{2}{7} \cdot \frac{1}{2} \cdot 168$. We have

$$\frac{1}{6} \cdot \frac{2}{7} \cdot \frac{1}{2} \cdot 168 = \frac{1 \cdot 2 \cdot 1}{6 \cdot 7 \cdot 2} \cdot 168 = \frac{1}{42} \cdot 168 = \boxed{4}.$$

4.59 One hour is 60 minutes, so $\frac{9}{10}$ hour is $\frac{9}{10} \cdot 60 = 9 \cdot \frac{60}{10} = 9 \cdot 6 = 54$ minutes. So, $5\frac{9}{10}$ hours is 5 hours and 54 minutes. 5 hours after 10:20 a.m. is 3:20 p.m., and 54 minutes after 3:20 p.m. is 6 minutes before one hour after 3:20 p.m., or $\boxed{4:14 \text{ p.m.}}$

4.60 $99 \div \frac{101}{102} = 99 \cdot \frac{102}{101} = 99 \cdot 1\frac{1}{101} = 99\left(1 + \frac{1}{101}\right) = 99 + \frac{99}{101}$. So, $\boxed{99 \div \frac{101}{102}}$ is greater than 99. We didn't have to compute the product $99 \cdot \frac{102}{101}$. All we really have to notice is that $\frac{102}{101}$ is greater than 1. The product of 99 and a number that is greater than 1 will always be greater than 99. (Similarly, the quotient of 99 and a positive number that is less than 1 will always be greater than 99.)

4.61 $2\frac{1}{5} + 3\frac{1}{3} + 5\frac{1}{2} = (2 + 3 + 5) + \left(\frac{1}{5} + \frac{1}{3} + \frac{1}{2}\right)$. The least common denominator of all three fractions is 30, and we have

$$(2 + 3 + 5) + \left(\frac{1}{5} + \frac{1}{3} + \frac{1}{2}\right) = 10 + \left(\frac{6}{30} + \frac{10}{30} + \frac{15}{30}\right) = 10 + \frac{31}{30} = 10 + 1 + \frac{1}{30} = \boxed{11\frac{1}{30}}.$$

4.62 The numerator of each fraction except the last one equals the denominator of the next fraction. This allows us to do a lot of canceling:

$$\frac{3}{2} \times \frac{4}{3} \times \frac{5}{4} \times \cdots \times \frac{2012}{2011} = \frac{3 \times 4 \times 5 \times \cdots \times 2011 \times 2012}{2 \times 3 \times 4 \times 5 \times \cdots \times 2011} = \frac{2012}{2} \times \frac{3}{3} \times \frac{4}{4} \times \frac{5}{5} \times \cdots \times \frac{2011}{2011}.$$

All of the fractions after the first equal 1, so the product is $\frac{2012}{2} = \boxed{1006}$.

4.63 We have $\frac{1}{2} \cdot \frac{2}{3} \cdot \frac{3}{4} = \frac{1 \cdot 2 \cdot 3}{2 \cdot 3 \cdot 4} = \frac{1}{4}$ and $\frac{6}{8} \cdot \frac{6}{9} \cdot \frac{1}{2} = \frac{3}{4} \cdot \frac{2}{3} \cdot \frac{1}{2} = \frac{3 \cdot 2 \cdot 1}{4 \cdot 3 \cdot 2} = \frac{1}{4}$. So, the product in the numerator of the original fraction equals the product in the denominator, which means the original fraction equals $\boxed{1}$.

4.64 After paying his income tax, Kory has $1 - \frac{1}{3} = \frac{2}{3}$ of his income remaining. He places $1 - \frac{4}{5} = \frac{1}{5}$ of this amount into a savings account. Since he places $\frac{1}{5}$ of $\frac{2}{3}$ of his income into a savings account, the fraction of his income that he puts in savings is $\frac{1}{5} \cdot \frac{2}{3} = \boxed{\frac{2}{15}}$.

4.65 We have $xy = \frac{3}{7} \cdot \frac{4}{3} = \frac{4}{7}$ and $7x + 3y = 7 \cdot \frac{3}{7} + 3 \cdot \frac{4}{3} = 3 + 4 = 7$, so $\frac{xy}{7x+3y} = \frac{4/7}{7} = \frac{4}{7} \cdot \frac{1}{7} = \boxed{\frac{4}{49}}$.

4.66

(a) Since $-\frac{13}{32}$ is the only negative number in the list, it is the smallest number. Writing the other three numbers with 64 as the denominator gives $\frac{3}{8} = \frac{24}{64}$, $\frac{7}{16} = \frac{28}{64}$, and $\frac{23}{64}$. So, in order from least to greatest, the numbers are $\boxed{-\frac{13}{32}, \frac{23}{64}, \frac{3}{8}, \frac{7}{16}}$.

(b) If we write all three fractions with a common denominator, our least common denominator is $7 \cdot 4 \cdot 11$. Rather than dealing with such large numbers, we'll compare the fractions two at a time. First, we compare $\frac{9}{7}$ and $\frac{5}{4}$. Writing these with common denominator 28 gives $\frac{9}{7} = \frac{36}{28}$ and $\frac{5}{4} = \frac{35}{28}$, so $\frac{9}{7}$ is greater than $\frac{5}{4}$. Next, we compare $\frac{5}{4}$ and $\frac{14}{11}$. Writing these with common denominator 44 gives $\frac{5}{4} = \frac{55}{44}$ and $\frac{14}{11} = \frac{56}{44}$, so $\frac{5}{4}$ is less than $\frac{14}{11}$. Therefore, $\frac{5}{4}$ is the smallest of the three numbers.

Finally, we compare $\frac{14}{11}$ and $\frac{9}{7}$ by writing them with the common denominator 77. We have $\frac{14}{11} = \frac{14 \cdot 7}{11 \cdot 7} = \frac{98}{77}$ and $\frac{9}{7} = \frac{99}{77}$, so $\frac{9}{7}$ is greater than $\frac{14}{11}$. Therefore, the three numbers in order from least to greatest are $\boxed{\frac{5}{4}, \frac{14}{11}, \frac{9}{7}}$.

(c) Since $\frac{1}{2} = \frac{200}{400}$, we know that $\frac{1}{2}$ is greater than $\frac{199}{400}$. (We might also have noted that 199 is less than half of 400, so $\frac{199}{400}$ is less than $\frac{1}{2}$.) Next we compare $\frac{100}{199}$ and $\frac{1}{2}$. We might write them with a common denominator: $\frac{100}{199} = \frac{200}{398}$ and $\frac{1}{2} = \frac{199}{398}$. Or, we could notice that 100 is greater than half of 199, so $\frac{100}{199}$ is greater than $\frac{1}{2}$. Since $\frac{1}{2}$ is greater than $\frac{199}{400}$ and less than $\frac{100}{199}$, the numbers in order from least to greatest are $\boxed{\frac{199}{400}, \frac{1}{2}, \frac{100}{199}}$.

4.67 Since $\frac{725}{60}$ is just a little more than $\frac{720}{60} = 12$ and $\frac{25}{6}$ is a little more than $\frac{24}{6} = 4$, we expect the quotient to be close to 3. Dividing gives us $\frac{725}{60} \div \frac{25}{6} = \frac{725}{60} \cdot \frac{6}{25} = \frac{725}{25} \cdot \frac{6}{60} = 29 \cdot \frac{1}{10} = \frac{29}{10} = 2\frac{9}{10}$. The number $2\frac{9}{10}$ is between 2 and 3. Since $2\frac{9}{10}$ is $\frac{1}{10}$ from 3 and $\frac{9}{10}$ from 2, the integer closest to $2\frac{9}{10}$ is indeed $\boxed{3}$.

4.68 *Solution 1: Subtractions first.* We have

$$2\left(1 - \tfrac{1}{2}\right) + 3\left(1 - \tfrac{1}{3}\right) + 4\left(1 - \tfrac{1}{4}\right) + \cdots + 10\left(1 - \tfrac{1}{10}\right) = 2 \cdot \tfrac{1}{2} + 3 \cdot \tfrac{2}{3} + 4 \cdot \tfrac{3}{4} + 5 \cdot \tfrac{4}{5} + \cdots + 10 \cdot \tfrac{9}{10}$$
$$= 1 + 2 + 3 + 4 + \cdots + 9 = \boxed{45}.$$

Solution 2: Apply the distributive property to each of the 9 products. This gives us

$$2\left(1 - \tfrac{1}{2}\right) + 3\left(1 - \tfrac{1}{3}\right) + \cdots + 10\left(1 - \tfrac{1}{10}\right) = 2 \cdot 1 - 2 \cdot \tfrac{1}{2} + 3 \cdot 1 - 3 \cdot \tfrac{1}{3} + 4 \cdot 1 - 4 \cdot \tfrac{1}{4} + \cdots + 10 \cdot 1 - 10 \cdot \tfrac{1}{10}$$
$$= (2 - 1) + (3 - 1) + (4 - 1) + \cdots + (10 - 1)$$
$$= 1 + 2 + 3 + 4 + 5 + 6 + 7 + 8 + 9 = \boxed{45}.$$

4.69 After using half of the cheese for a casserole, Sam has the other half remaining. Half of the initial $3\frac{1}{4}$ pounds is $\frac{1}{2} \cdot 3\frac{1}{4} = \frac{1}{2} \cdot \frac{13}{4} = \frac{13}{8}$ pounds. He then uses $\frac{1}{4}$ pound for sandwiches, which leaves him with $\frac{13}{8} - \frac{1}{4} = \frac{13}{8} - \frac{2}{8} = \frac{11}{8} = \boxed{1\frac{3}{8} \text{ pounds}}$.

4.70 For any numbers a and b, if a is the reciprocal of b, then b is the reciprocal of a. So, if the reciprocal of a fraction is an integer, then the fraction is the reciprocal of the integer. Therefore, all of the fractions that satisfy the problem must be of the form $\frac{1}{n}$, where n is an integer.

Since the fractions must be greater than $\frac{1}{6}$, they must all be positive. Any fraction of the form $\frac{1}{n}$ is in simplest form, so now we only have to determine which of these fractions is greater than $\frac{1}{6}$. We consider the number line:

If n is greater than 6, then $\frac{1}{n}$ is closer to 0 than $\frac{1}{6}$ is, so $\frac{1}{n}$ is less than $\frac{1}{6}$. If n is a positive number less than 6, then $\frac{1}{n}$ is farther from 0 than $\frac{1}{6}$ is, so $\frac{1}{n}$ is greater than $\frac{1}{6}$. So, the fractions that satisfy the problem are $\boxed{\frac{1}{1}, \frac{1}{2}, \frac{1}{3}, \frac{1}{4}, \frac{1}{5}}$.

4.71 Factoring gives us

$$\frac{3}{19} \cdot 95 - \frac{3}{19} \cdot 57 = \frac{3}{19}(95 - 57) = \frac{3}{19} \cdot 38 = 3 \cdot \frac{38}{19} = 3 \cdot 2 = \boxed{6}.$$

4.72 Since $\frac{15}{42} = \frac{5 \cdot 3}{14 \cdot 3} = \frac{5}{14}$ and $\left(\frac{3}{2}\right)^{-2} = \left(\left(\frac{3}{2}\right)^{-1}\right)^2 = \left(\frac{2}{3}\right)^2$, we have

$$\frac{15}{42}\left(-\frac{63}{55}\right)\left(\frac{3}{2}\right)^{-2}\left(\frac{11}{2}\right)^2 = \frac{5}{14}\left(-\frac{63}{55}\right)\left(\frac{2}{3}\right)^2\left(\frac{11}{2}\right)^2 = -\frac{5}{14} \cdot \frac{63}{55} \cdot \frac{2^2}{3^2} \cdot \frac{11^2}{2^2} = -\frac{5 \cdot 63 \cdot 2^2 \cdot 11^2}{14 \cdot 55 \cdot 3^2 \cdot 2^2}.$$

We can then use the prime factorization of each number to simplify the fraction:

$$-\frac{5 \cdot 63 \cdot 2^2 \cdot 11^2}{14 \cdot 55 \cdot 3^2 \cdot 2^2} = -\frac{5 \cdot (3^2 \cdot 7) \cdot 2^2 \cdot 11^2}{(2 \cdot 7) \cdot (5 \cdot 11) \cdot 3^2 \cdot 2^2} = -\frac{2^2 \cdot 3^2 \cdot 5 \cdot 7 \cdot 11^2}{2^3 \cdot 3^2 \cdot 5 \cdot 7 \cdot 11}$$

$$= -\frac{2^2}{2^3} \cdot \frac{3^2}{3^2} \cdot \frac{5}{5} \cdot \frac{7}{7} \cdot \frac{11^2}{11}$$

$$= -\frac{1}{2} \cdot 1 \cdot 1 \cdot 1 \cdot 11 = \boxed{-\frac{11}{2}}.$$

4.73 After Maya gives $\frac{3}{5}$ of her pennies to Mitch, she has $1 - \frac{3}{5} = \frac{2}{5}$ of her original 400 pennies, which is $\frac{2}{5} \cdot 400 = 2 \cdot \frac{400}{5} = 160$ pennies. She gives $\frac{3}{4}$ of these pennies to her mother, so she has $1 - \frac{3}{4} = \frac{1}{4}$ of the 160 pennies left, which is $\frac{1}{4} \cdot 160 = \boxed{40}$ pennies.

4.74 We have

$$\frac{9}{5}\left(3\frac{1}{3}\cdot\frac{1}{4}-\frac{10}{12}\cdot\frac{1}{8}\right)=\frac{9}{5}\left(\frac{10}{3}\cdot\frac{1}{4}-\frac{5}{6}\cdot\frac{1}{8}\right)$$

$$=\frac{9}{5}\left(\frac{10}{12}-\frac{5}{48}\right)=\frac{9}{5}\left(\frac{40}{48}-\frac{5}{48}\right)$$

$$=\frac{9}{5}\cdot\frac{35}{48}=\frac{35}{5}\cdot\frac{9}{48}=7\cdot\frac{3}{16}=\boxed{\frac{21}{16}}.$$

4.75 Since $\frac{7}{19}$ is a factor in each of the products, we can factor:

$$\frac{7}{19}\cdot\frac{13}{44}+\frac{7}{19}\cdot\frac{19}{44}+\frac{7}{19}\cdot\frac{25}{44}+\frac{7}{19}\cdot\frac{31}{44}=\frac{7}{19}\left(\frac{13}{44}+\frac{19}{44}+\frac{25}{44}+\frac{31}{44}\right)$$

$$=\frac{7}{19}\cdot\frac{88}{44}=\frac{7}{19}\cdot2-\boxed{\frac{14}{19}}.$$

4.76 We have $99\cdot2\frac{1}{49}=99\left(2+\frac{1}{49}\right)=99\cdot2+99\cdot\frac{1}{49}=198+\frac{99}{49}=198+2\frac{1}{49}=200\frac{1}{49}$. So, $\boxed{99\cdot2\frac{1}{49}}$ is greater than 200.

4.77 We know that $\frac{1}{3}+\frac{1}{3}+\frac{1}{3}=1$. So, if the reciprocals of three *different* positive integers sum to 1, at least one of these reciprocals must be greater than $\frac{1}{3}$. The only reciprocal of an integer between $\frac{1}{3}$ and 1 is $\frac{1}{2}$. So, one of the three reciprocals is $\frac{1}{2}$. We now need two other reciprocals that add to $\frac{1}{2}$ in order to have three reciprocals that sum to 1.

We have $\frac{1}{4}+\frac{1}{4}=\frac{1}{2}$. So, if the reciprocals of two *different* positive integers sum to $\frac{1}{2}$, one of the reciprocals is greater than $\frac{1}{4}$ and the other is less than $\frac{1}{4}$. The only reciprocal between $\frac{1}{4}$ and $\frac{1}{2}$ is $\frac{1}{3}$, so one of our reciprocals is $\frac{1}{3}$. This leaves $\frac{1}{2}-\frac{1}{3}=\frac{3}{6}-\frac{2}{6}=\frac{1}{6}$ for the other reciprocal.

Checking, we have $\frac{1}{2}+\frac{1}{3}+\frac{1}{6}=\frac{3}{6}+\frac{2}{6}+\frac{1}{6}=1$, so the three different positive integers whose reciprocals sum to 1 are $\boxed{2,3,\text{ and }6}$.

4.78 Each stool requires $12\frac{2}{3}$ feet of tubing, so 300 stools require

$$300\cdot12\frac{2}{3}=300\left(12+\frac{2}{3}\right)=300\cdot12+300\cdot\frac{2}{3}=3600+\frac{300}{3}\cdot2=3600+200=\boxed{3800\text{ feet of tubing}}.$$

4.79 The reciprocal of 5 plus the reciprocal of 7 is $\frac{1}{5}+\frac{1}{7}=\frac{7}{35}+\frac{5}{35}=\frac{12}{35}$. The reciprocal of $\frac{12}{35}$ is $\frac{35}{12}$, which is written as a mixed number as $\frac{35}{12}=\frac{24}{12}+\frac{11}{12}=\boxed{2\frac{11}{12}}$.

Challenge Problems

4.80

(a) Since $\frac{22}{44}=\frac{1}{2}$, the fraction $\frac{23}{44}$ is $\frac{1}{44}$ greater than $\frac{1}{2}$. Similarly, $\frac{32}{64}=\frac{1}{2}$, so $\frac{33}{64}$ is $\frac{1}{64}$ greater than $\frac{1}{2}$. Therefore, to compare $\frac{23}{44}$ and $\frac{33}{64}$, we only have to compare $\frac{1}{44}$ and $\frac{1}{64}$.

Since $\frac{1}{44} = \frac{64}{44 \cdot 64}$ and $\frac{1}{64} = \frac{44}{44 \cdot 64}$, we know that $\frac{1}{44}$ is greater than $\frac{1}{64}$. We also could have used the number line. Dividing the segment between 0 and 1 into 44 equal pieces produces larger pieces than dividing the segment into 64 equal pieces. Therefore, $\frac{1}{44}$ is greater than $\frac{1}{64}$. Since $\frac{1}{44}$ is greater than $\frac{1}{64}$, we know that $\boxed{\frac{23}{44}}$ is greater than $\frac{33}{64}$.

(b) We have $\frac{52}{53} = 1 - \frac{1}{53}$ and $\frac{97}{98} = 1 - \frac{1}{98}$. Just as we saw that $\frac{1}{44}$ is greater than $\frac{1}{64}$ in the previous part, we can see that $\frac{1}{53}$ is greater than $\frac{1}{98}$. Therefore, we subtract more from 1 to get $\frac{52}{53}$ than we do to get $\frac{97}{98}$. This means that $\boxed{\frac{97}{98}}$ is greater than $\frac{52}{53}$.

4.81 We could just multiply everything out, but that looks scary. Instead, we notice that $25 = 5^2$ and $36 = 6^2$, so we can use some exponent laws:

$$\frac{\left(\frac{6}{5}\right)^3 \left(\frac{25}{36}\right)^4}{\left(\frac{5}{6}\right)^4} = \frac{\left(\frac{6}{5}\right)^3 \left(\frac{5^2}{6^2}\right)^4}{\left(\frac{5}{6}\right)^4} = \frac{\left(\frac{6}{5}\right)^3 \left(\left(\frac{5}{6}\right)^2\right)^4}{\left(\frac{5}{6}\right)^4} = \frac{\left(\frac{6}{5}\right)^3 \left(\frac{5}{6}\right)^{2 \cdot 4}}{\left(\frac{5}{6}\right)^4} = \frac{\left(\frac{6}{5}\right)^3 \left(\frac{5}{6}\right)^8}{\left(\frac{5}{6}\right)^4}.$$

Now, we notice that we have $\frac{5}{6}$ in the numerator and denominator, so we can apply some more exponent laws:

$$\frac{\left(\frac{6}{5}\right)^3 \left(\frac{5}{6}\right)^8}{\left(\frac{5}{6}\right)^4} = \frac{\left(\frac{6}{5}\right)^3}{1} \cdot \frac{\left(\frac{5}{6}\right)^8}{\left(\frac{5}{6}\right)^4} = \left(\frac{6}{5}\right)^3 \left(\frac{5}{6}\right)^{8-4} = \left(\frac{6}{5}\right)^3 \left(\frac{5}{6}\right)^4.$$

Now we see that we can cancel:

$$\left(\frac{6}{5}\right)^3 \left(\frac{5}{6}\right)^4 = \frac{6^3}{5^3} \cdot \frac{5^4}{6^4} = \frac{6^3 \cdot 5^4}{5^3 \cdot 6^4} = \frac{5 \cdot (5^3 \cdot 6^3)}{6 \cdot (5^3 \cdot 6^3)} = \boxed{\frac{5}{6}}.$$

Note that we also could have used the fact that $\frac{6}{5} = \left(\frac{5}{6}\right)^{-1}$ to finish, or we could have cleverly used the exponent law $a^c b^c = (ab)^c$ like this:

$$\left(\frac{6}{5}\right)^3 \left(\frac{5}{6}\right)^4 = \left(\frac{6}{5}\right)^3 \left(\frac{5}{6}\right)^3 \left(\frac{5}{6}\right)^1 = \left(\frac{6}{5} \cdot \frac{5}{6}\right)^3 \left(\frac{5}{6}\right) = (1)^3 \left(\frac{5}{6}\right) = \frac{5}{6}.$$

4.82

(a) The fraction of the students at Central Middle School with blond hair equals the sum of the fraction of the students who are boys with blond hair and the fraction of the students who are girls with blond hair. Since $\frac{2}{5}$ of the students are boys, and $\frac{1}{4}$ of the boys have blond hair, the fraction of all students who are boys with blond hair is $\frac{2}{5} \cdot \frac{1}{4} = \frac{2}{20} = \frac{1}{10}$. Similarly, $1 - \frac{2}{5} = \frac{3}{5}$ of the students are girls, and $\frac{1}{3}$ of these girls have blond hair. Therefore, $\frac{3}{5} \cdot \frac{1}{3} = \frac{1}{5}$ of the students in the school are girls with blond hair.

Since $\frac{1}{10}$ of the students are boys with blond hair and $\frac{1}{5}$ of the students are girls with blond hair, the total fraction of students with blond hair is $\frac{1}{10} + \frac{1}{5} = \frac{1}{10} + \frac{2}{10} = \boxed{\frac{3}{10}}$.

(b) Since $\frac{3}{10}$ of the students have blond hair, and 36 students have blond hair, we have

$$\frac{3}{10} \cdot (\text{Total number of students}) = 36.$$

So, we divide to find the total number of students:

$$\text{Total number of students} = 36 \div \frac{3}{10} = 36 \cdot \frac{10}{3} = \frac{36}{3} \cdot 10 = \boxed{120}.$$

We also might have reasoned that if 36 students is $\frac{3}{10}$ of the students, then $36 \div 3 = 12$ students make are $\frac{1}{10}$ of the students, so there are $12 \cdot 10 = \boxed{120}$ students.

4.83 The positive factors of 30 are 1, 2, 3, 5, 6, 10, 15, and 30, so the desired sum is

$$\frac{1}{1} + \frac{1}{2} + \frac{1}{3} + \frac{1}{5} + \frac{1}{6} + \frac{1}{10} + \frac{1}{15} + \frac{1}{30}.$$

Since each of the denominators is a factor of 30, we can use 30 as a common denominator:

$$\frac{1}{1} + \frac{1}{2} + \frac{1}{3} + \frac{1}{5} + \frac{1}{6} + \frac{1}{10} + \frac{1}{15} + \frac{1}{30} = \frac{30 + 15 + 10 + 6 + 5 + 3 + 2 + 1}{30} = \frac{72}{30} = \boxed{\frac{12}{5}}.$$

Notice that our answer equals the sum of the divisors of 30 divided by 30. Is that a coincidence?

4.84 We could compute the sums in the numerator and the denominator, but that might take a while. Instead, we notice that all of the numbers in the numerator are multiples of 2, and all of the numbers in the denominator are multiples of 3. So, we factor:

$$\frac{2 + 4 + 6 + \cdots + 36}{3 + 6 + 9 + \cdots + 54} = \frac{2(1 + 2 + 3 + \cdots + 18)}{3(1 + 2 + 3 + \cdots + 18)} = \frac{2}{3} \cdot \frac{1 + 2 + 3 + \cdots + 18}{1 + 2 + 3 + \cdots + 18} = \frac{2}{3} \cdot 1 = \boxed{\frac{2}{3}}.$$

4.85 The distance between the two numbers on the number line is the greater number minus the lesser number, which equals

$$\frac{3}{5} - \left(-2\frac{5}{6}\right) = \frac{3}{5} + 2\frac{5}{6} = \frac{3}{5} + \frac{17}{6} = \frac{18}{30} + \frac{85}{30} = \frac{103}{30}.$$

So, to get the number that is halfway between the two numbers, we start with the lesser number, $-2\frac{5}{6}$, and add half the distance the two numbers, which is $\frac{1}{2} \cdot \frac{103}{30} = \frac{103}{60}$. This gives us

$$-2\frac{5}{6} + \frac{103}{60} = -\frac{17}{6} + \frac{103}{60} = -\frac{170}{60} + \frac{103}{60} = \frac{-170 + 103}{60} = \frac{-67}{60} = \boxed{-1\frac{7}{60}}.$$

4.86 The large container has $600 + 600 = 1200$ ml of liquid. The original first pitcher had $\frac{1}{3} \cdot 600 = \frac{600}{3} = 200$ ml vinegar and the original second pitcher had $\frac{2}{5} \cdot 600 = 2 \cdot \frac{600}{5} = 2 \cdot 120 = 240$ ml vinegar. So, the mixture has $200 + 240 = 440$ ml vinegar. Therefore, the fraction of the mixture that is vinegar is $\frac{440}{1200} = \frac{44}{120} = \frac{4 \cdot 11}{4 \cdot 30} = \boxed{\frac{11}{30}}.$

4.87 After I climb half the steps, I have half the staircase left. I then climb $\frac{1}{3}$ of this $\frac{1}{2}$ of the staircase, which means the amount of staircase remaining is $1 - \frac{1}{3} = \frac{2}{3}$ of the $\frac{1}{2}$ staircase that I haven't climbed. So, I have $\frac{2}{3} \cdot \frac{1}{2} = \frac{1}{3}$ of the staircase to go. I climb $\frac{1}{8}$ of this $\frac{1}{3}$, leaving $1 - \frac{1}{8} = \frac{7}{8}$ of this $\frac{1}{3}$ unclimbed, which means I have $\frac{7}{8} \cdot \frac{1}{3} = \frac{7}{24}$ of the staircase left.

So, if the staircase has n stairs, the number of stairs I have left is $\frac{7}{24} \cdot n = \frac{7n}{24}$. Since I must have a whole number of steps left, and n must be an integer, the smallest possible positive value of n is 24. When there are 24 steps in the staircase, then the number of steps I have left is $\frac{7 \cdot 24}{24} = 7$.

We need to make sure that 24 steps in the staircase gives us a whole number of steps at each point in the problem. After I climb the first half of stairs, I have $\frac{1}{2} \cdot 24 = 12$ steps to go. I climb $\frac{1}{3}$ of these, or $\frac{1}{3} \cdot 12 = 4$ steps, which leaves $12 - 4 = 8$ steps. I then climb $\frac{1}{8}$ of these steps, which is just 1 step, leaving 7 to go. So, there can indeed be $\boxed{24}$ steps in the staircase.

4.88

(a) We work from the bottom up:

$$1 + \frac{1}{2 + \frac{1}{2}} = 1 + \frac{1}{\frac{4}{2} + \frac{1}{2}} = 1 + \frac{1}{\frac{5}{2}} = 1 + \frac{2}{5} = \frac{7}{5}.$$

Squaring this fraction gives $\left(\frac{7}{5}\right)^2 = \frac{7^2}{5^2} = \frac{49}{25}$, and subtracting 2 gives $\frac{49}{25} - 2 = \frac{49}{25} - \frac{50}{25} = \boxed{-\frac{1}{25}}$.

(b) We can use the previous part to help us out a little:

$$1 + \frac{1}{2 + \frac{1}{2 + \frac{1}{2}}} = 1 + \frac{1}{1 + \left(1 + \frac{1}{2 + \frac{1}{2}}\right)} = 1 + \frac{1}{1 + \frac{7}{5}} = 1 + \frac{1}{\frac{12}{5}} = 1 + \frac{5}{12} = \frac{17}{12}.$$

Squaring this gives $\left(\frac{17}{12}\right)^2 = \frac{289}{144}$. Subtracting 2 gives $\frac{289}{144} - 2 = \frac{289}{144} - \frac{288}{144} = \boxed{\frac{1}{144}}$.

(c) Again, we can use the previous part to help:

$$1 + \frac{1}{2 + \frac{1}{2 + \frac{1}{2 + \frac{1}{2}}}} = 1 + \frac{1}{1 + \left(1 + \frac{1}{2 + \frac{1}{2 + \frac{1}{2}}}\right)} = 1 + \frac{1}{1 + \frac{17}{12}} = 1 + \frac{1}{\frac{12}{12} + \frac{17}{12}} = 1 + \frac{1}{\frac{29}{12}} = 1 + \frac{12}{29} = \frac{41}{29}.$$

Squaring this gives $\left(\frac{41}{29}\right)^2 = \frac{1681}{841}$. Subtracting 2 gives us $\frac{1681}{841} - 2 = \frac{1681}{841} - \frac{1682}{841} = \boxed{-\frac{1}{841}}$.

Notice that as we go from one answer to the next in the three parts, our results get closer and closer to 0. In other words, the squares of expressions in the three parts are closer and closer to 2. If we continue the pattern of the expressions in the three parts by adding more to the "tower" of fractions, the squares of the resulting expressions continue to get closer to 2. If we continue building the "tower" in this pattern forever, we form what's called a **continued fraction**.

4.89 Let x be the amount of money that each person gave Ott. Moe gave Ott $\frac{1}{5}$ of his money and kept the other $\frac{4}{5}$. So, Moe kept 4 times as much money as he gave Ott. Since Moe gave Ott x, Moe kept $4x$. Similarly, Loki gave Ott $\frac{1}{4}$ of his money and kept the other $\frac{3}{4}$. So, Loki kept 3 times as much money as he gave Ott. Loki also gave Ott x, so Loki kept $3x$. Finally, Nick gave Ott $\frac{1}{3}$ of his money and kept the other $\frac{2}{3}$. This means Nick kept 2 times as much as Nick gave Ott, so Nick kept $2x$. So, after the three others each gave Ott x, Moe has $4x$, Loki has $3x$, Nick has $2x$, and Ott has $3x$. Combined, they have $4x + 3x + 2x + 3x = 12x$, so the fraction of the total that Ott has is $\frac{3x}{12x} = \frac{3}{12} \cdot \frac{x}{x} = \boxed{\frac{1}{4}}$.

4.90

(a) We have $\frac{1}{4} + \frac{1}{4} = \frac{1}{2}$. So, if we have two different positive reciprocals of integers that sum to $\frac{1}{2}$, one of them is greater than $\frac{1}{4}$ and the other is less than $\frac{1}{4}$. The only reciprocal between $\frac{1}{4}$ and $\frac{1}{2}$ is $\frac{1}{3}$, so one of our reciprocals is $\frac{1}{3}$. The other is $\frac{1}{2} - \frac{1}{3} = \frac{3}{6} - \frac{2}{6} = \frac{1}{6}$. The desired integers are $\boxed{3 \text{ and } 6}$.

(b) We have $\frac{1}{6} + \frac{1}{6} = \frac{2}{6} = \frac{1}{3}$. So, if we have two different positive reciprocals of integers that sum to $\frac{1}{3}$, one of them is greater than $\frac{1}{6}$ and the other is less than $\frac{1}{6}$. The only reciprocals between $\frac{1}{6}$ and $\frac{1}{3}$ are $\frac{1}{4}$ and $\frac{1}{5}$.

We'll try $\frac{1}{4}$ first. If $\frac{1}{4}$ is one of the pair of reciprocals that sum to $\frac{1}{3}$, then the other must be $\frac{1}{3} - \frac{1}{4} = \frac{4}{12} - \frac{3}{12} = \frac{1}{12}$. We've found our two reciprocals: $\frac{1}{4} + \frac{1}{12} = \frac{3}{12} + \frac{1}{12} = \frac{4}{12} = \frac{1}{3}$. So, the desired integers are $\boxed{4 \text{ and } 12}$.

(c) We could reason through this part in the same way we went through the first two parts, but instead, let's look for a pattern. In the first two parts, we found
$$\frac{1}{2} = \frac{1}{3} + \frac{1}{6}, \qquad \frac{1}{3} = \frac{1}{4} + \frac{1}{12}.$$
In both cases, the denominator of the larger fraction on the right-hand side is 1 greater than the denominator on the left-hand side. Therefore, we guess that $\frac{1}{4}$ is the sum of $\frac{1}{5}$ and the reciprocal of an integer. The number that we must add to $\frac{1}{5}$ to get $\frac{1}{4}$ is $\frac{1}{4} - \frac{1}{5} = \frac{5}{20} - \frac{4}{20} = \frac{1}{20}$. Success! Since $\frac{1}{5} + \frac{1}{20} = \frac{1}{4}$, two integers that fit the problem are $\boxed{5 \text{ and } 20}$. However, this is not the only pair of numbers that works! The pair 6 and 12 also works; see if you can figure out how to use part (a) to find this pair.

(d) Our success in part (c) gives us a clear strategy. We guess that $\frac{1}{6}$ is one of the two reciprocals whose sum is $\frac{1}{5}$. The other then must be $\frac{1}{5} - \frac{1}{6} = \frac{6}{30} - \frac{5}{30} = \frac{1}{30}$. Success again! Since $\frac{1}{6} + \frac{1}{30} = \frac{1}{5}$, two positive integers whose reciprocals sum to $\frac{1}{5}$ are $\boxed{6 \text{ and } 30}$.

(e) One more time. We guess that $\frac{1}{7}$ is one of the two reciprocals whose sum is $\frac{1}{6}$. The other then must be $\frac{1}{6} - \frac{1}{7} = \frac{7}{42} - \frac{6}{42} = \frac{1}{42}$. Since $\frac{1}{7} + \frac{1}{42} = \frac{1}{6}$, two positive integers whose reciprocals sum to $\frac{1}{6}$ are $\boxed{7 \text{ and } 42}$. This isn't the only pair that works. The pair 9 and 18 works, as does the pair 8 and 24. See if you can figure out how we can use parts (a) and (b) to find these two pairs. Also, see if you can find one more pair that works.

(f) Let's consider the sums from the first five parts and look for a pattern:
$$\frac{1}{2} = \frac{1}{3} + \frac{1}{6}, \qquad \frac{1}{3} = \frac{1}{4} + \frac{1}{12}, \qquad \frac{1}{4} = \frac{1}{5} + \frac{1}{20}, \qquad \frac{1}{5} = \frac{1}{6} + \frac{1}{30}, \qquad \frac{1}{6} = \frac{1}{7} + \frac{1}{42}.$$
In each case, the denominator of the larger fraction on the right-hand side of the equation is 1 more than the denominator of the fraction on the left-hand side, and the denominator of the smaller fraction on the right-hand side is the product of the other two denominators. So, we guess that two fractions that sum to $\frac{1}{n}$ are $\frac{1}{n+1}$ and $\frac{1}{n(n+1)}$. We have to test that the sum of these fractions is indeed $\frac{1}{n}$. We write them with a common denominator by multiplying $\frac{1}{n+1}$ by $\frac{n}{n}$:
$$\frac{1}{n+1} + \frac{1}{n(n+1)} = \frac{n}{n} \cdot \frac{1}{n+1} + \frac{1}{n(n+1)} = \frac{n}{n(n+1)} + \frac{1}{n(n+1)} = \frac{n+1}{n(n+1)} = \frac{1}{n} \cdot \frac{n+1}{n+1} = \frac{1}{n} \cdot 1 = \frac{1}{n}.$$

Success! Since $\frac{1}{n+1} + \frac{1}{n(n+1)} = \frac{1}{n}$, the integers we seek are $\boxed{n+1 \text{ and } n(n+1)}$.

CHAPTER **5**

Equations and Inequalities

$24^{\circledR}$ Cards

First Card $5 \cdot 5 - 5/5$.

Second Card $(5 - 1/5) \cdot 5$.

Third Card $4/(1 - 5/6)$ or $6/(5/4 - 1)$.

Fourth Card $(13 + 5)/9 + 22$.

Exercises for Section 5.1

5.1.1

(a) $2r + 3r - 7r = (2 + 3 - 7)r = \boxed{-2r}$.

(b) $3y - 2y + 7y - 9y = (3 - 2 + 7 - 9)y = (-1)y = \boxed{-y}$.

(c) $6 - t + 3t - 4 + 2t = (-t + 3t + 2t) + (6 - 4) = (-1 + 3 + 2)t + 2 = \boxed{4t + 2}$.

(d) $-5z + \frac{3}{2} - 2 + 3z = -5z + 3z + \frac{3}{2} - 2 = \boxed{-2z - \frac{1}{2}}$.

(e) $-\frac{x}{2} + x + \frac{x}{3} = \left(-\frac{1}{2} + 1 + \frac{1}{3}\right)x = \boxed{\frac{5}{6}x}$.

(f) $5 - \frac{5}{2}r + 7 - \frac{7}{3}r = -\frac{5}{2}r - \frac{7}{3}r + 5 + 7 = \left(-\frac{5}{2} - \frac{7}{3}\right)r + 12 = \boxed{-\frac{29}{6}r + 12}$.

5.1.2

(a) $7(x - 2) + 5(2x + 3) = 7 \cdot x - 7 \cdot 2 + 5 \cdot (2x) + 5 \cdot 3 = 7x - 14 + 10x + 15 = \boxed{17x + 1}$.

(b) $4(3a - 4) - 6(2a - 1) = 4 \cdot 3a - 4 \cdot 4 - 6 \cdot 2a - 6(-1) = 12a - 16 - 12a + 6 = (12a - 12a) + (-16 + 6) = \boxed{-10}$.

(c) $-3(1 + 3t) - (t + 3)(1 + 4) = -3 \cdot 1 + (-3) \cdot 3t - (t + 3)(5) = -3 - 9t - (t \cdot 5 + 3 \cdot 5) = -3 - 9t - (5t + 15) = $
$-3 - 9t - 5t - 15 = \boxed{-14t - 18}$.

(d) $-5(22 - 31y) + 22(4y + 3) = -5 \cdot 22 - 5 \cdot (-31y) + 22 \cdot 4y + 22 \cdot 3 = -110 + 155y + 88y + 66 = \boxed{243y - 44}$.

5.1.3

(a) $\dfrac{12-4c}{4}+\dfrac{27+18c}{3}=\dfrac{12}{4}-\dfrac{4c}{4}+\dfrac{27}{3}+\dfrac{18c}{3}=3-c+9+6c=\boxed{5c+12}$.

(b) $\dfrac{1}{2}(6-4y)+\dfrac{3}{2}(6y+4)=\dfrac{1}{2}(6)-\dfrac{1}{2}(4y)+\dfrac{3}{2}(6y)+\dfrac{3}{2}(4)=3-2y+9y+6=\boxed{7y+9}$.

(c) $3r+7-\dfrac{24-16r}{8}=3r+7-\left(\dfrac{24}{8}-\dfrac{16r}{8}\right)=3r+7-(3-2r)=3r+7-3-(-2r)=3r+4+2r=\boxed{5r+4}$.

(d) We won't have any useful cancellation by breaking up each fraction, so we start by writing the fractions with a common denominator:

$$\frac{x-7}{3}-\frac{5-x}{2}=\frac{2(x-7)}{2(3)}-\frac{3(5-x)}{3(2)}=\frac{2x-14}{6}-\frac{15-3x}{6}.$$

Now, we can combine the fractions:

$$\frac{2x-14}{6}-\frac{15-3x}{6}=\frac{(2x-14)-(15-3x)}{6}=\frac{2x-14-15-(-3x)}{6}=\frac{2x-29+3x}{6}=\boxed{\frac{5x-29}{6}}.$$

5.1.4 $\boxed{\text{No}}$. Suppose $x=2$. Then, we have

$$\frac{(2/x)}{4}=\frac{(2/2)}{4}=\frac{1}{4}, \qquad \frac{2}{(x/4)}=\frac{2}{2/4}=\frac{2}{1/2}=2\cdot\frac{2}{1}=4.$$

Since the expressions have different values for the same value of x, the two expressions are not equivalent. We might also have noted that the first expression simplifies as

$$\frac{(2/x)}{4}=\frac{2}{x}\cdot\frac{1}{4}=\frac{2}{4x}=\frac{1}{2x},$$

while the second simplifies as

$$\frac{2}{x/4}=2\cdot\frac{4}{x}=\frac{8}{x}.$$

These simplified forms are clearly not equivalent.

5.1.5 $\boxed{\text{No}}$. Just as $3x+4$ cannot be simplified any further, we cannot simplify $3x+4y$.

5.1.6

(a) $\boxed{\text{Yes}}$. Just as we have $x+x=1x+1x=(1+1)x=2x$, we have $x^2+x^2=1x^2+1x^2=(1+1)x^2=2x^2$.

(b) $\boxed{\text{Not quite}}$. The expressions x^2 and x are not the same so we can't use the distributive property to simplify x^2+x the way we simplified $x+x$. We can, however, write $x^2+x=x\cdot x+1\cdot x=(x+1)\cdot x=(x+1)x$.

Exercises for Section 5.2

5.2.1

(a) Subtracting 235 from both sides gives $t = 137 - 235 = \boxed{-98}$.

(b) Subtracting $\frac{7}{9}$ from both sides gives $a = -\frac{2}{9} - \frac{7}{9} = \frac{-9}{9} = \boxed{-1}$.

(c) Adding 14 to both sides gives $14 - 6\frac{1}{10} = c$, so $c = \boxed{7\frac{9}{10}}$.

(d) Simplifying the left side gives $y + 2\frac{3}{5} = 1\frac{7}{10}$. Subtracting $2\frac{3}{5}$ from both sides gives

$$y = 1\frac{7}{10} - 2\frac{3}{5} = \frac{17}{10} - \frac{13}{5} = \frac{17}{10} - \frac{26}{10} = \boxed{-\frac{9}{10}}.$$

5.2.2

(a) Dividing both sides by -7 gives $y = 343/(-7) = \boxed{-49}$.

(b) We first write the right side as a fraction instead of a mixed number:

$$16x = \frac{10}{3}.$$

Multiplying both sides by $\frac{1}{16}$ (which is the same as dividing both sides by 16) gives

$$x = \frac{1}{16} \cdot \frac{10}{3} = \boxed{\frac{5}{24}}.$$

(c) Multiplying both sides by 5 cancels the 5 in the denominator on the left and gives $x = \frac{6}{7} \cdot 5 = \boxed{\frac{30}{7}}$.

(d) We can write the left side as $-\frac{5}{2}y$. We can eliminate the coefficient of y in the equation by multiplying both sides by the reciprocal of the coefficient:

$$\left(-\frac{2}{5}\right)\left(-\frac{5}{2}\right)y = \left(-\frac{2}{5}\right)\left(-\frac{14}{15}\right).$$

The product of the fractions on the left is 1, and we have $y = \left(-\frac{2}{5}\right)\left(-\frac{14}{15}\right) = \boxed{\frac{28}{75}}$.

5.2.3 Subtracting $5\frac{1}{4}$ from both sides gives $-y = 19\frac{3}{4} - 5\frac{1}{4} = 14\frac{1}{2}$. Since $-y = 14\frac{1}{2}$, we have $y = \boxed{-14\frac{1}{2}}$.

5.2.4 We don't have a single step that will get us x. However, since $x - 3$ divided by 7 equals 2, and 14 is the number we divide by 7 to get 2, we know that $x - 3$ equals 14. Therefore, $x = \boxed{17}$.

5.2.5 As in the last problem, we can't isolate r in one step. However, since 3 times $r - 7$ equals 24, and 8 is the number we multiply by 3 to get 24, we know that $r - 7$ equals 8. Therefore, $r = \boxed{15}$.

5.2.6 Just as we can multiply both sides of the equation $\frac{z}{2} = 6$ by 2 to get $z = 2 \cdot 6$, we can multiply both sides of $\frac{x}{c} = 3$ by c to get $x = 3c$. When $x = 2$, this equation becomes $2 = 3c$, and dividing both sides by 3 gives $c = \boxed{\frac{2}{3}}$.

We also could have substituted $x = 2$ into $\frac{x}{c} = 3$ right away. Since $x = 2$ is a solution to $\frac{x}{c} = 3$, we have $\frac{2}{c} = 3$. Multiplying both sides by c gives $2 = 3c$, and dividing by 3 gives $c = \boxed{\frac{2}{3}}$.

Exercises for Section 5.3

5.3.1

(a) Subtracting 5 from both sides gives $2x = 6$. Dividing both sides by 2 gives $x = \boxed{3}$.

(b) Adding $1\frac{1}{2}$ to both sides gives $-6a = \frac{1}{3} + 1\frac{1}{2} = \frac{1}{3} + \frac{3}{2} = \frac{2}{6} + \frac{9}{6} = \frac{11}{6}$. Now our equation is $-6a = \frac{11}{6}$. Dividing both sides by -6 gives $a = \boxed{-\frac{11}{36}}$.

(c) Subtracting 19 from both sides gives $-7t = 61 - 19 = 42$. Dividing both sides of $-7t = 42$ by -7 gives $t = \frac{42}{-7} = \boxed{-6}$.

5.3.2

(a) Subtracting $2y$ from both sides gives $y + 9 = 1$. Subtracting 9 from both sides gives $y = \boxed{-8}$.

(b) Simplifying both sides gives $4x - 3 = -3x + 25$. Adding $3x$ to both sides gives $7x - 3 = 25$. Adding 3 to both sides gives $7x = 28$, so $x = \boxed{4}$.

(c) Subtracting $998a$ from both sides gives $2a + 218 = 232$. Subtracting 218 from both sides gives $2a = 14$, so $a = \boxed{7}$.

5.3.3 Adding 2 to both sides gives $3x = 13$. We could solve for x by dividing by 3 to give $x = \frac{13}{3}$. However, we notice that we want $6x + 5$, so we multiply both sides of $3x = 13$ by 2 to give $6x = 26$. Adding 5 to both sides gives $6x + 5 = \boxed{31}$.

5.3.4

(a) Subtracting $\frac{4}{5}$ from both sides gives $\frac{2}{3}t = -\frac{1}{2} - \frac{4}{5} = -\frac{5}{10} - \frac{8}{10} = -\frac{13}{10}$. We solve $\frac{2}{3}t = -\frac{13}{10}$ by multiplying both sides by $\frac{3}{2}$, which gives

$$t = \frac{3}{2}\left(-\frac{13}{10}\right) = \boxed{-\frac{39}{20}}.$$

(b) We get rid of the fractions by multiplying both sides by 6. This gives us $6 \cdot \frac{1}{2}(z + 3) = 6 \cdot \frac{1}{3}(z - 7)$, so $3(z + 3) = 2(z - 7)$. Expanding both products gives $3z + 9 = 2z - 14$. Subtracting $2z$ from both sides gives $z + 9 = -14$. Subtracting 9 from both sides gives $z = \boxed{-23}$.

(c) We start by multiplying both sides by a number that gets rid of all of the fractions. The common denominator of all of the fractions is $7 \cdot 4 \cdot 5 = 140$. Multiplying the left side by 140 gives

$$140\left(\frac{4x}{7} - \frac{1}{2}\right) = 140 \cdot \frac{4x}{7} - 140 \cdot \frac{1}{2} = \frac{(140)(4x)}{7} - \frac{140}{2} = \frac{140}{7} \cdot 4x - 70 = 20 \cdot 4x - 70 = 80x - 70.$$

Multiplying the right side of the original equation by 140 gives

$$140\left(-\frac{3}{4} - \frac{2x}{5}\right) = 140\left(-\frac{3}{4}\right) - 140\left(\frac{2x}{5}\right)$$

$$= -105 - \frac{(140)(2x)}{5} = -105 - \frac{140}{5}(2x) = -105 - 28(2x) = -105 - 56x.$$

Therefore, multiplying both sides of the original equation by 140 gives $80x - 70 = -105 - 56x$. Adding $56x$ to both sides gives $136x - 70 = -105$. Adding 70 to both sides gives $136x = -35$. Finally, dividing by 136 gives $x = \boxed{-\frac{35}{136}}$.

5.3.5 We multiply both sides by 40 to get rid of the denominators. This gives $40 \cdot \frac{2x+7}{5} = -40 \cdot \frac{1-3x}{8}$, or $\frac{(40)(2x+7)}{5} = -\frac{40(1-3x)}{8}$. We can now divide on each side to give $8(2x+7) = -5(1-3x)$. Expanding both sides gives $8(2x) + 8(7) = -5(1) - 5(-3x)$, so $16x + 56 = -5 + 15x$. Subtracting $15x$ from both sides gives $x + 56 = -5$, so $x = \boxed{-61}$.

5.3.6

(a) First, we expand all of the products:

$$2(z) + 2(3) - 5(6) - 5(-z) = 8(3z) + 8(3) - 4(1) - 4(-2z).$$

Computing the products gives $2z + 6 - 30 + 5z = 24z + 24 - 4 + 8z$. Simplifying both sides gives $7z - 24 = 32z + 20$. Subtracting $7z$ from both sides produces $-24 = 25z + 20$. Subtracting 20 from both sides gives $-44 = 25z$, and dividing both sides by 25 gives $z = \boxed{-\frac{44}{25}}$.

(b) We multiply both sides by 12, which is the common denominator of the three fractions. This gives us $12 \left(\frac{m+11}{3} + \frac{m-2}{6} \right) = 12 \cdot \frac{2m-1}{12}$, or

$$12 \cdot \frac{m+11}{3} + 12 \cdot \frac{m-2}{6} = 12 \cdot \frac{2m-1}{12}.$$

Thus, we have $4(m+11)+2(m-2) = 1(2m-1)$. Expanding the products gives $4m+44+2m-4 = 2m-1$. Simplifying the left side gives $6m+40 = 2m-1$. Subtracting $2m$ from both sides gives $4m+40 = -1$, and subtracting 40 from both sides gives $4m = -41$. Dividing both sides by 4 gives $m = \boxed{-\frac{41}{4}}$.

(c) Multiplying both sides by 8 gives $8 \cdot \frac{p-2}{4} = 8 \cdot \frac{2p-3}{8}$, so $2(p-2) = 2p-3$. Expanding the product on the left gives $2p-4 = 2p-3$. Subtracting $2p$ from both sides gives $-4 = -3$, which is never true. Therefore, the original equation has $\boxed{\text{no solutions}}$.

Exercises for Section 5.4

5.4.1 Let k be Kellie's number. Double her number is $2k$, and triple this is $6k$. Since this is 65 greater than her original number, we have $6k = k+65$. Subtracting k from both sides gives $5k = 65$, and dividing by 5 gives $k = \boxed{13}$.

5.4.2 Let n be the number. Adding 5 to $\frac{1}{3}$ the number gives $5 + \frac{1}{3}n$, so we must have $5 + \frac{1}{3}n = \frac{1}{2}n$. Subtracting $\frac{1}{3}n$ from both sides gives $5 = \frac{1}{2}n - \frac{1}{3}n = \left(\frac{1}{2} - \frac{1}{3} \right) n - \frac{1}{6}n$. Multiplying both sides of $5 - \frac{1}{6}n$ by 6 gives $n = \boxed{30}$.

5.4.3 Let w be the weight of the heavier dog in pounds, so the lighter dog weighs $w - 25$ pounds. Since the two dogs together weigh 137 pounds, we have $w + (w - 25) = 137$, or $2w - 25 = 137$. Adding 25 to both sides gives $2w = 162$. Dividing both sides by 2 gives $w = 81$, so the heavier dog weighs $\boxed{81 \text{ pounds}}$.

5.4.4 Let n be the integer. Triple the integer is $3n$. The sum of 9 and three-fourths of the integer is $9 + \frac{3}{4}n$, so we have $3n = 9 + \frac{3}{4}n$. Subtracting $\frac{3}{4}n$ from both sides gives $3n - \frac{3}{4}n = 9$, so $\frac{9}{4}n = 9$. Multiplying both sides by $\frac{4}{9}$ gives $n = \boxed{4}$.

5.4.5 Let the age of the youngest child be y. The oldest is twice as old as the youngest, so the oldest is $2y$ years old. The two older children differ by three years, so the middle child is $2y - 3$ years old. The sum of their ages is 32, so $y + 2y + (2y - 3) = 32$. Simplifying the left side gives $5y - 3 = 32$, so $5y = 35$, which means $y = 7$. Therefore, the youngest child is $\boxed{7 \text{ years}}$ old.

5.4.6 Let the smallest of the integers be n, so the sum of the six integers is $n + (n + 2) + (n + 4) + (n + 6) + (n + 8) + (n + 10)$. Simplifying this expression gives $6n + 30$. We are given that the numbers add to 282, so $6n + 30 = 282$. Subtracting 30 from both sides gives $6n = 252$. Dividing by 6 gives $n = 42$. Therefore, the largest of the six integers is $n + 10 = \boxed{52}$.

5.4.7 Let t be the number of two-wheel bikes they have. Since they have 47 bikes total, they must have $47 - t$ three-wheel bikes. The two-wheel bikes need $2t$ tires total and the three-wheel bikes need $3(47 - t)$ tires total, so we must have $2t + 3(47 - t) = 112$. Expanding the product on the left gives $2t + 141 - 3t = 112$. Simplifying the left side gives $141 - t = 112$. Subtracting 141 from both sides gives $-t = -29$, so $t = \boxed{29}$.

We also could have noted that if each of the 47 bikes has 2 wheels, then there are 94 wheels total. That's $112 - 94 = 18$ wheels too few. For each two-wheel bike that we replace with a three-wheel bike, we gain one wheel. So, if we start with 47 two-wheel bikes, which have 94 wheels total, and then replace 18 of them with three-wheel bikes, then we'll have $94 + 18 = 112$ wheels total and $47 - 18 = \boxed{29}$ two-wheel bikes.

5.4.8 If we let c be the number of cents in the cost of the comic book, then $\frac{c}{25}$ is the number of quarters I use to buy the comic book and $\frac{c}{10}$ is the number of dimes I would have used to buy the comic book. Since I would need 9 more dimes than quarters, we have

$$\frac{c}{25} + 9 = \frac{c}{10}.$$

Multiplying both sides by 50 gives

$$50\left(\frac{c}{25} + 9\right) = 50 \cdot \frac{c}{10},$$

so $2c + 450 = 5c$. Solving this equation gives $c = 150$, which means the comic book cost $\boxed{\$1.50}$.

5.4.9 Let g be the number of games in a row the Phillies win. After these games, they have won $3 + g$ games total out of their $21 + g$ games. Since they must win $\frac{3}{5}$ of their games, we must have $3 + g = \frac{3}{5}(21 + g)$. Multiplying both sides by 5 gives $5(3 + g) = 3(21 + g)$. Expanding both sides gives $15 + 5g = 63 + 3g$. Subtracting $3g$ from both sides gives $15 + 2g = 63$, so $2g = 48$, which gives $g = \boxed{24}$.

Exercises for Section 5.5

5.5.1 Since z cannot be greater than y, and x must be greater than y, we know that x must be greater than z. So, x cannot be equal to z.

5.5.2

(a) The values of t that satisfy the inequality are all numbers from 4 to 7, including 7 but excluding 4. We graph them on the number line below:

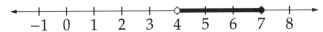

(b) Adding 41 to both sides gives $3x \le 2x + 4$, and subtracting $2x$ from both sides gives $x \le 4$. All numbers less than or equal to 4 satisfy the original inequality; these are graphed below:

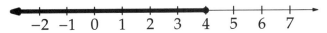

(c) Subtracting 14 from both sides, and adding $2y$ to both sides, gives $-5y < -10$. Dividing both sides by -2, and remembering to reverse the inequality, gives $y > 2$. So, all numbers greater than 2 satisfy the original inequality. These solutions are graphed below:

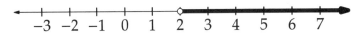

5.5.3 Let my favorite number be n, so we must have $\frac{n}{2} > 6 + n$. Multiplying both sides by 2 gives $n > 2(6 + n)$, or $n > 12 + 2n$. Subtracting n from both sides gives $0 > 12 + n$. Subtracting 12 from both sides gives $-12 > n$. So, my favorite number could be $\boxed{\text{any number less than } -12}$.

5.5.4 For the first two parts, we can consider the location of the reciprocals on the number line:

(a) $\boxed{\frac{1}{2}}$ is greater.

(b) $\boxed{\frac{1}{5}}$ is greater.

(c) We have
$$\frac{1}{1/2} = 1 \cdot \frac{2}{1} = 2.$$

Similarly, we have $\frac{1}{1/3} = 3$, so $\boxed{\frac{1}{1/3}}$ is greater.

(d) In each of the first three parts, we compared expressions of the form $\frac{1}{a}$ and $\frac{1}{b}$ in which a and b are both positive. In each case, we found that if $a > b$, then $\frac{1}{a} < \frac{1}{b}$. To see why this is always true, we start by dividing both sides of $a > b$ by a, assuming that a and b are positive. This gives us $1 > \frac{b}{a}$. Then, we divide both sides by b (or multiply both sides by $\frac{1}{b}$). This gives us $\frac{1}{b} > \frac{1}{a}$. So, we see that if a and b are positive and a is greater than b, then $\frac{1}{a}$ is less than $\frac{1}{b}$. Therefore, $\boxed{\frac{1}{b}}$ is greater.

(e) We proceed as in the previous part. We start with $a > b$, where a and b are both negative. We divide both sides by a, but we have to reverse the direction of the inequality because a is negative. This gives us $1 < \frac{b}{a}$. Then, we divide both sides by b, again reversing the direction of the inequality symbol because b is negative. This gives us $\frac{1}{b} > \frac{1}{a}$. So, again $\boxed{\frac{1}{b}}$ is greater.

(f) In the previous two cases, we saw that if $a > b$ and a and b have the same sign (positive or negative), then $\frac{1}{a} < \frac{1}{b}$. However, if a is positive and b is negative, then the reciprocal of a is positive and the reciprocal of b is negative. Any positive number is greater than any negative number. So, we have $\frac{1}{a} > \frac{1}{b}$, which means $\boxed{\frac{1}{a}}$ is greater.

We also could have followed the same process of the previous two parts. Dividing both sides of $a > b$ by a gives $1 > \frac{b}{a}$. Since b is negative, when we divide both sides by b, we get $\frac{1}{b} < \frac{1}{a}$.

5.5.5 Since x and y are positive, we can multiply both sides of $x > y$ by x to get $x^2 > xy$, and multiply $x > y$ by y to get $xy > y^2$. Combining $x^2 > xy$ and $xy > y^2$ gives us $x^2 > xy > y^2$, so $x^2 > y^2$.

5.5.6 Let w be the number of weeks my team plays after the end of Week 3. Since my team wins 3 games each week after the end of Week 3, the total number of wins my team has w weeks after Week 3 is $5 + 3w$. Similarly, the total number of losses my team has w weeks after Week 3 is $7 + w$. Since the number of my team's wins must be at least twice as many games as it has lost, we seek the smallest value of w such that

$$5 + 3w \geq 2(7 + w).$$

Expanding the right side gives $5 + 3w \geq 14 + 2w$. Subtracting $2w$ and 5 from both sides gives $w \geq 9$. So, my team will first have at least twice as many wins as losses 9 weeks after Week 3, which is the end of $\boxed{\text{Week 12}}$.

Review Problems

5.33

(a) $4(2 - 3r) - \frac{1}{2}(4 + 24r) = 4(2) - 4(3r) - \frac{1}{2}(4) - \frac{1}{2}(24r) = 8 - 12r - 2 - 12r = \boxed{-24r + 6}$.

(b) $\frac{24x}{21} + \frac{35x}{49} - \frac{x}{2} = \left(\frac{24}{21} + \frac{35}{49} - \frac{1}{2}\right)x = \left(\frac{8}{7} + \frac{5}{7} - \frac{1}{2}\right)x = \left(\frac{13}{7} - \frac{1}{2}\right)x = \boxed{\frac{19}{14}x}$.

(c) $3y + \frac{y - 8}{2} + \frac{6y}{4} = 3y + \frac{y}{2} - \frac{8}{2} + \frac{3y}{2} = 3y + \frac{y + 3y}{2} - 4 = 3y + 2y - 4 = \boxed{5y - 4}$.

(d)

$$\begin{aligned}
\frac{20z - 1}{3} - \frac{8z + 4}{12} &= \frac{20z}{3} - \frac{1}{3} - \left(\frac{8z}{12} + \frac{4}{12}\right) \\
&= \frac{20z}{3} - \frac{1}{3} - \left(\frac{2z}{3} + \frac{1}{3}\right) \\
&= \frac{20z}{3} - \frac{1}{3} - \frac{2z}{3} - \frac{1}{3} = \frac{20z}{3} - \frac{2z}{3} - \frac{1}{3} - \frac{1}{3} = \frac{18z}{3} - \frac{2}{3} = \boxed{6z - \frac{2}{3}}.
\end{aligned}$$

5.34

(a) Subtracting 133 from both sides gives $w = \boxed{-138}$.

(b) Adding $12\frac{7}{8}$ to both sides and subtracting y from both sides gives

$$3y - y = 3\frac{1}{4} + 12\frac{7}{8} = 3\frac{2}{8} + 12\frac{7}{8} = 15 + \frac{9}{8} = 16\frac{1}{8},$$

so $2y = 16\frac{1}{8}$. Dividing both sides by 2 gives

$$y = \frac{16\frac{1}{8}}{2} = \frac{16 + \frac{1}{8}}{2} = \frac{16}{2} + \frac{(1/8)}{2} = 8 + \frac{1}{16} = \boxed{8\frac{1}{16}}.$$

(c) Multiplying both sides by $\frac{3}{2}$ gives $t = \frac{3}{2}(-18) = \boxed{-27}$.

(d) Simplifying the right-hand side gives $168 + 76a = 118a$. Subtracting $76a$ from both sides gives $168 = 42a$, and dividing both sides by 42 gives $a = \frac{168}{42} = \boxed{4}$.

(e) Expanding the product on the right-hand side and simplifying gives

$$4r - 5 = 7 - 3r + 3(2) - 3(r) = 7 - 3r + 6 - 3r = -6r + 13.$$

Adding $6r$ to both sides of $4r - 5 = -6r + 13$ gives $10r - 5 = 13$, and adding 5 to both sides gives $10r = 18$. Dividing both sides by 10 gives $r = \frac{18}{10} = \boxed{\frac{9}{5}}$.

(f) Expanding the products on both sides gives

$$6 - 4(2) - 4(-3x) = 74 - 2(3) - 2(-x),$$

so $6 - 8 + 12x = 74 - 6 + 2x$. Simplifying both sides gives $12x - 2 = 2x + 68$. Subtracting $2x$ from both sides gives $10x - 2 = 68$, and adding 2 to both sides gives $10x = 70$. Dividing by 10 gives $x = \boxed{7}$.

(g) Multiplying both sides by 6 gives $6\left(\frac{z}{3} - 4\right) = 6 \cdot \frac{2z - 9}{6}$. Expanding the product on the left and canceling on the right side gives

$$6 \cdot \frac{z}{3} - (6)(4) = 2z - 9.$$

Since $6 \cdot \frac{z}{3} = \frac{6z}{3} = 2z$, we have $2z - 24 = 2z - 9$. Subtracting $2z$ from both sides gives $-24 = -9$, which is never true. Therefore, there are $\boxed{\text{no solutions}}$ to the original equation.

(h) Multiplying both sides by 28 gives $28 \cdot \frac{3p+4}{7} = 28 \cdot \frac{2p-7}{4}$, so $4(3p + 4) = 7(2p - 7)$. Expanding both products gives $4(3p) + 4(4) = 7(2p) - 7(7)$, so $12p + 16 = 14p - 49$. Subtracting $12p$ from both sides and adding 49 to both sides gives $65 = 2p$. Dividing both sides by 2 gives $p = \boxed{\frac{65}{2}}$.

(i) Applying the distributive property on both sides gives

$$\frac{12y}{6} - \frac{8}{6} + \frac{9y}{3} + \frac{1}{3} = 5y - 5 \cdot \frac{1}{5},$$

so $2y - \frac{4}{3} + 3y + \frac{1}{3} = 5y - 1$. Simplifying the left side gives $5y - 1 = 5y - 1$. This equation is always true because both sides are the same! Therefore, $\boxed{\text{all values of } y}$ satisfy the equation.

(j)   We start by simplifying the right-hand side as $\frac{x}{7} - \frac{2x}{3} = \left(\frac{1}{7} - \frac{2}{3}\right)x = -\frac{11}{21}x$. Expanding the first product on the left and simplifying the second product then gives

$$3(4) - 3(2x) - 3x = -\frac{11}{21}x.$$

Simplifying the left side gives $-9x + 12 = -\frac{11}{21}x$. Adding $9x$ to both sides gives $12 = -\frac{11}{21}x + 9x = \frac{178}{21}x$. Multiplying both sides by $\frac{21}{178}$ gives $x = 12 \cdot \frac{21}{178} = \boxed{\frac{126}{89}}$.

5.35   First, we simplify the left side as $\frac{t/4}{16} = \frac{t}{4(16)} = \frac{t}{64}$, so the equation is $\frac{t}{64} = \frac{1}{6}$. Multiplying both sides by 64 gives $t = 64 \cdot \frac{1}{6} = \boxed{\frac{32}{3}}$.

We also could have solved this problem by first multiplying $\frac{t/4}{16} = \frac{t}{6}$ by 48 (the least common denominator of 16 and 6):

$$48 \cdot \frac{t/4}{16} = 48 \cdot \frac{1}{6} = 8.$$

Since $\frac{48}{16} = 3$, simplifying the left side of $48 \cdot \frac{t/4}{16} = 8$ gives $3(t/4) = 8$, so $\frac{3}{4}t = 8$. Multiplying both sides of $\frac{3}{4}t = 8$ by $\frac{4}{3}$ gives $t = \boxed{\frac{32}{3}}$.

5.36   Multiplying both sides by 10^6 gives

$$\begin{aligned}
x &= 10^6 \left(\frac{1}{10^1} + \frac{1}{10^2} + \frac{1}{10^3} + \frac{1}{10^4} + \frac{1}{10^5} + \frac{1}{10^6} \right) \\
&= \frac{10^6}{10^1} + \frac{10^6}{10^2} + \frac{10^6}{10^3} + \frac{10^6}{10^4} + \frac{10^6}{10^5} + \frac{10^6}{10^6} \\
&= 10^5 + 10^4 + 10^3 + 10^2 + 10^1 + 10^0 \\
&= \boxed{111111}.
\end{aligned}$$

5.37

(a)   When $x = 0$, we have

$$\frac{x-2}{2x+7} = \frac{0-2}{2 \cdot 0 + 7} = \frac{-2}{7} = \boxed{-\frac{2}{7}}.$$

(b)   When $x = -3$, we have

$$\frac{x-2}{2x+7} = \frac{-3-2}{2 \cdot (-3) + 7} = \frac{-5}{1} = \boxed{-5}.$$

(c)   When $x = \frac{1}{2}$, we have

$$\frac{x-2}{2x+7} = \frac{\frac{1}{2}-2}{2 \cdot \frac{1}{2} + 7} = \frac{-\frac{3}{2}}{8} = -\frac{3}{2} \cdot \frac{1}{8} = \boxed{-\frac{3}{16}}.$$

(d)   We can't divide by 0, so we cannot have the denominator equal to zero. Otherwise, the expression is defined. So, the only values of x for which the expression is not defined are the values of x such that $2x + 7 = 0$. Subtracting 7 from both sides of $2x + 7 = 0$ gives $2x = -7$, and dividing by 2 gives $x = \boxed{-\frac{7}{2}}$.

5.38   When $y = 7$, the expression on the left, $2y - 7$, is an integer, but the expression on the right, $y/3 + 9$, is not. Therefore, the two sides cannot be equal.

5.39   Subtracting 2 from both sides gives $-7g = 21$. Dividing both sides by -7 gives $g = -3$. Substituting this into the given expression gives $\frac{g+7}{g+4} = \frac{-3+7}{-3+4} = \boxed{4}$.

5.40   Multiplying both sides of $\frac{8}{x} = -3$ by x gives $\frac{8x}{x} = -3x$, so $8 = -3x$. Dividing by -3 gives $x = \boxed{-\frac{8}{3}}$.

5.41   We'll start with the first two quantities: $\frac{2}{3} = \frac{x}{24}$. Multiplying both sides by 24 gives $x = \frac{2}{3} \cdot 24 = 16$. We also have $\frac{2}{3} = \frac{84}{y}$. Multiplying both sides by 3 gives $2 = \frac{84 \cdot 3}{y}$. Multiplying both sides by y gives $2y = 84 \cdot 3 = 252$. Dividing both sides of $2y = 252$ by 2 gives $y = 126$. So $x + y = 16 + 126 = \boxed{142}$.

5.42   Let the number be n, so we have $6 + 2n = \frac{n}{2} - 12$. Adding 12 to both sides gives $18 + 2n = \frac{n}{2}$. Multiplying both sides by 2 gives $2(18 + 2n) = n$. Expanding the product on the left gives $36 + 4n = n$. Subtracting $4n$ from both sides gives $36 = -3n$, and dividing by -3 gives $n = \boxed{-12}$.

5.43   First, we find the number that Tom multiplied by $2\frac{1}{2}$. Let n be the number, so $n \cdot \frac{5}{2} = 50$ (since $2\frac{1}{2} = \frac{5}{2}$). Multiplying both sides by $\frac{2}{5}$ gives $n = 50 \cdot \frac{2}{5} = 20$. He was supposed to divide this number by $2\frac{1}{2}$, which would have given him

$$\frac{20}{2\frac{1}{2}} = \frac{20}{5/2} = 20 \cdot \frac{2}{5} = \boxed{8}.$$

5.44   If all 60 tickets were worth two points, he'd have 120 points total. That's 111 too few. Each time we replace a two-point ticket with a five-point ticket, we gain three points. So, we must replace $111/3 = 37$ of the two-point tickets with five-point tickets. Therefore, there must be $60 - 37 = \boxed{23}$ two-point tickets.

We could also have used a variable to solve the problem. Let t be the number of two-point tickets Jay had. Since he had 60 tickets total, he must have had $60 - t$ five-point tickets. So, the total number of points the tickets are worth is $2t + 5(60 - t)$, which means we have

$$2t + 5(60 - t) = 231.$$

Expanding the product on the left gives $2t + 300 - 5t = 231$, and simplifying gives $300 - 3t = 231$. Subtracting 300 from both sides gives $-3t = -69$, and dividing by -3 gives $t = \boxed{23}$ tickets.

5.45   Let x be the number of dollars she has. If I give her 5 dollars, then she'll have $x + 5$ dollars. Since we must have the same amount of money if I give her \$5, I must have $x + 5$ dollars after giving her \$5. So, I must have had $x + 10$ dollars before giving her any money.

If instead she gives me 8 dollars, I'll have $x + 18$ dollars and she'll have $x - 8$ dollars. Since I'll then have twice as much money as she has, we know that $x + 18 = 2(x - 8)$. Expanding the product on the right gives $x + 18 = 2x - 16$. Adding 16 to both sides gives $x + 34 = 2x$, and subtracting x from both sides gives $x = 34$. So, she has $\boxed{34 \text{ dollars}}$.

5.46   Let a be the number of apples on the tree, so Jenny picked $\frac{a}{4}$ apples and Lenny picked $\frac{a}{3}$. Since Lenny picked 7 more apples than Jenny, we have $\frac{a}{3} - \frac{a}{4} = 7$. This means $\left(\frac{1}{3} - \frac{1}{4}\right) a = 7$, so $\frac{1}{12}a = 7$. Multiplying both sides by 12 gives $a = 84$. Therefore, Jenny picked $\frac{84}{4} = 21$ apples and Lenny picked $21 + 7 = 28$ apples. This leaves $84 - 21 - 28 = \boxed{35}$ apples for Penny to pick.

5.47 Let x be the cost of the computer in dollars. Initially, the five members in the club each paid $\frac{x}{5}$ dollars. After the three new members joined and paid their fair share, the eight members of the club must have each paid $\frac{x}{8}$ dollars. Since this amount is \$15 less than what the five members initially paid, we have $\frac{x}{5} - \frac{x}{8} = 15$. Multiplying both sides by 40 to get rid of the fractions gives $40\left(\frac{x}{5} - \frac{x}{8}\right) = 15 \cdot 40$, so $8x - 5x = 600$. Simplifying the right-hand side gives $3x = 600$, so $x = 200$. Therefore the price of the computer was $\boxed{\$200}$.

5.48 Let the length of the road be r yards. The crew paved $\frac{2}{5}r$ yards in the first day, $\frac{1}{3}r$ yards on the second day, and 1500 yards on the third day. Since they paved all r yards of the road in these three days, we must have $\frac{2}{5}r + \frac{1}{3}r + 1500 = r$. Simplifying the left side gives $\frac{11}{15}r + 1500 = r$. Subtracting $\frac{11}{15}r$ from both sides gives $1500 = \frac{4}{15}r$. Multiplying both sides by $\frac{15}{4}$ gives $r = 1500 \cdot \frac{15}{4} = 5625$. The road is $\boxed{5625 \text{ yards}}$ long.

5.49 Let n be the number of employees in the company. The manager's original plan would have cost $50n$ dollars. Since this is \$5 more than the fund had, the fund must have had $50n - 5$ dollars. When the manager gave each employee 45 dollars, he paid $45n$ dollars out of the fund. Since there were 95 dollars left in the fund, the fund must have had $45n + 95$ dollars. We now have two expressions for the original amount of money in the fund, so we set these equal: $50n - 5 = 45n + 95$. Adding 5 to both sides and subtracting $45n$ from both sides gives $5n = 100$, so $n = 20$. So, there are 20 employees. The question asks for the amount of money that was originally in the fund, which is $50n - 5 = 1000 - 5 = 995$ dollars. The fund had $\boxed{\$995}$ before any bonuses were paid.

5.50 Let the number my teacher told me be n, so I computed $(n - 8) \cdot 5$, which means $(n - 8) \cdot 5 = 70$. Dividing both sides by 5 gives $n - 8 = 14$, so $n = 22$. If I had done what I was supposed to do, I would have subtracted 5 to get 17, and then multiplied by 8 to get $\boxed{136}$.

5.51

(a) $\boxed{\text{False}}$. If $a = b = c = 5$, then we have $a \leq b$ and $b \leq c$, but it is not true that $a < c$.

(b) $\boxed{\text{True}}$. Since b must be no greater than a ($a \geq b$) and b must be no less than a ($b \geq a$), we must have $b = a$.

(c) $\boxed{\text{False}}$. If $c = 0$, then $ac = bc$. If c is negative, then $ac < bc$.

(d) $\boxed{\text{True}}$. If c is 0, then $ac = bc$. If $c < 0$, then $ac < bc$, so $ac \leq bc$.

(e) $\boxed{\text{True}}$. Subtracting a from both sides of $x + a \geq y + a$ gives $x \geq y$.

(f) $\boxed{\text{False}}$. Suppose $x + a \geq y + b$ is $5 + 1 \geq 0 + 2$. Here, we have $5 \geq 0$ and $1 < 2$, so we don't have $x \geq y$ and $a \geq b$. So, just because $x + a \geq y + b$, it is not necessarily true that $x \geq y$ and $a \geq b$.

5.52

(a) $\boxed{\text{Yes}}$. Since $a > b > c > d$, we have $a > b$ and $c > d$. So, it seems like $a + c$ must be greater than $b + d$ because $a + c$ is the sum of the greater numbers in these two inequalities and $b + d$ is the sum of the lesser numbers. We can use the facts we discovered in the chapter to be sure this intuition is correct. Adding c to both sides of $a > b$ gives $a + c > b + c$, and adding b to both sides of $c > d$ gives $b + c > b + d$. Combining $a + c > b + c$ and $b + c > b + d$ gives $a + c > b + c > b + d$, so $a + c > b + d$.

(b) $\boxed{\text{No}}$. Suppose $a > b > c > d$ is $8 > 7 > 6 > 1$. Then, we have $a + d = 9$ and $b + c = 13$, so $a + d$ is not greater than $b + c$.

(c) $\boxed{\text{No}}$. We have to be careful about negatives when multiplying inequalities. Suppose $a > b > c > d$ is $2 > -1 > -2 > -3$. Then, $ac = -4$ but $bd = 3$, so ac is not greater than bd.

(d) $\boxed{\text{No}}$. Again, we have to be careful about negatives. If $a > b > c > d$ is $2 > -1 > -2 > -3$, then $ab = -2$ and $cd = 6$, and ab is not greater than cd.

5.53

(a) The numbers between $-2\frac{1}{10}$ and $4\frac{1}{2}$ satisfy the inequality. A graph of these values is shown below:

(b) Simplifying both sides gives $-3t + 5 \geq 3t + 17$. Subtracting $3t$ from both sides gives $-6t + 5 \geq 17$. Subtracting 5 from both sides gives $-6t \geq 12$. Dividing by -6 (and reversing the direction of the inequality symbol) gives $t \leq -2$. Therefore, all numbers less than or equal to -2 satisfy the original inequality. These values are graphed below:

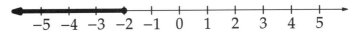

(c) Multiplying both sides by 12 gives $12 \cdot \frac{3}{4}(3-x) \leq -12 \cdot \frac{2}{3}(2+x)$, so $9(3-x) \leq -8(2+x)$. Expanding both products gives $27 - 9x \leq -16 - 8x$. Adding $9x$ to both sides gives $27 \leq x - 16$, so $43 \leq x$. Therefore, all numbers greater than or equal to 43 satisfy the inequality. These numbers are graphed below:

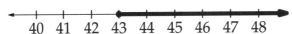

5.54

(a) Simplifying both sides gives $2 - 3x \geq -3x - 3$. Adding $3x$ to both sides gives $2 \geq -3$. This inequality is always true, so $\boxed{\text{all values of } x}$ satisfy the original inequality.

(b) Simplifying both sides gives $9 - 3x \geq 12 - 3x$. Adding $3x$ to both sides gives $9 \geq 12$, which is never true. Therefore, there are $\boxed{\text{no values of } x}$ that satisfy this inequality.

5.55 Let d be the number of dollars in the pile of money. If she puts \$100 in her left pocket, gives away $\frac{2}{3}$ of the rest of the pile, and then puts the remaining money in her right pocket, she keeps \$100 plus one-third of the remaining $d - 100$ dollars in the pile. Therefore, she has $100 + \frac{1}{3}(d - 100)$ dollars. If she instead gives away \$500 and keeps the rest, she has $d - 500$ dollars. In order for her to have more money in the first case than in the second, we must have $100 + \frac{1}{3}(d - 100) > d - 500$. Subtracting 100 from both sides gives $\frac{1}{3}(d - 100) > d - 600$. Multiplying both sides by 3 gives $3 \cdot \frac{1}{3}(d - 100) > 3(d - 600)$, so $d - 100 > 3d - 1800$. Subtracting d from both sides and adding 1800 to both sides gives $1700 > 2d$. Dividing by 2 gives $850 > d$. Combining this with the given fact that there is at least 500 dollars in the pile, there must be $\boxed{\text{less than 850 dollars and at least 500 dollars}}$ in the original pile of money.

Challenge Problems

5.56 First, we write the mixed numbers as fractions, which gives

$$\frac{\frac{25}{4}}{\frac{5}{2}} = \frac{\frac{3}{2}}{x}.$$

Therefore, we have $\frac{25}{4} \cdot \frac{2}{5} = \frac{3}{2} \cdot \frac{1}{x}$, so $\frac{5}{2} = \frac{3}{2x}$. Multiplying both sides by $2x$ gives $\frac{5}{2} \cdot 2x = \frac{3}{2x} \cdot 2x$, or $5 \cdot \frac{2x}{2} = 3 \cdot \frac{2x}{2x}$. This gives us $5x = 3$, and dividing by 5 gives $x = \boxed{\frac{3}{5}}$.

5.57 We might be tempted to simply multiply both sides by n and write $1 \geq 6n$, so $\frac{1}{6} \geq n$. However, we have to be careful. First, n cannot be 0, since we cannot divide by 0. Second, if n is negative, then we have to reverse the direction of the inequality symbol. But looking back at the original inequality, we can see that n cannot be negative, since this would make the greater side of $\frac{1}{n} \geq 6$ negative while the other side is positive. Therefore, n must be positive, which means we can indeed multiply by n to produce $1 \geq 6n$. Dividing by 6 gives $\frac{1}{6} \geq n$, but we must remember that n has to be positive. So, the full solution is $0 < n \leq \frac{1}{6}$, or $\boxed{\text{all positive numbers less than or equal to } \frac{1}{6}}$.

5.58 Let n be Kayla's number, so we must have $\frac{1+n}{10+n} = \frac{2}{3}$. Multiplying both sides by 3 and by $10 + n$ gives

$$3 \cdot (10 + n) \cdot \frac{1 + n}{10 + n} = 3 \cdot \frac{2}{3} \cdot (10 + n).$$

The right-hand side is simply $2(10 + n)$, while the left side simplifies as

$$3 \cdot (10 + n) \cdot \frac{1 + n}{10 + n} = 3 \cdot \frac{(10 + n)(1 + n)}{10 + n} = 3 \cdot \frac{10 + n}{10 + n} \cdot (1 + n) = 3(1 + n).$$

So, our equation is now $3(1 + n) = 2(10 + n)$. Expanding both sides gives $3 + 3n = 20 + 2n$. Subtracting $2n$ and 3 from both sides gives $n = \boxed{17}$.

5.59 Let x be Douglas's favorite number. Next, we have to figure out how to represent the number Douglas makes when he writes a 7 at the end of his favorite number. We know how to represent a new number that results when we write a 0 at the end of a number. That's just 10 times the original number. So, if Douglas put a 0 at the end of his favorite number, he would make the number $10x$. The number he makes when he puts a 7 at the end of his favorite number is 7 greater than the number he would make by putting 0 at the end of his favorite number. So, when he places a 7 on the end of his favorite number, the new number's value equals $10x + 7$. This number is 385 more than Douglas's favorite number, so we must have $10x + 7 = x + 385$. Subtracting x and 7 from both sides gives $9x = 378$, and dividing by 9 gives $x = \boxed{42}$.

5.60 We start with the first two fractions, $\frac{5}{6} = \frac{n}{72}$. Multiplying both sides by 72 gives $n = \frac{5}{6} \cdot 72 = 60$. Substituting this value in for n gives

$$\frac{5}{6} = \frac{60}{72} = \frac{m + 60}{84} = \frac{p - m}{120}.$$

So, now we have $\frac{5}{6} = \frac{m+60}{84}$. Multiplying both sides by 84 gives $m + 60 = \frac{5}{6} \cdot 84 = 70$, so $m = 10$. Substituting this value in for m gives

$$\frac{5}{6} = \frac{60}{72} = \frac{10 + 60}{84} = \frac{p - 10}{120}.$$

So, now we have $\frac{5}{6} = \frac{p-10}{120}$. Multiplying both sides by 120 gives $p - 10 = \frac{5}{6} \cdot 120 = 100$, so $p = \boxed{110}$.

5.61

(a) The denominators of the fractions are the same, so we can apply the distributive property:
$$\frac{1}{x} + \frac{3}{x} = \frac{1+3}{x} = \boxed{\frac{4}{x}}.$$

(b) If we make the denominators the same, we can apply the distributive property. The second denominator is twice the first, so we multiply the numerator and denominator of the second fraction by 2:
$$\frac{5}{4x} - \frac{2}{2x} = \frac{5}{4x} - \frac{2 \cdot 2}{2x \cdot 2} = \frac{5}{4x} - \frac{4}{4x} = \frac{5-4}{4x} = \boxed{\frac{1}{4x}}.$$

(c) The least common multiple of 16 and 10 is 80, and we have
$$\frac{7}{16x} - \frac{3}{10x} = \frac{7}{16x} \cdot \frac{5}{5} - \frac{3}{10x} \cdot \frac{8}{8} = \frac{35}{80x} - \frac{24}{80x} = \frac{35-24}{80x} = \boxed{\frac{11}{80x}}.$$

5.62 We notice two of the fractions have the same denominator, so they will be easy to add or subtract. Subtracting $\frac{1}{z-1}$ from both sides gives
$$\frac{5}{3} = \frac{3}{z-1} - \frac{1}{z-1} = \frac{3-1}{z-1} = \frac{2}{z-1}.$$

Therefore, we have $\frac{5}{3} = \frac{2}{z-1}$. Multiplying both sides by 3 gives $5 = \frac{6}{z-1}$, and multiplying both sides of this equation by $z - 1$ gives $5(z - 1) = 6$. Expanding the left side gives $5z - 5 = 6$, so $5z = 11$ and $z = \boxed{\frac{11}{5}}$.

5.63 If the squares of two numbers are equal, then either the two numbers are equal or they are opposites. So, we must have either $12 - t = 3 + 2t$ or $12 - t = -(3 + 2t)$. Solving the first equation gives $t = 3$. Turning to the second equation, we can write $12 - t = -(3 + 2t)$ as $12 - t = -3 - 2t$. Solving this equation gives $t = -15$. So, the two solutions to the equation is $\boxed{3 \text{ and } -15}$.

5.64 Since a must equal both $2b + 21$ and $b - 28$, we must have $2b + 21 = b - 28$. Subtracting b and 21 from both sides gives $b = -49$. Substituting this into $a = b - 28$, we have $a = -77$. So, the integers are $\boxed{a = -77 \text{ and } b = -49}$.

5.65 Simplifying the right side gives $7y + 3c = 7y + 15$. Subtracting $7y$ from both sides gives $3c = 15$. In order for the original equation to have infinitely many solutions for y, the equation $3c = 15$ must always be true. Therefore, we must have $c = \boxed{5}$.

5.66 First, we solve the equation Paula wrote down. Adding 7 to both sides of $2x - 7 = 23$ gives $2x = 30$, so $x = 15$. Therefore, the correct answer must have been $15 - 5 = 10$. Let a be the number that was supposed to have been the coefficient of x, so Paula should have written $ax - 7 = 23$. Since the answer to the problem was supposed to be $x = 10$, we must have $a(10) - 7 = 23$. Therefore, we must have $10a = 30$, so $a = \boxed{3}$.

5.67 In order to have $7 + x \geq 2x + 3 > 12 - x$, we must have both $7 + x \geq 2x + 3$ and $2x + 3 > 12 - x$. So, we tackle these two inequalities separately and then combine the results. First, we solve $7 + x \geq 2x + 3$.

Subtracting x and 3 from both sides gives $4 \geq x$. Next, we solve $2x + 3 > 12 - x$. Adding x to both sides and subtracting 3 from both sides gives $3x > 9$, so $x > 3$. We then combine the results of these two inequalities. We must have both $4 \geq x$ and $x > 3$, so x can be any number that is greater than 3 and less than or equal to 4. We can write this as $3 < x \leq 4$, and graph the solutions as shown below:

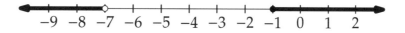

5.68 If $\frac{2x+2}{x+7} \geq 0$, then the fraction must either be positive or equal 0. If the fraction equals 0, then we must have $2x + 2 = 0$, so $x = -1$. If the fraction is positive, then the numerator and denominator have the same sign (positive or negative). We investigate these cases separately.

Case 1: Both are positive. If $2x + 2 > 0$, then $2x > -2$, so $x > -1$. If $x + 7 > 0$, then $x > -7$. So, in order for both $2x + 2$ and $x + 7$ to be positive, we must have both $x > -1$ and $x > -7$. Therefore, all values of x greater than -1 make both expressions positive.

Case 2: Both are negative. If $2x + 2 < 0$, then $2x < -2$, so $x < -1$. If $x + 7 < 0$, then $x < -7$. So, in order for both $2x + 2$ and $x + 7$ to be negative, we must have both $x < -1$ and $x < -7$. Therefore, all values of x less than -7 make both expressions negative.

So, the inequality $\frac{2x+2}{x+7} \geq 0$ is satisfied if $x = -1$, if $x > -1$, or if $x < -7$. We can combine the first two as $x \geq -1$, and graph the solutions on the number line as shown below:

5.69 Since the number x is twice its reciprocal, we must have $x = 2 \cdot \frac{1}{x}$. Multiplying both sides by x gives $x \cdot x = 2 \cdot \frac{1}{x} \cdot x$, so $x^2 = 2$. There's no integer whose square is 2, but we can still answer the question! We want the value of x^6, and $x^6 = x^{2 \cdot 3} = (x^2)^3$. Since $x^2 = 2$, we have $x^6 = (x^2)^3 = 2^3 = \boxed{8}$.

5.70

(a) Subtracting $\frac{1}{3}$ from both sides gives $\frac{1}{n} = \frac{1}{2} - \frac{1}{3} = \frac{1}{6}$, so $n = \boxed{6}$.

(b) Subtracting $\frac{1}{4}$ from both sides gives $\frac{1}{n} = \frac{1}{3} - \frac{1}{4} = \frac{4}{3 \cdot 4} - \frac{3}{4 \cdot 3} = \frac{1}{12}$, so $n = \boxed{12}$.

(c) Subtracting $\frac{1}{5}$ from both sides gives $\frac{1}{n} = \frac{1}{4} - \frac{1}{5} = \frac{5}{4 \cdot 5} - \frac{4}{5 \cdot 4} = \frac{1}{20}$, so $n = \boxed{20}$.

(d) Subtracting $\frac{1}{1001}$ from both sides gives $\frac{1}{n} = \frac{1}{1000} - \frac{1}{1001} = \frac{1001}{1000 \cdot 1001} - \frac{1000}{1001 \cdot 1000} = \frac{1}{1001000}$, so $n = \boxed{1001000}$.

(e) We see a pattern in our first four parts. Specifically, it appears that if $\frac{1}{m} = \frac{1}{m+1} + \frac{1}{n}$, then $n = m(m+1)$. For example, in the first part, the solution to $\frac{1}{2} = \frac{1}{3} + \frac{1}{n}$ is $n = 2(2 + 1)$, and in the second part, the solution to $\frac{1}{3} = \frac{1}{4} + \frac{1}{n}$ is $n = 3(3 + 1)$. Let's see why this is true. Subtracting $\frac{1}{m+1}$ from both sides of $\frac{1}{m} = \frac{1}{m+1} + \frac{1}{n}$ gives

$$\frac{1}{n} = \frac{1}{m} - \frac{1}{m+1} = \frac{1}{m} \cdot \frac{m+1}{m+1} - \frac{1}{m+1} \cdot \frac{m}{m} = \frac{m+1}{m(m+1)} - \frac{m}{m(m+1)} = \frac{m+1-m}{m(m+1)} = \frac{1}{m(m+1)}.$$

Since $\frac{1}{n} = \frac{1}{m(m+1)}$, we have $n = m(m + 1)$.

Therefore, we can write any fraction of the form $\frac{1}{m(m+1)}$ as $\frac{1}{m} - \frac{1}{m+1}$. When we do this with the fractions in our sum, we have

$$\frac{1}{1\cdot 2} = \frac{1}{1} - \frac{1}{2},$$
$$\frac{1}{2\cdot 3} = \frac{1}{2} - \frac{1}{3},$$
$$\frac{1}{3\cdot 4} = \frac{1}{3} - \frac{1}{4},$$
$$\vdots$$
$$\frac{1}{98\cdot 99} = \frac{1}{98} - \frac{1}{99},$$
$$\frac{1}{99\cdot 100} = \frac{1}{99} - \frac{1}{100}.$$

When we add all the left sides of these equations, we get our desired sum. This must equal the sum of all the right sides. In the sum of all the right sides, every fraction except $\frac{1}{1}$ and $\frac{1}{100}$ is added once and subtracted once, so they all cancel. We are left with $\frac{1}{1} - \frac{1}{100} = \boxed{\frac{99}{100}}$.

We call the sum in this part a **telescoping** sum, since we can write the terms in the sum such that part of each term cancels with part of another term.

CHAPTER 6

Decimals

24® Cards

First Card $6 + 6 + 6 + 6$.

Second Card $6 \cdot 10 - 6 \cdot 6$.

Third Card $6/(1 - 3/4)$.

Fourth Card $(9 - 6)(21 - 13)$.

Exercises for Section 6.1

6.1.1   All five numbers have the same tenths digit, so we move on to the hundredths digit. Only 0.99 has a nonzero hundredths digit, so 0.99 is the largest of the numbers. Moving on to the thousandths digit, both 0.9099 and 0.909 have 9 as the thousandths digit while 0.9 and 0.9009 have 0 as the thousandths digit. So, 0.9099 and 0.909 are both larger than 0.9 and 0.9009. Going to the ten-thousandths digit, we see that 0.9099 is larger than 0.909, and 0.9009 is larger than 0.9. Therefore, in order from smallest to largest, the numbers are

$$\boxed{0.9, 0.9009, 0.909, 0.9099, 0.99}.$$

6.1.2   We are asked to find the largest of the numbers

$$0.94321, 0.59321, 0.54921, 0.54391, 0.54329.$$

Since 0.94321 has the largest tenths digit, it is the largest number. So we should change the digit $\boxed{5}$ of 0.54321 to get 0.94321.

6.1.3

(a) $0.4 + 0.02 + 0.006 = \boxed{0.426}$.

(b)   Our addition is shown to the right. Notice that we have to carry from the hundredths place to the tenths place, and from the tenths place to the ones place. The resulting sum is $0.92 + 0.093 = \boxed{1.013}$.

$$\begin{array}{r} 0.920 \\ + 0.093 \\ \hline 1.013 \end{array}$$

(c) Our subtraction is shown at the right, where we see that $1.28 - 0.377 = \boxed{0.903}$.

$$\begin{array}{r} 1.280 \\ -\,0.377 \\ \hline 0.903 \end{array}$$

(d) Writing 1.001 as $1 + 0.001$ gives us

$$8 - 1.001 = 8 - (1 + 0.001) = 8 - 1 - 0.001 = 7 - 0.001 = \boxed{6.999}.$$

(e) We note that 0.002 is greater than 0.0006, so the result must be negative:

$$0.0006 - 0.002 = -(0.002 - 0.0006) = -(0.0020 - 0.0006) = \boxed{-0.0014}.$$

(f) Rearranging the terms makes the computation easier:

$$1.1 - 0.11 + 0.011 = 1.1 + 0.011 - 0.11 = 1.111 - 0.11 = \boxed{1.001}.$$

6.1.4

(a) Multiplying by 100 moves the decimal point two places to the right: $23.879 \cdot 100 = \boxed{2387.9}$.

(b) Dividing by 10^5 moves the decimal point five places to the left:

$$2 \div 10^5 = 2.0 \div 10^5 = \boxed{0.00002}.$$

(c) We have

$$1.6 \div 400 = \frac{1.6}{400} = \frac{16 \cdot 10^{-1}}{4 \cdot 10^2} = \frac{16}{4} \cdot \frac{10^{-1}}{10^2} = 4 \cdot 10^{-1-2} = 4 \cdot 10^{-3} = \boxed{0.004}.$$

(d) Multiplying by 10^6 moves the decimal point 6 places to the right, so $0.0031 \cdot 10^6 = \boxed{3100}$.

(e) We write the quotient as a fraction, and then multiply the numerator and denominator by the appropriate power of 10 to make the denominator an integer:

$$3.6 \div 0.09 = \frac{3.6}{0.09} = \frac{3.6 \cdot 10^2}{0.09 \cdot 10^2} = \frac{360}{9} = \boxed{40}.$$

(f) We use the distributive property, since it's easy to multiply by 1 and by 0.01:

$$1.01 \cdot 3.03 = (1 + 0.01) \cdot 3.03 = 1 \cdot 3.03 + 0.01 \cdot 3.03 = 3.03 + 0.0303 = \boxed{3.0603}.$$

6.1.5 We factor 25 out of the first sum and 5 out of the second, and find that the remaining factors are the same, so they cancel:

$$\frac{250 + 25 + 2.5 + 0.25 + 0.025}{50 + 5 + 0.5 + 0.05 + 0.005} = \frac{25(10 + 1 + 0.1 + 0.01 + 0.001)}{5(10 + 1 + 0.1 + 0.01 + 0.001)} = \frac{25}{5} = \boxed{5}.$$

6.1.6 First, we note that x is a tiny bit larger than 0. So, $2 + x$ is a little larger than 2 and $2 - x$ is a little smaller than 2, which means $2 + x$ is greater than $2 - x$. Both $2x$ and $\frac{x}{2}$ are less than 1, so both are less than $2 - x$. Doubling x to get $2x$ produces a number that is greater than x, while halving x to get $\frac{x}{2}$ produces a number that is smaller than x. So, $2x$ is greater than $\frac{x}{2}$.

Finally, we note that $\frac{2}{x} = \frac{2}{10^{-5001}} = 2 \cdot 10^{5001}$, which is much greater than all of the other numbers. So, from least to greatest, the numbers are

$$\boxed{\frac{x}{2}, 2x, 2-x, 2+x, \frac{2}{x}}.$$

6.1.7 Betty computed $75 \cdot 256$. We have

$$0.075 \cdot 2.56 = (75 \cdot 10^{-3}) \cdot (256 \cdot 10^{-2}) = (75 \cdot 256) \cdot (10^{-3} \cdot 10^{-2}) = (75 \cdot 256) \cdot 10^{-5}.$$

So, to get the correct answer, we must multiply Betty's result by 10^{-5}, which moves the decimal point 5 places to the left: $19200 \cdot 10^{-5} = \boxed{0.192}$. Notice that we don't need to include the trailing zeros in 0.19200. Also, note that our answer makes sense; if we multiply 0.075 by a number that is between 2 and 3, then we will get a number that is between 0.15 and 0.225.

6.1.8 We separate factors of 1993 from the powers of 10:

$$1000 \cdot 1993 \cdot 0.1993 \cdot 10 = 10^3 \cdot 1993 \cdot (1993 \cdot 10^{-4}) \cdot 10$$
$$= 1993^2 \cdot (10^3 \cdot 10^{-4} \cdot 10) = 1993^2 \cdot 10^{3+(-4)+1} = 1993^2 \cdot 10^0 = \boxed{1993^2}.$$

6.1.9 Writing 0.0481 as 4.81 times a power of 10, we have

$$0.0481 \cdot 10^{-4} = (4.81 \cdot 10^{-2}) \cdot 10^{-4} = 4.81 \cdot (10^{-2} \cdot 10^{-4}) = 4.81 \cdot 10^{-2+(-4)} = 4.81 \cdot 10^{-6},$$

so $N = \boxed{10^{-6}}$.

6.1.10 Let x be the number that Curt multiplied by 10, so he computed $10x$. He should have divided by 10, which means that he should have computed $0.1x$. Since the number he computed was 33.66 greater than the number he should have found, we have $10x - 0.1x = 33.66$. Simplifying the left-hand side gives $(10 - 0.1)x = 33.66$, or $9.9x = 33.66$. Dividing both sides by 9.9 gives

$$x = \frac{33.66}{9.9} = \frac{33.66 \cdot 100}{9.9 \cdot 100} = \frac{3366}{990} = \frac{11 \cdot 306}{11 \cdot 90} = \frac{306}{90} = \frac{9 \cdot 34}{9 \cdot 10} = \frac{34}{10} = \boxed{3.4}.$$

Exercises for Section 6.2

6.2.1

(a) 28.2508 is between 20 and 30, and is closer to 30 than to 20, so 28.2508 rounded to the nearest ten is $\boxed{30}$.

(b) 28.2508 is between 28.2 and 28.3, and is closer to 28.3 than to 28.2, so 28.2508 rounded to the nearest tenth is $\boxed{28.3}$.

(c) 28.2508 is between 28.25 and 28.26, and is closer to 28.25 than to 28.26, so 28.2508 rounded to the nearest hundredth is $\boxed{28.25}$.

6.2.2

(a) -0.155 is between -1 and 0, and is closer to 0 than to -1, so -0.155 rounded to the nearest integer is $\boxed{0}$.

(b) -0.155 is between -0.1 and -0.2. It is 0.055 from -0.1 and 0.045 from -0.2, so -0.155 is closer to -0.2 than to -0.1. Therefore, -0.155 rounded to the nearest tenth is $\boxed{-0.2}$.

(c) -0.155 is exactly halfway between -0.16 and -0.15, so by rule, it rounds to the larger quantity, which is $\boxed{-0.15}$.

6.2.3 7.6397 is between 7.639 and 7.640, and is closer to 7.640 than to 7.639, so 7.6397 rounded to the nearest thousandth is $\boxed{7.640}$, which is the same as 7.64.

6.2.4 Because x rounded to the nearest tenth is 1.8, we have $1.75 \le x < 1.85$. Because x rounded to the nearest hundredth is 1.82, we have $1.815 \le x < 1.825$. Because x rounded to the nearest thousandth is 1.819, we have $1.8185 \le x < 1.8195$.

Both bounds of $1.8185 \le x < 1.8195$ are inside the bounds of $1.75 \le x < 1.85$ and $1.815 \le x < 1.825$. So, any number x that satisfies $1.8185 \le x < 1.8195$ will automatically satisfy the other two. For example, $\boxed{1.819}$ satisfies the problem. Any other number x that satisfies $1.8185 \le x < 1.8195$ is also a valid answer.

6.2.5 We multiply the numerator and the denominator by 10^3, so that we aren't dividing by a decimal:

$$\frac{401}{0.205} = \frac{401 \cdot 10^3}{0.205 \cdot 10^3} = \frac{401 \cdot 10^3}{205} = \frac{401}{205} \cdot 10^3.$$

Since 401 is very close to $2 \cdot 205 = 410$, we know that $\frac{401}{205}$ is nearly 2, which means that among the choices given, $\frac{401}{.205}$ is closest to $\boxed{2000}$.

To be sure that $\frac{401}{205} \cdot 10^3$ is closer to 2000 than to 200, we note that $\frac{401}{205} = 1\frac{196}{205}$ is greater than 1.5, so $\frac{401}{205} \cdot 10^3$ is greater than $1.5 \cdot 10^3 = 1500$. This means that $\frac{401}{205} \cdot 10^3$ is between 1500 and 2000, which means it is closer to 2000 than to 200.

Exercises for Section 6.3

6.3.1

(a) Multiplying 25 by 4 produces a power of 10, and we have $\frac{2}{25} = \frac{2 \cdot 4}{25 \cdot 4} = \frac{8}{100} = \boxed{0.08}$.

(b) Since $16 = 2^4$, we see that multiplying 16 by 5^4 produces a power of 10:

$$\frac{5}{16} = \frac{5 \cdot 5^4}{16 \cdot 5^4} = \frac{5 \cdot 625}{2^4 \cdot 5^4} = \frac{3125}{10^4} = \boxed{0.3125}.$$

(c) Multiplying 4 by 25 gives a power of 10, and we have $-\frac{11}{4} = -\frac{11 \cdot 25}{4 \cdot 25} = -\frac{275}{100} = \boxed{-2.75}$.

(d) The denominator is already a power of 10: $\frac{81}{1000} = 81 \cdot 0.001 = \boxed{0.081}$.

(e) Multiplying 4 by 25 gives a power of 10, so multiplying 40 by 25 gives a power of 10:

$$\frac{17}{40} = \frac{17 \cdot 25}{40 \cdot 25} = \frac{425}{1000} = \boxed{0.425}.$$

We might also have noted that $\frac{17}{4} = 4\frac{1}{4} = 4.25$, and $\frac{17}{40} = \frac{17}{4} \cdot \frac{1}{10} = (4.25) \cdot \frac{1}{10} = \boxed{0.425}$.

(f) The denominator is already a power of 10. Dividing by 10000 is the same as moving the decimal point 4 places to the left. We are careful to include zeros in the appropriate places after the decimal point: $\frac{3}{10000} = \boxed{0.0003}$.

6.3.2 We convert each fraction to a decimal separately, and then add:

$$\frac{2}{10} + \frac{4}{100} + \frac{6}{1000} = 0.2 + 0.04 + 0.006 = \boxed{0.246}.$$

6.3.3

(a) $-0.7 = \boxed{-\frac{7}{10}}$.

(b) There are four places past the decimal point, so the denominator of our fraction is 10^4:

$$0.0138 = \frac{138}{10^4} = \frac{2 \cdot 69}{2 \cdot 5000} = \boxed{\frac{69}{5000}}.$$

(c) $0.375 = \frac{375}{1000} = \frac{3 \cdot 5^3}{2^3 \cdot 5^3} = \frac{3}{2^3} = \boxed{\frac{3}{8}}$.

(d) $1.11 = 1\frac{11}{100} = 1 + \frac{11}{100} = \frac{100}{100} + \frac{11}{100} = \boxed{\frac{111}{100}}$.

(e) There are three places past the decimal point, so the denominator of our fraction is 10^3:

$$0.002 = \frac{2}{10^3} = \frac{2}{1000} = \boxed{\frac{1}{500}}.$$

(f) $2.6 = 2\frac{6}{10} = 2\frac{3}{5} = 2 + \frac{3}{5} = \frac{10}{5} + \frac{3}{5} = \boxed{\frac{13}{5}}$. We also could have reasoned: $2.6 = 26 \cdot \frac{1}{10} = \frac{26}{10} = \boxed{\frac{13}{5}}$.

6.3.4 We write each decimal as a fraction, and we have

$$8 \times .25 \times 2 \times .125 = 8 \times \frac{1}{4} \times 2 \times \frac{125}{1000} = 2 \times 2 \times \frac{125}{1000} = 4 \times \frac{125}{8 \times 125} = 4 \times \frac{1}{8} = \boxed{\frac{1}{2}}.$$

6.3.5 We have $3.2 = 32 \cdot \frac{1}{10} = \frac{32}{10} = \frac{16}{5}$, so the reciprocal of 3.2 is the reciprocal of $\frac{16}{5}$, which is $\boxed{\frac{5}{16}}$.

6.3.6 We have $5.75 = 5\frac{3}{4} = 5 + \frac{3}{4}$. There are $5 \cdot 8 = 40$ eighths in 5. Since $\frac{3}{4} = \frac{6}{8}$, there are 6 eighths in $\frac{3}{4}$. So, there are $40 + 6 = \boxed{46}$ eighths in 5.75.

6.3.7 We list the heights (in meters) that the ball reaches on each bounce for the first several bounces by repeatedly multiplying by $\frac{2}{3}$:

$$2, \frac{4}{3}, \frac{8}{9}, \frac{16}{27}, \frac{32}{81}, \frac{64}{243}.$$

Since $0.5 = \frac{1}{2}$, we seek the first time that the ball's height is less than $\frac{1}{2}$ meters. A positive fraction is less than $\frac{1}{2}$ if its denominator is more than twice its numerator. The first of the fractions above whose denominator is more than twice its numerator is $\frac{32}{81}$, so the $\boxed{5^{\text{th}}}$ bounce is the first on which the ball will not reach 0.5 meters.

6.3.8 We are given that $\frac{y}{x} = 1.0625$, so we convert the decimal 1.0625 to a fraction. There are four places past the decimal point, so we have $1.0625 = \frac{10625}{10000}$. We simplify by finding the prime factorizations of the numerator and the denominator: $\frac{10625}{10000} = \frac{5^4 \cdot 17}{2^4 \cdot 5^4} = \frac{17}{2^4} = \frac{17}{16}$. So, Julian could have divided 16 into 17.

However, this isn't the only possibility. There are other fractions equivalent to $\frac{17}{16}$, such as $\frac{17 \cdot 2}{16 \cdot 2} = \frac{34}{32}$ and $\frac{17 \cdot 3}{16 \cdot 3} = \frac{51}{48}$. Both $\frac{17}{16}$ and $\frac{34}{32}$ have numerator and denominator less than 50. If we multiply the numerator and denominator of $\frac{17}{16}$ by any integer greater than 2, then the numerator is greater than 50. So, the only possibilities for the division Julian performed are $17 \div 16$ and $34 \div 32$. Therefore, the desired sum is $17 + 16 + 34 + 32 = \boxed{99}$.

Exercises for Section 6.4

6.4.1

(a) We divide 11 into 2 as shown at the right. In the next step of the long division at the bottom, we will divide 11 into 90. This repeats the second step of the long division, so we know that the decimal will repeat at that point. Therefore, we have $\frac{2}{11} = \boxed{0.\overline{18}}$.

```
       0.181....
   11 | 2.0000...
        1.1
        0.90
        0.88
        0.020
        0.011
        0.0090
```

(b) We divide 11 into 21 as shown at the right. In the next step of the long division at the bottom, we will divide 11 into 100. This repeats the second step of the long division, so we know that the decimal will repeat with alternating 9's and 0's from there on. Therefore, we have $\frac{21}{11} = \boxed{1.\overline{90}}$. Notice that the bar does not go over the 1, since the 1 is not part of the repeating block. Also notice that $1.\overline{90}$ is not the same number as $1.\overline{9}$; indeed, we have $1.\overline{9} = 2$.

```
       1.9090....
   11 | 21.00000...
        11.
        10.0
        9.9
        0.100
        0.099
        0.00100
```

(c) In the text, we found that $\frac{1}{3} = 0.\overline{3}$. Since $\frac{1}{30} = \frac{1}{3} \cdot \frac{1}{10}$, we have $\frac{1}{30} = (0.\overline{3}) \cdot \frac{1}{10}$. Multiplying by $\frac{1}{10}$ is the same as moving the decimal point one place to the left, so we have $\frac{1}{30} = \boxed{0.0\overline{3}}$. Notice that the bar does not cover the 0; only the 3 repeats.

(d) We divide 33 into 5 as shown at the right. In the next step of the long division at the bottom, we will divide 33 into 170. This repeats the second step of the long division, so we know that the decimal will repeat with alternating 1's and 5's from there on. Therefore, we have $\frac{5}{33} = \boxed{0.\overline{15}}$.

```
        .151....
   33 | 5.0000...
        3.3
        1.70
        1.65
        0.050
        0.033
        0.0170
```

(e) We start by noting that $\frac{71}{90} = \frac{71}{9} \cdot \frac{1}{10}$. So we just need to write the decimal for $\frac{71}{9}$ and then move the decimal point one place to the left. Rather than dividing 9 into 71, we can note that $\frac{71}{9} = 7\frac{8}{9}$. So, we just need to find the decimal for $\frac{8}{9}$. We can use the long division at the right to see that $\frac{8}{9} = 0.\overline{8}$, so we have

$$\frac{71}{90} = \frac{71}{9} \cdot \frac{1}{10} = (7.\overline{8}) \cdot \frac{1}{10} = \boxed{0.7\overline{8}}.$$

```
    .8....
9 | 8.00...
    7.2
    0.80
```

(f) We start by noting that $\frac{118}{55} = 2\frac{8}{55}$, so now we just have to write $\frac{8}{55}$ as a decimal. From the division at the right, we have $\frac{8}{55} = 0.1\overline{45}$. Make sure you see why the bar does not extend over the 1; only the alternating 4 and 5 repeat. So, we have $\frac{118}{55} = 2\frac{8}{55} = \boxed{2.1\overline{45}}$.

```
     0.145....
55 | 8.0000...
     5.5
     2.50
     2.20
     0.300
     0.275
     0.0250
```

6.4.2 From the (very) long division at the right, we see that the decimal for $\frac{1}{13}$ consists of a repeating block of 6 digits, $\frac{1}{13} = 0.\overline{076923}$. So, to get to the 14th digit after the decimal point, we write this block twice, and then write 2 more digits. Thus, the 14th digit after the decimal point is the 2nd of this repeating block, which is $\boxed{7}$.

```
      0.07692307....
13 | 1.000000000...
      0.91
      0.090
      0.078
      0.0120
      0.0117
      0.00030
      0.00026
      0.000040
      0.000039
      0.00000100
      0.00000091
      0.000000090
```

6.4.3 If $\frac{a}{b}$ is in simplest form, then when we write $\frac{a}{b}$ as a decimal, the decimal will be infinitely repeating if b has prime factors other than 2 and 5, and the decimal will be finite if b has no prime factors besides 2 and 5. Since 16 is the smallest integer greater than 10 that has no prime factors besides 2 and 5, the smallest positive integer x such that $\frac{1}{10+x}$ has a finite decimal form is $x = \boxed{6}$.

6.4.4

(a) Letting $x = 0.\overline{7}$, we have $10x = 7.\overline{7}$. Subtracting x from $10x$ gives

$$10x - x = 7.\overline{7} - 0.\overline{7} = 7 + (0.\overline{7} - 0.\overline{7}) = 7.$$

Therefore, we have $9x = 7$, so $x = \boxed{\frac{7}{9}}$.

(b) Let $x = 0.\overline{12}$. Here, the repeating portion has two digits, so we multiply by 100, getting $100x = 12.\overline{12}$, and

$$100x - x = 12.\overline{12} - 0.\overline{12} = 12 + (0.\overline{12} - 0.\overline{12}) = 12.$$

Therefore, we have $99x = 12$, so $x = \frac{12}{99} = \boxed{\frac{4}{33}}$.

(c) Let $x = 0.\overline{16}$. Here, the repeating portion has two digits, so we multiply by 100, getting $100x = 16.\overline{16}$, and

$$100x - x = 16.\overline{16} - 0.\overline{16} = 16 + (0.\overline{16} - 0.\overline{16}) = 16.$$

Therefore, we have $99x = 16$, so $x = \boxed{\frac{16}{99}}$.

(d) Let $x = 0.\overline{45}$. Here, the repeating portion has two digits, so we multiply by 100, getting $100x = 45.\overline{45}$, and

$$100x - x = 45.\overline{45} - 0.\overline{45} = 45 + (0.\overline{45} - 0.\overline{45}) = 45.$$

Therefore, we have $99x = 45$, so $x = \frac{45}{99} = \boxed{\frac{5}{11}}$.

(e) Let $x = 0.\overline{912}$. The repeating portion has three digits, so we multiply by 1000, getting $1000x = 912.\overline{912}$. We then have

$$1000x - x = 912.\overline{912} - 0.\overline{912} = 912 + (0.\overline{912} - 0.\overline{912}) = 912.$$

Therefore, we have $999x = 912$, so $x = \frac{912}{999} = \boxed{\frac{304}{333}}$.

(f) There's only one digit repeating, so we let $x = 0.00\overline{1}$, and multiply by 10 to get $10x = 0.0\overline{1}$. Then, we have

$$10x - x = 0.0\overline{1} - 0.00\overline{1} = 0.01\overline{1} - 0.00\overline{1} = 0.01 + (0.00\overline{1} - 0.00\overline{1}) = 0.01.$$

Therefore, we have $9x = 0.01$, so $900x = 1$ and $x = \boxed{\frac{1}{900}}$.

We might also have recognized $0.\overline{1}$ as $\frac{1}{9}$, so $0.00\overline{1} = \frac{1}{100} \cdot (0.\overline{1}) = \frac{1}{100} \cdot \frac{1}{9} = \boxed{\frac{1}{900}}$.

(g) Again, we only have one digit repeating. We let $x = 0.3\overline{6}$, and multiplying by 10 gives $10x = 3.6\overline{6}$. So, we have

$$10x - x = 3.6\overline{6} - 0.3\overline{6} = 3.6 - 0.3 + (0.0\overline{6} - 0.0\overline{6}) = 3.3.$$

This gives us $9x = 3.3$, so $90x = 33$ and $x = \frac{33}{90} = \boxed{\frac{11}{30}}$.

We might also have recognized $0.\overline{6}$ as $\frac{2}{3}$, so

$$0.3\overline{6} = 0.3 + 0.0\overline{6} = \frac{3}{10} + \frac{1}{10} \cdot (0.\overline{6}) = \frac{3}{10} + \frac{1}{10} \cdot \frac{2}{3} = \frac{3}{10} + \frac{2}{30} = \frac{9}{30} + \frac{2}{30} = \boxed{\frac{11}{30}}.$$

(h) We recognize $0.\overline{9}$ as 1, so

$$0.0\overline{9} = \frac{1}{10} \cdot (0.\overline{9}) = \frac{1}{10} \cdot 1 = \boxed{\frac{1}{10}}.$$

(i) We let $x = 2.\overline{02}$. Two digits repeat, so we multiply by 100 to get $100x = 202.\overline{02}$, and we have

$$100x - x = 202.\overline{02} - 2.\overline{02} = 202 - 2 + (0.\overline{02} - 0.\overline{02}) = 200.$$

This produces $99x = 200$, so $x = \boxed{\frac{200}{99}}$.

6.4.5 Continuing each number one more digit allows us to order them:

Number	First 6 digits
1.2345	1.23450
1.234$\overline{5}$	1.23455
1.23$\overline{45}$	1.23454
1.2$\overline{345}$	1.23453
1.$\overline{2345}$	1.23452

The numbers have the same first 5 digits, so their order matches the order of the 6th digits. From smallest to largest, the numbers are

$$\boxed{1.2345, 1.\overline{2345}, 1.2\overline{345}, 1.23\overline{45}, 1.234\overline{5}}.$$

6.4.6 We let $x = 0.\overline{63}$. Two digits repeat, so we multiply to get $100x = 63.\overline{63}$. Subtracting gives

$$100x - x = 63.\overline{63} - 0.\overline{63} = 63 + 0.\overline{63} - 0.\overline{63} = 63,$$

so $99x = 63$. Dividing by 99 gives $x = \frac{63}{99} = \frac{7}{11}$, so

$$0.\overline{63} - 0.63 = 0.63\overline{63} - 0.63 = (0.63 + 0.00\overline{63}) - 0.63 = (0.63 - 0.63) + 0.00\overline{63} = \frac{1}{100}(0.\overline{63}) = \frac{1}{100} \cdot \frac{7}{11} = \boxed{\frac{7}{1100}}.$$

6.4.7 We let $x = 0.\overline{48}$. Two digits repeat, so we multiply to get $100x = 48.\overline{48}$. Subtracting gives

$$100x - x = 48.\overline{48} - 0.\overline{48} = 48 + 0.\overline{48} - 0.\overline{48} = 48,$$

so $99x = 48$. Dividing by 99 gives $x = \frac{48}{99} = \frac{16}{33}$.

Similarly, we have $0.\overline{15} = \frac{15}{99} = \frac{5}{33}$, so

$$\frac{0.\overline{48}}{0.\overline{15}} = \frac{16/33}{5/33} = \frac{16}{33} \cdot \frac{33}{5} = \frac{16}{5} = \boxed{3\frac{1}{5}}.$$

We might also have reasoned as follows:

$$\frac{0.\overline{48}}{0.\overline{15}} = \frac{0.48484848\ldots}{0.15151515\ldots} = \frac{48(0.01010101\ldots)}{15(0.01010101\ldots)} = \frac{48}{15} = \frac{16}{5} = \boxed{3\frac{1}{5}}.$$

Review Problems

6.21 Writing the first four digits after the decimal point for each number makes them easy to compare:

$$0.9700, 0.9790, 0.9709, 0.9070, 0.9089.$$

Now, comparing the tenths digit, then the hundredths, then the thousandths, and finally the ten thousandths, we get the following order from smallest to largest:

$$\boxed{0.907, 0.9089, 0.97, 0.9709, 0.979}.$$

6.22

(a) Our addition is shown to the right. Notice that we have to carry from the hundredths place to the tenths place, and from the tenths place to the ones place. The resulting sum is $8.97 + 0.254 = \boxed{9.224}$.

$$\begin{array}{r} 8.970 \\ + 0.254 \\ \hline 9.224 \end{array}$$

(b) Since 1.006 is greater than 0.27, our result is negative: $0.27 - 1.006 = -(1.006 - 0.270)$. We perform the subtraction $1.006 - 0.270$ as shown at right, and we have $0.27 - 1.006 = -(1.006 - 0.27) = \boxed{-0.736}$.

$$\begin{array}{r} 1.006 \\ -\ 0.270 \\ \hline 0.736 \end{array}$$

(c) Multiplying by 10000 is the same as moving the decimal point 4 places to the right, so we have $0.902 \cdot 10000 = \boxed{9020}$.

(d) $25.5 \div 0.05 = \dfrac{25.5}{0.05} = \dfrac{25.5 \cdot 100}{0.05 \cdot 100} = \dfrac{2550}{5} = \boxed{510}$.

(e) We have $0.025 \cdot 0.042 = (25 \cdot 10^{-3})(42 \cdot 10^{-3}) = (25 \cdot 42)(10^{-3} \cdot 10^{-3}) = (1050)(10^{-3+(-3)}) = 1050(10^{-6})$. Multiplying by 10^{-6} is the same as moving the decimal point 6 places to the left, so $1050(10^{-6}) = 0.001050 = \boxed{0.00105}$.

(f) $(0.11)^3 = (11 \cdot 10^{-2})^3 = (11^3) \cdot (10^{-2})^3 = 1331 \cdot 10^{(-2)(3)} = 1331 \cdot 10^{-6} = \boxed{0.001331}$.

6.23 We write the decimals as 3367 times a power of 10, and then group the powers of 10:

$$100 \times 33.67 \times 3.367 \times 1000 = 10^2 \times (3367 \times 10^{-2}) \times (3367 \times 10^{-3}) \times 10^3$$
$$= 3367^2 \times (10^2 \times 10^{-2} \times 10^{-3} \times 10^3)$$
$$= 3367^2 \times 10^{2+(-2)+(-3)+3} = 3367^2 \times 10^0 = 3367^2.$$

So, the given product is the square of $\boxed{3367}$.

6.24 We have $\frac{6}{.3} = \frac{6 \cdot 10}{.3 \cdot 10} = \frac{60}{3} = 20$ and $\frac{.3}{.06} = \frac{.3 \cdot 100}{.06 \cdot 100} = \frac{30}{6} = 5$, so

$$\frac{6}{.3} + \frac{.3}{.06} = 20 + 5 = \boxed{25}.$$

6.25 Note that $3.5 = 3\frac{1}{2} = \frac{7}{2}$. So the reciprocal of 3.5 is $\frac{1}{\frac{7}{2}} = \frac{2}{7}$, and our answer is

$$\frac{7}{2} - \frac{2}{7} = \frac{7 \cdot 7}{2 \cdot 7} - \frac{2 \cdot 2}{7 \cdot 2} = \frac{49}{14} - \frac{4}{14} = \boxed{\frac{45}{14}}.$$

6.26 $\dfrac{(.2)^3}{(.02)^2} = \dfrac{(2/10)^3}{(2/100)^2} = \dfrac{2^3/10^3}{2^2/100^2} = \dfrac{2^3}{10^3} \cdot \dfrac{100^2}{2^2} = \dfrac{2^3}{2^2} \cdot \dfrac{100^2}{10^3} = 2 \cdot \dfrac{10000}{1000} = 2 \cdot 10 = \boxed{20}$.

6.27 10.68494 is between 10.68 and 10.69, and is closer to 10.68 than to 10.69, so 10.68494 rounded to the nearest hundredth is 10.68. This is Nanette's number. 10.68494 is between 10 and 11, and is closer to 11 than to 10, so 10.68494 rounded to the nearest whole number is 11. This is Duane's number. Therefore, the positive difference between Duane's number and Nanette's number is

$$11 - 10.68 = 0.32 = \frac{32}{100} = \boxed{\frac{8}{25}}.$$

6.28 20 dollars is $20 \cdot 100 = 2000$ cents. Since each copy costs 2.5¢, the number of copies you can make is

$$\frac{2000}{2.5} = \frac{2000 \cdot 2}{2.5 \cdot 2} = \frac{4000}{5} = \boxed{800}.$$

We also could have reasoned in steps. If 1 copy costs 2.5¢, then 2 copies cost 5¢, which means $20 \cdot (5¢) = 100¢ = \$1$ will get you $20 \cdot 2 = 40$ copies. So, 20 dollars will get you $20 \cdot 40 = \boxed{800}$ copies.

6.29

(a) Since $8 = 2^3$, we multiply the numerator and denominator by 5^3 to make the denominator a power of 10: $\frac{11}{8} = \frac{11 \cdot 5^3}{2^3 \cdot 5^3} = \frac{11 \cdot 125}{10^3} = \frac{1375}{1000} = \boxed{1.375}$. We also might have noted that $\frac{11}{8} = 1\frac{3}{8}$, and recognized $\frac{3}{8} = 0.375$, so $1\frac{3}{8} = \boxed{1.375}$. (Since the decimal representations of fractions with denominator of 8 appear frequently, it's probably not a bad idea to just memorize them.)

(b) From the long division on the right, we have $\frac{10}{7} = \boxed{1.\overline{428571}}$.

$$
\begin{array}{r}
1.428571\ldots \\
7\,\overline{\big)\,10.0000000\ldots} \\
7. \\
\overline{3.0} \\
2.8 \\
\overline{0.20} \\
0.14 \\
\overline{0.060} \\
0.056 \\
\overline{0.0040} \\
0.0035 \\
\overline{0.00050} \\
0.00049 \\
\overline{0.000010} \\
0.000007 \\
\overline{0.0000030}
\end{array}
$$

We also could have recalled that $\frac{1}{7} = 0.\overline{142857}$, so $\frac{10}{7} = 10 \cdot \frac{1}{7} = 10 \cdot (0.\overline{142857})$. Multiplying by 10 is the same as moving the decimal point one place to the right, so the repeating block after the decimal point will start at 4, not 1. But we have to remember that 1 is still part of the repeating block:

$$10 \cdot (0.\overline{142857}) = 10 \cdot (0.142857142857\ldots)$$
$$= 1.428571428571\ldots$$
$$= \boxed{1.\overline{428571}}.$$

Or, we could note that $\frac{10}{7} = 1\frac{3}{7}$, and recall that $\frac{3}{7} = 0.\overline{428571}$, so $\frac{10}{7} = 1\frac{3}{7} = \boxed{1.\overline{428571}}$.

(c) The long division is shown at the right. In the next step of the long division, we will divide 15 into 100. This repeats the second step of the long division, so we know that the decimal will repeat. We find $\frac{7}{15} = \boxed{0.4\overline{6}}$.

$$
\begin{array}{r}
0.46\ldots \\
15\,\overline{\big)\,7.000\ldots} \\
6.0 \\
\overline{1.00} \\
0.90 \\
\overline{0.100}
\end{array}
$$

(d) Multiplying 20 by 5 gives a power of 10, so we have $\frac{39}{20} = \frac{39 \cdot 5}{20 \cdot 5} = \frac{195}{100} = \boxed{1.95}$.

We also might have noted

$$\frac{39}{20} = \frac{40}{20} - \frac{1}{20} = 2 - \frac{5}{100} = 2 - 0.05 = \boxed{1.95}.$$

(e) From the long division at the right, we see that $\frac{25}{33}$ as a decimal consists of infinitely alternating 7's and 5's: $\frac{25}{33} = \boxed{0.\overline{75}}$.

$$
\begin{array}{r}
0.757\ldots \\
33\,\overline{\big)\,25.0000\ldots} \\
23.1 \\
\overline{1.90} \\
1.65 \\
\overline{0.250} \\
0.231 \\
\overline{0.0190}
\end{array}
$$

(f) In the next step of the long division, we will divide 21 into 40. This repeats the first step of the long division, so the decimal will repeat infinitely and we have $\frac{4}{21} = \boxed{0.\overline{190476}}$.

```
        0.190476....
   21 | 4.0000000...
        2.1
        ———
        1.90
        1.89
        ————
        0.0100
        0.0084
        ——————
        0.00160
        0.00147
        ———————
        0.000130
        0.000126
        ————————
        0.0000040
```

6.30 Our long division is at the right. In the next step of the long division, we will divide 37 into 40. This repeats the first step of the long division, so the decimal will repeat infinitely. So, we have $\frac{4}{37} = 0.\overline{108}$. We must find the 100th digit after the decimal point. We write $3 \cdot 33 = 99$ digits when writing the repeating pattern "108" the first 33 times. The next digit, the 100th after the decimal point, is the start of the repeating block, $\boxed{1}$.

```
        0.108....
   37 | 4.0000...
        3.7
        —————
        0.300
        0.296
        —————
        0.0040
```

6.31

(a) Letting $x = 0.\overline{6}$, we have $10x = 6.\overline{6}$, so

$$10x - x = 6.\overline{6} - 0.\overline{6} = 6 + 0.\overline{6} - 0.\overline{6} = 6.$$

Therefore, we have $9x = 6$, so $x = \frac{6}{9} = \boxed{\frac{2}{3}}$.

(b) Let $x = 0.\overline{97}$. The repeating portion has two digits, so we multiply by 100, getting $100x = 97.\overline{97}$, and

$$100x - x = 97.\overline{97} - 0.\overline{97} = 97 + (0.\overline{97} - 0.\overline{97}) = 97.$$

Therefore, we have $99x = 97$, so $x = \boxed{\frac{97}{99}}$.

(c) Let $x = 0.0\overline{8}$. The repeating portion has one digit, so we multiply by 10, getting $10x = 0.8\overline{8}$. Then we have

$$10x - x = 0.8\overline{8} - 0.0\overline{8} = 0.8 + 0.0\overline{8} - 0.0\overline{8} = 0.8.$$

This gives us $9x = 0.8$. Multiplying both sides by 10 gives $90x = 8$, and dividing by 90 gives $x = \frac{8}{90} = \boxed{\frac{4}{45}}$.

We might also recognize that $0.\overline{8} = \frac{8}{9}$, so $0.0\overline{8} = \frac{1}{10} \cdot (0.\overline{8}) = \frac{1}{10} \cdot \frac{8}{9} = \frac{8}{90} = \boxed{\frac{4}{45}}$.

(d) Let $x = 0.\overline{36}$. The repeating portion has two digits, so we multiply by 100, getting $100x = 36.\overline{36}$, and

$$100x - x = 36.\overline{36} - 0.\overline{36} = 36 + (0.\overline{36} - 0.\overline{36}) = 36.$$

Therefore, we have $99x = 36$, so $x = \frac{36}{99} = \boxed{\frac{4}{11}}$.

(e) Let $x = 0.3\overline{21}$. The repeating portion has two digits, so we multiply by 100, getting $100x = 32.1\overline{21}$, and

$$100x - x = 32.1\overline{21} - 0.3\overline{21} = 32.1 + 0.0\overline{21} - (0.3 + 0.0\overline{21}) = 32.1 - 0.3 + 0.0\overline{21} - 0.0\overline{21} = 31.8.$$

Therefore, we have $99x = 31.8$. Multiplying both sides by 10 gives $990x = 318$, and dividing both sides by 990 gives $x = \frac{318}{990} = \frac{6 \cdot 53}{6 \cdot 165} = \boxed{\frac{53}{165}}$.

(f) We recognize $0.\overline{9}$ as 1, so $0.00\overline{9} = 0.01 \cdot 0.\overline{9} = 0.01 \cdot 1 = 0.01$. This means

$$0.46\overline{9} = 0.46 + 0.00\overline{9} = 0.46 + 0.01 = 0.47 = \boxed{\frac{47}{100}}.$$

Challenge Problems

6.32 We have $\frac{1}{2} = 0.5$, so $\frac{1}{2} = 0.1 \cdot \frac{1}{2} = 0.1 \cdot 0.5 = 0.05$. We also have $\frac{1}{.2} = \frac{1 \cdot 10}{.2 \cdot 10} = \frac{10}{2} = 5$, so $\frac{1}{2} + \frac{1}{2} + \frac{1}{.2} = 0.5 + 0.05 + 5 = \boxed{5.55}$.

6.33

(a) Any positive number that is greater than or equal to 1 and less than 10 is essentially already in scientific notation. For example, $5.2 = 5.2 \cdot 10^0$.

If the number is greater than 10, we can repeatedly factor out powers of 10. Each time we do so, we move the decimal point one place to the left. For example:

$$56739 = 5673.9 \cdot 10 = 567.39 \cdot 10^2 = 56.739 \cdot 10^3 = 5.6739 \cdot 10^4.$$

If a number has n digits to the left of its decimal point, then factoring out 10^{n-1} leaves the product of 10^{n-1} and a number with only one digit, which is nonzero, to the left of its decimal point. A number with only one digit, which is nonzero, to the left of its decimal point must be greater than or equal to 1 and less than 10. So, factoring 10^{n-1} out of a number with n digits to the left of its decimal point leaves a number in scientific notation.

Similarly, if the number is between 0 and 1, then we repeatedly factor out powers of 10^{-1} until we finally get a single digit, which is nonzero, to the left of the decimal point. For example:

$$0.00043 = 0.0043 \cdot 10^{-1} = 0.043 \cdot 10^{-2} = 0.43 \cdot 10^{-3} = 4.3 \cdot 10^{-4}.$$

Therefore, any positive number can be expressed in scientific notation.

(b) We have $(3 \cdot 10^5) \cdot (4 \cdot 10^6) = (3 \cdot 4) \cdot (10^5 \cdot 10^6) = 12 \cdot 10^{11}$. However, 12 is greater than 10. We write 12 in scientific notation as $1.2 \cdot 10^1$, so $12 \cdot 10^{11} = 1.2 \cdot 10^1 \cdot 10^{11} = \boxed{1.2 \cdot 10^{12}}$.

(c) We express the division as a fraction, and separate the powers of 10 from the 3 and the 5:

$$(3 \cdot 10^2) \div (5 \cdot 10^{-3}) = \frac{3 \cdot 10^2}{5 \cdot 10^{-3}} = \frac{3}{5} \cdot \frac{10^2}{10^{-3}} = 0.6 \cdot 10^{2-(-3)} = 0.6 \cdot 10^5.$$

But 0.6 is less than 1, so $0.6 \cdot 10^5$ is not in scientific notation. We write 0.6 in scientific notation as $0.6 = 6 \cdot 10^{-1}$, and we have $0.6 \cdot 10^5 = 6 \cdot 10^{-1} \cdot 10^5 = 6 \cdot 10^{-1+5} = \boxed{6 \cdot 10^4}$.

(d) Since n is a positive integer, the exponent of 10 when n is written in scientific notation must be nonnegative. So, in the scientific notation expression $a \cdot 10^b$ for n, either $b = 0$ and a is a single digit, or b is positive and $1 \le a < 10$.

Suppose b is positive. Then, in the expression $a \cdot 10^b$, we start with a number a that has exactly one digit to the left of the decimal point. We then multiply a by 10 exactly b times. Each multiplication by 10 moves the decimal point one place to the right, thereby increasing the number of digits to the left of the decimal point by 1. So, multiplying a by 10 exactly b times increases the number of digits to the left of the decimal point by b, to a total of $\boxed{b + 1}$ digits.

Note that if $b = 0$, then n is a one-digit integer, so it also has $b + 1$ digits.

6.34 A fraction can be expressed as a terminating decimal if the denominator of the fraction in simplest form does not have any prime factors besides 2 and 5. So, we seek the smallest positive integer k such that writing $\frac{k}{660}$ in simplest form produces a fraction whose denominator does not have any prime factors besides 2 and 5. Since we can't have any prime factors besides 2 and 5, we must choose a k that allows us to cancel out the prime factors of 660 that are not 2 or 5.

The prime factorization of 660 is $2^2 \cdot 3 \cdot 5 \cdot 11$, so the prime factors we must cancel are 3 and 11. We therefore must include 3 and 11 in the prime factorization of k. The smallest positive integer with both 3 and 11 in its prime factorization is $3 \cdot 11 = 33$, so $\boxed{33}$ is the smallest value of k such that $\frac{k}{660}$ can be expressed as a terminating decimal. Checking, we see that $\frac{33}{660} = \frac{33}{33 \cdot 20} = \frac{1}{20} = \frac{5}{100} = 0.05$.

6.35 Multiplying x by 10^n moves the decimal point n places to the right. But since K is an n-digit number, the resulting number is $10^n x = K.\overline{K}$. Subtracting x from $10^n x$ gives

$$10^n x - x = K.\overline{K} - 0.\overline{K} = (K + 0.\overline{K}) - 0.\overline{K} = K.$$

Thus

$$x = \frac{K}{10^n - 1} = \frac{K}{\underbrace{9\ldots9}_{n \text{ nines}}}.$$

6.36

(a) The number 10^{30} is a 1 followed by 30 zeros, so it has $\boxed{31}$ digits.

(b) Multiplying out 2^{30} is quite a chore. However, we can take a bit of a shortcut by noticing that $2^{30} = 2^{5 \cdot 6} = (2^5)^6 = 32^6$. We then note that $32^6 = 32^{2 \cdot 3} = (32^2)^3 = 1024^3$. We don't need to compute 1024^3; we only need to count its digits. We know that 1024^3 is greater than 1000^3, which equals 1,000,000,000. We also know that 1024^3 is less than 2000^3, which equals 8,000,000,000. Because 1024^3 is between 1,000,000,000 and 8,000,000,000, we know that 1024^3 has $\boxed{10}$ digits.

(c) Multiplying out 5^{30} isn't much fun, and it isn't particularly easy to find a slick method to count the digits like we counted the digits of 2^{30} in part (b). However, we do have

$$10^{30} = (2 \cdot 5)^{30} = 2^{30} \cdot 5^{30}.$$

Moreover, we know that 2^{30} has 10 digits. Using our result from Problem 6.33, we know that 2^{30}, which has 10 digits, can be written in scientific notation as $a \cdot 10^9$ for some number a with $1 < a < 10$ (we know that a cannot be 1 because 2^{30} does not equal a power of 10). So, now we have

$$10^{30} = (a \cdot 10^9) \cdot 5^{30}.$$

Suppose we also write 5^{30} in scientific notation, as $c \cdot 10^d$, where $1 < c < 10$ (we know that c is not 1 since 5^{30} is not a power of 10). Then, we have

$$10^{30} = (a \cdot 10^9) \cdot (c \cdot 10^d) = (ac) \cdot (10^9 \cdot 10^d) = (ac)(10^{9+d}).$$

Next, we note that because a and c are each greater than 1 but less than 10, we have $1 < ac < 100$. On the other hand, since $(ac)(10^{a+d}) = 10^{30}$, we have

$$ac = \frac{10^{30}}{10^{9+d}} = 10^{30-(9+d)} = 10^{21-d}.$$

In particular, this means that ac must be a power of 10. The only power of 10 between 1 and 100 (but not equal to 1 or 100) is 10. So we must have $ac = 10 = 10^1$, and hence $10^1 = 10^{21-d}$. This gives $1 = 21 - d$, and thus $d = 20$.

Therefore, we know that 5^{30} in scientific notation is $c \cdot 10^{20}$ for some number c. Again applying our result from Problem 6.33, this tells us that 5^{30} has $\boxed{21}$ digits.

6.37 Let n be a positive integer that is less than 100. The reciprocal $\frac{1}{n}$ has a terminating decimal representation if n has no prime factors besides 2 and 5. Otherwise, $\frac{1}{n}$ does not have a terminating decimal representation. So, our problem is to count the number of positive integers less than 100 that have no prime factors besides 2 and 5. We will organize our counting based on the number of 5's in n's prime factorization.

Case 1: No 5's. This means that the only prime factor of n is 2. So, n is a power of 2. We have to be careful to include 1. The powers of 2 less than 100 are 1, 2, 4, 8, 16, 32, 64. So, there are 7 values of n that satisfy this case.

Case 2: One 5. The numbers less than 100 with one 5 in their prime factorizations, and no other primes besides 2, are 5, 10, 20, 40, 80. (We can generate these quickly by starting with 5 and then repeatedly multiplying by 2 to add 2's to the prime factorization of n.) There are 5 values of n that satisfy this case.

Case 3: Two 5's. The only numbers less than 100 that have two 5's in their prime factorizations and no other primes besides 2 are 25 and 50. There are 2 values of n that satisfy this case.

Case 4: More than two 5's. The smallest such number is $5^3 = 125$, which is greater than 100, so there are no values of n that satisfy this case.

Combining our cases, we find $7 + 5 + 2 = \boxed{14}$ positive integers less than 100 whose reciprocals have terminating decimal representations.

6.38

(a) The fractions look daunting, but all of the denominators are powers of 10, so we can easily write each term as a decimal:

$$\frac{7}{10} + \frac{7}{100} + \frac{7}{1000} + \frac{7}{10000} + \cdots = 0.7 + 0.07 + 0.007 + 0.0007 + \cdots = 0.7777\ldots.$$

Aha! The right-hand side is simply $0.\overline{7}$, which equals $\boxed{\frac{7}{9}}$.

(b) Writing each term as a decimal, we have

$$\frac{6}{10} + \frac{3}{100} + \frac{6}{1000} + \frac{3}{10000} + \cdots = 0.6 + 0.03 + 0.006 + 0.0003 + \cdots.$$

Combining pairs of terms makes the sum $0.63 + 0.0063 + 0.000063 + \cdots$. We have another repeating decimal:

$$0.63 + 0.0063 + 0.000063 + \cdots = 0.\overline{63} = \frac{63}{99} = \boxed{\frac{7}{11}}.$$

(c) We handled repeating decimals by assigning them to a variable and then multiplying by an appropriate power of 10. For example, we express $0.\overline{1}$ as a fraction by letting $x = 0.\overline{1}$ and multiplying by 10 to get $10x = 1.\overline{1}$. Let's see what this looks like if we write out $0.1111\ldots$ as a sum of fractions:

$$x = \frac{1}{10} + \frac{1}{100} + \frac{1}{1000} + \frac{1}{10000} + \cdots.$$

Multiplying both sides by 10 gives

$$10x = 1 + \frac{1}{10} + \frac{1}{100} + \frac{1}{1000} + \cdots.$$

When we subtract our expression for x from our expression from $10x$, everything cancels except 1:

$$10x - x = \left(1 + \frac{1}{10} + \frac{1}{100} + \frac{1}{1000} + \cdots\right) - \left(\frac{1}{10} + \frac{1}{100} + \frac{1}{1000} + \cdots\right)$$
$$= 1 + \left(\frac{1}{10} + \frac{1}{100} + \frac{1}{1000} + \cdots\right) - \left(\frac{1}{10} + \frac{1}{100} + \frac{1}{1000} + \cdots\right)$$
$$= 1.$$

Then we have $9x = 1$, so $x = \frac{1}{9}$.

We just computed the sum of reciprocals of powers of 10. However, in this part of the problem, we are asked to compute the sum of reciprocals of powers of 3. Maybe the same strategy will work. We start by setting the sum equal to x:

$$x = \frac{1}{3} + \frac{1}{9} + \frac{1}{27} + \frac{1}{81} + \cdots.$$

When working with the sum of reciprocals of powers of 10, our next step was multiplying by 10. Here, we're working with powers of 3, so we'll multiply by 3:

$$3x = 3 \cdot \frac{1}{3} + 3 \cdot \frac{1}{9} + 3 \cdot \frac{1}{27} + 3 \cdot \frac{1}{81} + \cdots$$
$$= 1 + \frac{1}{3} + \frac{1}{9} + \frac{1}{27} + \cdots.$$

When we subtract our expression for x from this expression for $3x$, everything cancels except 1:

$$3x - x = \left(1 + \frac{1}{3} + \frac{1}{9} + \frac{1}{27} + \cdots\right) - \left(\frac{1}{3} + \frac{1}{9} + \frac{1}{27} + \cdots\right)$$
$$= 1 + \left(\frac{1}{3} + \frac{1}{9} + \frac{1}{27} + \cdots\right) - \left(\frac{1}{3} + \frac{1}{9} + \frac{1}{27} + \cdots\right)$$
$$= 1.$$

So, we have $3x - x = 1$, which means $2x = 1$ and $x = \boxed{\frac{1}{2}}$.

CHAPTER **7**

Ratios, Conversions, and Rates

24® Cards

First Card $7 \cdot (3 + 3/7)$.

Second Card $7 \cdot (4 - 4/7)$.

Third Card $(13 - 7) \cdot 7 - 18$.

Fourth Card $7 \cdot (2 + 10/7)$.

Exercises for Section 7.1

7.1.1

(a) The greatest common factor of 20 and 8 is 4. Dividing out this factor, we have $20 : 8 = \frac{20}{4} : \frac{8}{4} = \boxed{5 : 2}$.

(b) Since $6^3 = (2 \cdot 3)^3 = 2^3 \cdot 3^3$ and $8^3 = (2 \cdot 4)^3 = 2^3 \cdot 4^3$, the greatest common factor of 6^3 and 8^3 is 2^3. Dividing by this common factor, we have

$$6^3 : 8^3 = \frac{2^3 \cdot 3^3}{2^3} : \frac{2^3 \cdot 4^3}{2^3} = 3^3 : 4^3 = \boxed{27 : 64}.$$

(c) Multiplying both fractions by 10 turns each into an integer:

$$\frac{3}{5} : \frac{1}{10} = 10 \cdot \frac{3}{5} : 10 \cdot \frac{1}{10} = \boxed{6 : 1}.$$

(d) Dividing both 100 and 500 by 100 gives $100 : 500 = \frac{100}{100} : \frac{500}{100} = \boxed{1 : 5}$.

(e) First, we write both mixed numbers as fractions, which gives us $2\frac{1}{4} : 3\frac{5}{8} = \frac{9}{4} : \frac{29}{8}$. Then, multiplying both by 8 turns both into integers:

$$\frac{9}{4} : \frac{29}{8} = 8 \cdot \frac{9}{4} : 8 \cdot \frac{29}{8} = \boxed{18 : 29}.$$

(f) Dividing both parts of the ratio by 672 gives a simplified ratio of $\boxed{1 : 0}$.

7.1.2 There are $25 - 10 = 15$ girls, so the ratio of girls to boys is $15 : 10$. We can divide by 5 to simplify the ratio as $15 : 10 = \boxed{3 : 2}$.

7.1.3 The ratio of the number of revolutions A makes to the number of revolutions B makes is $2 : 5$. We multiply both parts of the ratio by 18 because A makes 36 revolutions:

$$2 : 5 = 2 \cdot 18 : 5 \cdot 18 = 36 : 90.$$

So, B makes $\boxed{90}$ revolutions when A makes 36 revolutions.

We also could have used the $2 : 5$ ratio to note that B makes $\frac{5}{2}$ as many revolutions as A. So, when A makes 36 revolutions, B makes $\frac{5}{2} \cdot 36 = \boxed{90}$ revolutions.

7.1.4 Since the ratio of girls to boys is $7 : 4$, the ratio of girls to the total number of students is $7 : (7+4) = 7 : 11$. Multiplying both parts of this ratio by 6 to produce 42 girls gives us $7 : 11 = 6 \cdot 7 : 6 \cdot 11 = 42 : 66$, so there are $\boxed{66}$ total students.

We also could have used the $7 : 11$ ratio to see that the total number of students is $\frac{11}{7}$ times the number of girls. Therefore, there are $42 \cdot \frac{11}{7} = \boxed{66}$ total students.

7.1.5 Since the ratio of the smaller number to the larger number is $3 : 8$, the numbers are $3n$ and $8n$ for some value of n. Therefore, we must have $3n + 8n = 44$, so $11n = 44$. Dividing by 11 gives $n = 4$, so the larger number is $8 \cdot 4 = \boxed{32}$.

7.1.6 Since the ratio of the longer piece to the shorter piece is $7 : 5$, the shorter piece is $\frac{5}{7+5} = \frac{5}{12}$ of the whole sandwich. Since the sandwich is 8 inches long, the length of the shorter piece is $\frac{5}{12} \cdot 8 = \boxed{\frac{10}{3} \text{ inches}} = \boxed{3\frac{1}{3} \text{ inches}}$.

7.1.7 Al receives $\frac{4}{4+3} = \frac{4}{7}$ of the land, so he gets $\frac{4}{7} \cdot 280 = \boxed{160 \text{ acres}}$.

7.1.8 Since the numbers are in the ratio $4 : 9$, the numbers are $4n$ and $9n$ for some value of n. Therefore, we have $9n - 4n = 30$, so $5n = 30$. Dividing by 5 gives $n = 6$, so the sum of the numbers is $4n + 9n = 13n = 13(6) = \boxed{78}$.

7.1.9 First, we find the total number of students. Since the ratio of female students to total number of students is $4 : 9$, the total number of students is $\frac{9}{4}$ the number of female students. Therefore, the total number of students is $\frac{9}{4} \cdot 396$. Noticing that $396 = 400 - 4$ allows us to compute this very quickly:

$$\frac{9}{4} \cdot 396 = \frac{9}{4}(400 - 4) = \frac{9}{4} \cdot 400 - \frac{9}{4} \cdot 4 = 900 - 9 = 891.$$

Next, we count the teachers. Since the ratio of teachers to students is $1 : 11$, the number of teachers is $\frac{1}{11}$ the number of students. There are 891 students, so there are $\frac{1}{11} \cdot 891 = \boxed{81}$ teachers.

7.1.10 The ratio of losses to wins for Kyle's team is $3 : 2$, which means that the team won 2 out of every 5 games. If it had won twice as many games, but played the same number of games, then it would have won 4 out of every 5 games. This leaves 1 out of every 5 games to be a loss, so the ratio of losses to wins is $\boxed{1 : 4}$.

7.1.11 Since the ratio of pennies to dimes is $2 : 5$, the pennies are $\frac{2}{2+5} = \frac{2}{7}$ of the pennies and dimes in the jar. There are 245 pennies and dimes total, so there are $\frac{2}{7} \cdot 245 = 70$ pennies and $245 - 70 = 175$ dimes.

We wish to add pennies until the ratio of pennies to dimes is 3 : 7, at which point the number of pennies is $\frac{3}{7}$ of the number of dimes. We'll still have 175 dimes after adding pennies, so we need $\frac{3}{7} \cdot 175 = 75$ pennies total. This means we need $75 - 70 = \boxed{5}$ more pennies.

Exercises for Section 7.2

7.2.1 The shortest piece is $\frac{1}{1+3+5} = \frac{1}{9}$ of the whole. So, the length of the shortest piece is $\frac{1}{9} \cdot 60 = \boxed{\frac{20}{3} \text{ inches}} = \boxed{6\frac{2}{3} \text{ inches}}$.

7.2.2 The greatest of the numbers is $\frac{3}{1+2+3} = \frac{1}{2}$ of the sum of the numbers. Since the sum of the numbers is 48, the largest number is $\frac{1}{2} \cdot 48 = \boxed{24}$.

7.2.3 The white paint is $\frac{16}{16+3+1} = \frac{4}{5}$ of the mixture, so $\frac{4}{5} \cdot 1 = \boxed{\frac{4}{5} \text{ gallon}}$ is needed to make 1 gallon of purple paint.

7.2.4 The ratio of Akira's, Bruno's, and Carmela's investments is 25 : 20 : 35. Dividing all parts of this ratio by 5 gives 5 : 4 : 7. Therefore, Bruno invested $\frac{4}{5+4+7} = \frac{4}{16} = \frac{1}{4}$ of the money, so he should receive $\frac{1}{4}$ of the earnings. Therefore, Bruno will receive $\frac{1}{4} \cdot 2000 = \boxed{500}$ dollars.

7.2.5 The top winner receives $\frac{9}{9+5+2+1} = \frac{9}{17}$ of the total, so the ratio of the top winner to the total is 9 : 17. Multiplying this ratio by 5000 to make the top winner equal to 45000 gives 45000 : 85000, so the total prize pool is $\boxed{\$85,000}$.

7.2.6 From the information in the problem, we have

brown	:	red		black	:	brown		white	:	red
3	:	1		2	:	1		4	:	1

Conveniently, red's number is the same in the first and third ratios. So, we can combine these two ratios and write

brown	:	red	:	white		black	:	brown
3	:	1	:	4		2	:	1

To combine these two ratios, we need the number corresponding to brown shoes to be the same in both. So, we multiply both parts of the second ratio by 3:

brown	:	red	:	white		black	:	brown
3	:	1	:	4		6	:	3

Now, we can combine the ratios:

black	:	brown	:	red	:	white
6	:	3	:	1	:	4

So, we see that the ratio of white shoes to black shoes is 4 : 6, which equals $\boxed{2 : 3}$.

7.2.7 In order to make the ratio easier to work with, we start by simplifying it. The least common denominator of the fractions in the ratio is 12, so we multiply all parts of the ratio by 12. That gives

$$12 \cdot \frac{1}{2} : 12 \cdot \frac{1}{3} : 12 \cdot \frac{1}{4} = 6 : 4 : 3.$$

Therefore, the sibling who gets the most money receives $\frac{6}{6+4+3} = \frac{6}{13}$ of the money. Since there is $169 to divide, the one who gets the most money receives $\frac{6}{13} \cdot (\$169) = \boxed{\$78}$.

Exercises for Section 7.3

7.3.1 Let there be x sheets in the 7.5 cm stack, so we have $500 : 5 = x : 7.5$. Writing this using fractions, we have $\frac{500}{5} = \frac{x}{7.5}$, so $\frac{x}{7.5} = 100$. Multiplying both sides by 7.5 gives $x = \boxed{750}$.

7.3.2 Let x be the number of dollars the traveler needs. Since the ratio of dollars to yen is $1 : 80$, we have $1 : 80 = x : 10000$. Writing this using fractions gives $\frac{1}{80} = \frac{x}{10000}$. Multiplying both sides by 10000 gives $x = \frac{10000}{80} = \frac{1000}{8} = \boxed{125 \text{ dollars}}$.

7.3.3 The width of the website logo is 4 times as large as Alexia's design, so the height of the website logo must also be 4 times as large as Alexia's design, or $4 \cdot 1.5 = \boxed{6 \text{ inches}}$.

7.3.4 Considering the taller object, we have

$$\text{Object height : Shadow height} = 30 : 20 = 3 : 2.$$

So, we have

$$\text{Short pole height : 15 feet} = 3 : 2.$$

Multiplying both parts of the ratio on the right by 7.5, we have

$$\text{Short pole height : 15 feet} = 22.5 : 15.$$

So, the height of the short pole is $\boxed{22.5 \text{ feet}}$.

We might also have reasoned as follows. The shadow of the shorter pole is $\frac{15}{20}$ of the shadow of the taller pole. So, the height of the shorter pole is $\frac{15}{20}$ of the height of the taller pole. Therefore, the height of the shorter pole is $\frac{15}{20} \cdot (30 \text{ feet}) = \boxed{22.5 \text{ feet}}$.

7.3.5 We have

$$\text{Map : Actual distance} = \frac{1}{4} \text{ inch : 50 miles.}$$

Multiplying by 4 makes this ratio 1 inch : 200 miles. So, each inch on the map represents 200 miles Therefore, $2\frac{7}{8}$ inches represents $200 \cdot \left(2\frac{7}{8}\right) = 200 \left(\frac{23}{8}\right) = \boxed{575 \text{ miles}}$.

7.3.6 Let x cm be the smaller dimension in the drawing. Since we must have

$$\text{Drawing : Building} = 1 \text{ cm : 2.5 m,}$$

we have x cm : 30 m $= 1$ cm : 2.5 m. Writing this as fractions, we have $\frac{x}{30} = \frac{1}{2.5}$. Multiplying by 30 gives $x = \frac{30}{2.5} = \frac{300}{25} = 12$. So, the smaller dimension of the drawing is $\boxed{12 \text{ cm}}$.

7.3.7 The 12 meals they ordered were enough for 18 people, so we have the ratio

$$\text{Meals needed : People} = 12 : 18 = 2 : 3.$$

So, we need $\frac{2}{3}$ as many meals as we have people. Therefore, to feed 12 people, we need $\frac{2}{3} \cdot 12 = \boxed{8}$ meals.

7.3.8 As the temperature rose from 20° to 32°, it rose by 3° exactly 4 times. Each time the temperature rises 3°, the volume of the gas increases by 4 cubic centimeters. So, as the temperature rose 4 times by 3°, the volume expanded by $4 \cdot 4 = 16$ cubic centimeters. After this expansion, the volume of the gas is 24 cubic centimeters, so the volume of the gas before the expansion was $24 - 16 = \boxed{8 \text{ cubic centimeters}}$.

Exercises for Section 7.4

7.4.1 I need 1 US dollar for every 1.25 Canadian dollars, so my 15 Canadian dollars convert to $15/1.25 = 12$ US dollars. We can also do this using a conversion factor:

$$\text{C\$15} = \text{C\$15} \cdot \frac{\text{US\$1}}{\text{C\$1.25}} = \text{US\$}\frac{15}{1.25} = \boxed{\text{US\$12}}.$$

7.4.2 Each foot is 12 inches, so 7 feet is $7 \cdot 12 = 84$ inches. Adding the 2 extra inches gives $84 + 2 = \boxed{86}$ inches.

7.4.3 First, since there are 12 inches in a foot, there are $12 \cdot \frac{1}{2} = 6$ inches in $\frac{1}{2}$ foot. Thus the square is 6 inches per side.

We could first convert the side length to centimeters, giving

$$6 \text{ in} = 6 \text{ in} \cdot \frac{2.5 \text{ cm}}{1 \text{ in}} = (6 \cdot 2.5) \text{ cm} = 15 \text{ cm}.$$

Thus the area of the square is $15 \cdot 15 = \boxed{225}$ cm^2.

Alternatively, we could use the fact (from the text) that 1 in^2 = 6.25 cm^2, and convert the 36 square inches of area to square centimeters:

$$36 \text{ in}^2 = 36 \text{ in}^2 \cdot \frac{6.25 \text{ cm}^2}{1 \text{ in}^2} = (36 \cdot 6.25) \text{ cm}^2 = \boxed{225} \text{ cm}^2.$$

7.4.4 We line up the conversion factors:

$$\frac{1}{4} \text{ pounds} \cdot \frac{16 \text{ ounces}}{1 \text{ pounds}} \cdot \frac{28.35 \text{ grams}}{1 \text{ ounces}} \approx \boxed{113 \text{ grams}}.$$

7.4.5 We line up the conversion factors:

$$2.5 \text{ cups} = 2.5 \text{ cups} \cdot \frac{8 \text{ oz}}{1 \text{ cup}} \cdot \frac{1 \text{ tbsp}}{0.5 \text{ oz}} \cdot \frac{3 \text{ tsp}}{1 \text{ tbsp}} = \frac{2.5 \cdot 8 \cdot 3}{0.5} \text{ tsp} = 120 \text{ tsp}.$$

Thus Natalya needs $\boxed{120}$ teaspoons for her recipe.

7.4.6 We have (note ha is short for hectare):

$$1 \text{ ha} = 1 \text{ ha} \cdot \frac{1 \text{ km}^2}{100 \text{ ha}} \cdot \left(\frac{1000 \text{ m}}{1 \text{ km}}\right)^2 = \frac{1000^2}{100} \text{ m}^2 = \boxed{10000 \text{ m}^2}.$$

7.4.7 Despite the foreign words, we can still use conversion factors!

$$500 \text{ piquat} = 500 \text{ piquat} \cdot \frac{20 \text{ stuun}}{1 \text{ piquat}} \cdot \frac{400 \text{ Ploktars}}{1 \text{ stuun}} = (500 \cdot 20 \cdot 400) \text{ Ploktars} = \boxed{4,000,000 \text{ Ploktars}}.$$

Exercises for Section 7.5

7.5.1

(a) A car that travels 50 miles in each hour for $2\frac{3}{4}$ hours goes $50 \cdot \left(2\frac{3}{4}\right) = 50 \cdot \left(\frac{11}{4}\right) = \boxed{137.5 \text{ miles}}$.

(b) In order to cover 320 miles by traveling 60 miles per hour, the car needs to travel for

$$\frac{320 \text{ miles}}{60 \frac{\text{miles}}{\text{hours}}} = \frac{320}{60} \text{ hours} = \boxed{\frac{16}{3} \text{ hours}}.$$

Since $\frac{16}{3} = 5\frac{1}{3}$, and $\frac{1}{3}$ of an hour is $\frac{1}{3} \cdot 60 = 20$ minutes, a better way to write our answer is $\boxed{5 \text{ hours and } 20 \text{ minutes}}$.

(c) The speed of the car is the distance it travels divided by the time it travels. If the car must travel 280 miles at a constant speed in $3\frac{1}{2}$ hours, it must travel $\frac{280}{3\frac{1}{2}}$ miles in each hour. Therefore, its speed is

$$\frac{280 \text{ miles}}{3\frac{1}{2} \text{ hours}} = \frac{280}{3\frac{1}{2}} \frac{\text{miles}}{\text{hour}} = \frac{280}{\frac{7}{2}} \frac{\text{miles}}{\text{hour}} = \boxed{80} \frac{\text{miles}}{\text{hour}}.$$

7.5.2 The car travels for 6 hours, and it goes $27816 - 27289 = 527$ kilometers. The speed of the car is the distance it travels divided by the time it travels, so the car's speed is

$$\frac{527 \text{ km}}{6 \text{ hr}} = \boxed{87\frac{5}{6}} \frac{\text{km}}{\text{hr}}.$$

7.5.3 Altogether, the car traveled $80 + 100 = 180$ miles in 4 hours, so its average speed is

$$\frac{180 \text{ miles}}{4 \text{ hours}} = \frac{180}{4} \frac{\text{miles}}{\text{hour}} = \boxed{45} \frac{\text{miles}}{\text{hour}}.$$

7.5.4 Peter drove 60 miles at 40 miles per hour. So, the amount of time he drove was

$$\frac{60 \text{ miles}}{40 \frac{\text{miles}}{\text{hour}}} = \frac{60}{40} \text{ hours} = \frac{3}{2} \text{ hours}.$$

Since Peter arrived 15 minutes late, he arrived at 12:15 p.m. Therefore, since he drove $\frac{3}{2} = 1\frac{1}{2}$ hours, he left his house $1\frac{1}{2}$ hours before 12:15 p.m., at $\boxed{10:45 \text{ a.m.}}$

7.5.5 Had I kept walking, the second half of my trip would have taken 10 more minutes. By doubling my speed for the second half of my trip, I halved the amount of time it took me to finish. So, the second half of my trip took 5 minutes, for a total trip time of $10 + 5 = \boxed{15 \text{ minutes}}$.

To see why doubling my speed resulted in halving my time, recall that

$$\text{time} = \frac{\text{distance}}{\text{speed}}.$$

So, if the distance remains the same and we double "speed," then "time" must be halved.

7.5.6 *Solution 1: Scale the traveling time.* First, we convert the 75 seconds to minutes. We have $75 \text{ seconds} = (75 \text{ seconds}) \cdot \frac{1 \text{ minute}}{60 \text{ seconds}} = \frac{75}{60} \text{ minutes} = \frac{5}{4} \text{ minutes}$. So, the train goes 1 mile every $\frac{5}{4}$ minutes. Multiplying by 4, we see that the train goes 4 miles in 5 minutes. Multiplying by 12, the train goes 48 miles in 60 minutes, which is one hour. So, we multiply by 2 to see that the train goes $\boxed{96 \text{ miles}}$ in two hours.

Solution 2: Multiply the train's speed by the time it travels. The train travels at a speed of 1 mile per 75 seconds. That is, its speed is $\frac{1 \text{ mile}}{75 \text{ seconds}}$. It travels at this speed for 2 hours, so the distance it travels is $\frac{1 \text{ mile}}{75 \text{ seconds}} \cdot (2 \text{ hours})$. The problem now is that the hours and the seconds don't cancel. We can use a couple of conversion factors to convert the hours to seconds:

$$\frac{1 \text{ mile}}{75 \text{ seconds}} \cdot (2 \text{ hours}) = \frac{1 \text{ mile}}{75 \text{ seconds}} \cdot (2 \text{ hours}) \cdot \frac{60 \text{ minutes}}{1 \text{ hour}} \cdot \frac{60 \text{ seconds}}{1 \text{ minute}}$$
$$= \frac{1 \text{ mile}}{75 \text{ seconds}} \cdot \frac{(2 \text{ hours})(60 \text{ minutes})(60 \text{ seconds})}{(1 \text{ hour})(1 \text{ minute})}$$
$$= \frac{1 \text{ mile}}{75 \text{ seconds}} \cdot (7200 \text{ seconds})$$
$$= \frac{7200}{75} \text{ miles} = \boxed{96 \text{ miles}}.$$

7.5.7 We compute separately how long it takes each person to travel to the factory. Jason drives 60 miles per hour for 25 miles, so his trip takes him

$$\frac{25 \text{ miles}}{60 \frac{\text{miles}}{\text{hour}}} = \frac{25}{60} \text{ hours} = \frac{5}{12} \text{ hours}.$$

Jeremy drives 70 miles per hour for 35 miles, so his trip takes him

$$\frac{35 \text{ miles}}{70 \frac{\text{miles}}{\text{hour}}} = \frac{35}{70} \text{ hours} = \frac{1}{2} \text{ hours}.$$

Since $\frac{1}{2} > \frac{5}{12}$, $\boxed{\text{Jason}}$ arrives before Jeremy. We convert both times to minutes to see how many minutes Jason arrives before Jeremy. An hour is 60 minutes, so Jeremy's $\frac{1}{2}$ hour is 30 minutes. Jason takes

$$\frac{5}{12} \text{ hours} \cdot \frac{60 \text{ minutes}}{1 \text{ hour}} = \frac{5}{12} \cdot 60 \text{ minutes} = 25 \text{ minutes}.$$

So, Jeremy arrives $30 - 25 = \boxed{5 \text{ minutes}}$ after Jason.

7.5.8 The total distance for the round trip is $2 \cdot 20 = 40$ miles. Next, we must compute how long the round trip took. On the way to the mall, he goes 30 miles per hour for 20 miles. So, the amount of time he takes is

$$\frac{20 \text{ miles}}{30 \frac{\text{miles}}{\text{hour}}} = \frac{20}{30} \text{ hours} = \frac{2}{3} \text{ hours.}$$

On the way back, he goes 12 miles per hour for 20 miles. So, the amount of time the return trip takes is

$$\frac{20 \text{ miles}}{12 \frac{\text{miles}}{\text{hour}}} = \frac{20}{12} \text{ hours} = \frac{5}{3} \text{ hours.}$$

Combining these, the whole round trip takes $\frac{2}{3} + \frac{5}{3} = \frac{7}{3}$ hours. Since the round trip covers 40 miles in $\frac{7}{3}$ hours, the average speed is

$$\frac{40 \text{ miles}}{\frac{7}{3} \text{ hours}} = \frac{40}{7/3} \frac{\text{miles}}{\text{hour}} = \frac{120}{7} \frac{\text{miles}}{\text{hour}}.$$

Since $\frac{120}{7} = 17\frac{1}{7}$, his speed rounded to the nearest mile per hour is $\boxed{17 \text{ miles per hour}}$.

7.5.9 Since one dog is 3 feet per second faster than the other, the faster dog's lead grows by 3 feet every second. The two dogs will be back at the same point when this lead grows to 300 feet, which is a full lap around the track. Since the lead grows by 3 feet every second, the lead reaches 300 feet in

$$\frac{300 \text{ feet}}{3 \frac{\text{feet}}{\text{second}}} = \frac{300}{3} \text{ seconds} = \boxed{100 \text{ seconds}}.$$

7.5.10 First, we determine how many train lengths the train travels during the 2 minutes it takes to clear the tunnel. In order for the train to clear the tunnel, the front of the train must travel the entire 9 train-lengths of the tunnel, and then the train must go 1 more train-length in order for the rest of the train to leave the tunnel. So, the train travels 10 train-lengths to clear the tunnel. This means that the train's speed is 10 train-lengths per two minutes, or

$$\frac{10 \text{ train-lengths}}{2 \text{ minutes}} = 5 \frac{\text{train-lengths}}{\text{minute}}.$$

We know that the train travels 30 miles per hour. So, if we convert our train-lengths per minute to train-lengths per hour, then we know how many train-lengths the train covers in order to go 30 miles:

$$5 \frac{\text{train-lengths}}{\text{minute}} = 5 \frac{\text{train-lengths}}{\text{minute}} \cdot \frac{60 \text{ minutes}}{1 \text{ hour}} = 300 \frac{\text{train-lengths}}{\text{hour}}.$$

Therefore, in each hour that the train goes 30 miles, it covers 300 train-lengths. This means that the train is $\frac{30}{300} = \frac{1}{10}$ mile long. Converting this to feet gives

$$\frac{1}{10} \text{ miles} \cdot \frac{5280 \text{ feet}}{1 \text{ mile}} = \boxed{528 \text{ feet}}.$$

Exercises for Section 7.6

7.6.1 Since it takes Casey 15 minutes to build each foot of the fence, it takes her $15 \cdot 100 = 1500$ minutes to build all 100 feet of the fence. We can then convert the minutes to hours:

$$1500 \text{ minutes} \cdot \frac{1 \text{ hours}}{60 \text{ minutes}} = \frac{1500}{60} \text{ hours} = \boxed{25 \text{ hours}}.$$

We also could have used conversion factors to do the whole problem from the beginning. Since she builds 1 foot per 15 minutes, we have

$$100 \text{ feet} \cdot \frac{15 \text{ minutes}}{1 \text{ feet}} \cdot \frac{1 \text{ hours}}{60 \text{ minutes}} = \frac{100 \cdot 15}{60} \text{ hours} = \boxed{25 \text{ hours}}.$$

7.6.2 Phil produces 1 page every 20 minutes. So in $3 \cdot 20 = 60$ minutes (which is 1 hour), he produces 3 pages. Therefore, in 8 hours, he produces $3 \cdot 8 = \boxed{24 \text{ pages}}$.

Again, we can use conversion factors:

$$8 \text{ hours} \cdot \frac{60 \text{ minutes}}{1 \text{ hour}} \cdot \frac{1 \text{ page}}{20 \text{ minutes}} = \frac{8 \cdot 60}{20} \text{ pages} = \boxed{24 \text{ pages}}.$$

7.6.3 The kangaroo gains 5 feet on every leap. The kangaroo must gain 150 feet total, so it must jump $\frac{150}{5} = \boxed{30}$ times to catch the rabbit.

7.6.4 We first determine how much money Maria makes per disk. Since she buys 4 disks for \$5, the per-disk rate she pays is $\frac{5 \text{ dollars}}{4 \text{ disks}} = \frac{5}{4} \frac{\text{dollars}}{\text{disk}}$. She sells 3 disks for \$5, so the per-disk rate she receives is $\frac{5 \text{ dollars}}{3 \text{ disks}} = \frac{5}{3} \frac{\text{dollars}}{\text{disk}}$. Therefore, her profit per disk is

$$\frac{5}{3} \frac{\text{dollars}}{\text{disk}} - \frac{5}{4} \frac{\text{dollars}}{\text{disk}} = \frac{5}{12} \frac{\text{dollars}}{\text{disk}}.$$

Since this is the amount she makes for each disk, the number of disks she needs to make \$100 is

$$\frac{100 \text{ dollars}}{\frac{5}{12} \frac{\text{dollars}}{\text{disk}}} = \frac{100 \cdot 12}{5} \text{ disks} = \boxed{240 \text{ disks}}.$$

7.6.5 The entire trip was $57{,}060 - 56{,}200 = 860$ miles. The tank was full after the driver added the 6 gallons of gas at the beginning. After that, the driver had to add a total of $12 + 20 = 32$ gallons of gas to have the gas tank completely filled at the end. So, the car used 32 gallons of gas while traveling the 860 miles. That gives an average miles-per-gallon rate of

$$\frac{860 \text{ miles}}{32 \text{ gallons}} = \frac{860}{32} \frac{\text{miles}}{\text{gallon}} \approx \boxed{26.9 \frac{\text{miles}}{\text{gallon}}}.$$

7.6.6 The clock will again show the correct time when it has lost a total of 12 hours. Every hour, the clock loses 1 minute. We want the clock to lose 12 hours, which is $12 \cdot 60 = 720$ minutes. It will take 720 hours to lose 720 minutes, so it will take $\boxed{720 \text{ hours}}$ to lose 12 hours.

As with many of our other problems, we also could have used conversion factors to solve the problem:

$$\frac{1 \text{ hours time}}{1 \text{ minute lost}} \cdot (12 \text{ hours lost}) \cdot \frac{60 \text{ minutes lost}}{1 \text{ hours lost}} = 12 \cdot 60 \text{ hours time} = \boxed{720 \text{ hours}}.$$

7.6.7 Homer peels 3 potatoes a minute, so in the four minutes he works alone, he peels $3 \cdot 4 = 12$ potatoes. This leaves $44 - 12 = 32$ potatoes for them to peel together. Together, they peel $3 + 5 = 8$ potatoes per minute, so they need

$$\frac{32 \text{ potatoes}}{8 \frac{\text{potatoes}}{\text{minute}}} = 4 \text{ minutes}$$

to peel the rest of the potatoes. Christen peels 5 potatoes a minute for 4 minutes, so she peels $5 \cdot 4 = \boxed{20}$ potatoes.

7.6.8 Since the faucet can fill the tub in 15 minutes, it fills $\frac{1}{15}$ of the tub every minute. Similarly, the drain empties $\frac{1}{20}$ of the tub every minute. Therefore, the fraction of the tub that is filled each minute when the faucet and drain are both open is

$$\frac{1}{15} - \frac{1}{20} = \frac{4}{60} - \frac{3}{60} = \frac{1}{60}.$$

Since $\frac{1}{60}$ of the tub is filled each minute, it will take $\boxed{60 \text{ minutes}}$ for the tub to fill.

7.6.9 Since Roger can shovel the driveway in an hour, he can shovel $\frac{1}{60}$ of the driveway every minute. Similarly, Alexis can shovel the driveway in 30 minutes, so she shovels $\frac{1}{30}$ of the driveway each minute. Together, each minute they shovel

$$\frac{1}{60} + \frac{1}{30} = \frac{3}{60} = \frac{1}{20}$$

of the driveway. Therefore, they need $\boxed{20 \text{ minutes}}$ to shovel the entire driveway.

7.6.10 The first pipe fills $\frac{1}{8}$ of the pool in 1 hour, the second pipe fills $\frac{1}{12}$ of the pool in 1 hour, and the third pipe fills $\frac{1}{24}$ of the pool in 1 hour. So together the pipes fill

$$\frac{1}{8} + \frac{1}{12} + \frac{1}{24} = \frac{3}{24} + \frac{2}{24} + \frac{1}{24} = \frac{6}{24} = \frac{1}{4}$$

of the pool in 1 hour. Therefore, the three pipes together will take $\boxed{4 \text{ hours}}$ to fill the pool.

Review Problems

7.34 Multiplying both parts of the ratio by 9 gives

$$\text{cats} : \text{dogs} = 18 : 27,$$

so there are $\boxed{27}$ dogs at the pound.

7.35 Since the ratio of boys to girls at the camp is 4 : 5, the boys are $\frac{4}{4+5} = \frac{4}{9}$ of the students at the camp. There are 108 students, so there are $\frac{4}{9} \cdot 108 = \boxed{48}$ boys.

7.36 We express both weights in ounces, so that we can compare them. 1 pound, 4 ounces equals $1 \cdot 16 + 4 = 20$ ounces. 3 pounds, 10 ounces equals $3 \cdot 16 + 10 = 58$ ounces. So, the ratio of the given weights is 20 : 58, which is $\boxed{10 : 29}$ in simplest form.

7.37 The longer piece is $\frac{5}{1+5} = \frac{5}{6}$ of the whole board, so the length of the longer piece is $\frac{5}{6} \cdot (12 \text{ meters}) = \boxed{10 \text{ meters}}$.

7.38 *Solution 1.* The number of girls is $\frac{3}{4+3} = \frac{3}{7}$ of the total number of students, so the total number of students is $\frac{7}{3}$ times the number of girls. Therefore, there are $\frac{7}{3} \cdot 87 = \boxed{203}$ students.

Solution 2. For some value of x, the number of boys is $4x$ and the number of girls is $3x$. Since there are 87 girls, we have $3x = 87$, so $x = \frac{87}{3} = 29$. Therefore, the total number of students is $4x + 3x = 7x = 7(29) = \boxed{203}$.

7.39 The ratio of the amount Marisa spends to the amount Andie spends is 3 : 5. So, for some value of x, Marisa spends $3x$ dollars and Andie spends $5x$ dollars. Andie spends \$120 more than Marisa spends, so $5x - 3x = 120$. Simplifying gives $2x = 120$, so $x = 60$. Therefore, Andie spends $5x = \boxed{300 \text{ dollars}}$.

7.40 Initially, the fraction of fish that are guppies is $\frac{3}{3+2} = \frac{3}{5}$, so there are $\frac{3}{5} \cdot 20 = 12$ guppies in the tank.

After adding the new fish, there are $20 + 20 = 40$ fish in the tank. Since the ratio of guppies to angelfish is then 2 : 3, we know that $\frac{2}{2+3} = \frac{2}{5}$ of these fish are guppies. So, there are $\frac{2}{5} \cdot 40 = 16$ guppies in the tank, which means $16 - 12 = \boxed{4}$ guppies were added.

7.41 The person who receives the largest amount gets $\frac{5}{2+3+3+5} = \frac{5}{13}$ of the total profit. So, this person receives $\frac{5}{13} \cdot (\$26{,}000) = \boxed{\$10{,}000}$.

7.42 The statue's height and the Memorial's height are in proportion. On the back of the \$5 bill, the Memorial is 25 cm and the statue is 5 cm, so the Memorial is 5 times as tall as the statue. Therefore, the real Memorial is 5 times as tall as the real statue, or $5 \cdot (6 \text{ meters}) = \boxed{30 \text{ meters}}$ tall.

7.43 *Solution 1.* Since $\frac{1}{2}$ inch represents 80 miles, we know that $2 \cdot \frac{1}{2} = 1$ inch represents $2 \cdot 80 = 160$ miles. Therefore, $5\frac{7}{8}$ inches represents

$$5\frac{7}{8} \cdot 160 = \frac{47}{8} \cdot 160 = 47 \cdot 20 = \boxed{940 \text{ miles}}.$$

Solution 2. We use conversion factors. Since $\frac{1}{2}$ inch corresponds to 80 miles, $5\frac{7}{8}$ inches corresponds to

$$\left(5\frac{7}{8} \text{ inches}\right) \cdot \frac{80 \text{ miles}}{\frac{1}{2} \text{ inches}} = \frac{5\frac{7}{8} \cdot 80}{\frac{1}{2}} \text{ miles} = \left(\frac{47}{8} \cdot 80\right) \cdot 2 \text{ miles} = \boxed{940 \text{ miles}}.$$

7.44 We set up conversion factors:

$$10 \, \frac{\text{miles}}{\text{day}} = 10 \, \frac{\text{miles}}{\text{day}} \cdot \frac{8 \text{ furlongs}}{1 \text{ mile}} \cdot \frac{14 \text{ days}}{1 \text{ fortnight}} = (10 \cdot 8 \cdot 14) \frac{\text{furlongs}}{\text{fortnight}} = \boxed{1120} \, \frac{\text{furlongs}}{\text{fortnight}}.$$

7.45 The ratio of the shorter dimension to the larger dimension is $3 : 5$. The smaller dimension of the enlarged picture is $12 + 3 = 15$ inches. Multiplying both parts of $3 : 5$ by 5 gives $15 : 25$, so the larger dimension of the picture is $\boxed{25}$ inches.

7.46 The ratio of the distance on the map to the actual distance is $\frac{4}{9}$ mm : 1 km. Integers are easier to work with than fractions, so we multiply both parts of the ratio by 9 to give 4 mm : 9 km. Next, we notice that the distance on the map between Montreal and Quebec City is in centimeters, not mm. In order to use our ratio, we convert the distance on the map to mm. Since there are 10 mm in every 1 cm, 10 cm equals 100 mm. Multiplying both parts of the ratio 4 mm : 9 km by 25 gives us 100 mm : 225 km, so Montreal and Quebec City are $\boxed{225 \text{ km}}$ apart.

7.47 Dale's trip from A to B is 120 miles at 60 mph, which takes

$$\frac{120 \text{ miles}}{60 \frac{\text{miles}}{\text{hour}}} = \frac{120}{60} \text{ hours} = 2 \text{ hours.}$$

Her trip from B to C is 120 miles at 40 mph, which takes

$$\frac{120 \text{ miles}}{40 \frac{\text{miles}}{\text{hour}}} = \frac{120}{40} \text{ hours} = 3 \text{ hours.}$$

Her trip from C to A is 120 miles at 24 mph, which takes

$$\frac{120 \text{ miles}}{24 \frac{\text{miles}}{\text{hour}}} = \frac{120}{24} \text{ hours} = 5 \text{ hours.}$$

Combining these, she travels $120 + 120 + 120 = 360$ miles in $2 + 3 + 5 = 10$ hours, so her average speed is $\frac{360 \text{ miles}}{10 \text{ hours}} = \frac{360}{10} \frac{\text{miles}}{\text{hour}} = \boxed{36 \frac{\text{miles}}{\text{hour}}}$.

7.48 30 gallons of liquid A weighs $30 \cdot 8 = 240$ pounds. Since each gallon of liquid B weighs 6 pounds, we need $240/6 = \boxed{40 \text{ gallons}}$ of liquid B to weigh the same amount.

We could also set this up using conversion factors as:

$$30 \text{ gal of } A = 30 \text{ gal of } A \cdot \frac{8 \text{ lb of } A}{1 \text{ gal of } A} \cdot \frac{1 \text{ lb of } B}{1 \text{ lb of } A} \cdot \frac{1 \text{ gal of } B}{6 \text{ lb of } B} = \frac{30 \cdot 8}{6} \text{ gal of } B = \boxed{40 \text{ gal}} \text{ of } B.$$

7.49 Since 1 minute is 60 seconds, Ike covers 1 mile in 80 seconds. We can use conversion factors to convert "miles per second" to "miles per hour":

$$\frac{1 \text{ mile}}{80 \text{ seconds}} \cdot \frac{60 \text{ seconds}}{1 \text{ minute}} \cdot \frac{60 \text{ minutes}}{1 \text{ hour}} = \frac{60 \cdot 60}{80} \frac{\text{mile}}{\text{hour}} = \boxed{45 \text{ miles per hour}}.$$

7.50 When Elisa started, it took her 25 minutes to cover 10 laps, so each lap took her $\frac{25}{10} = \frac{5}{2}$ minutes. Now, she swims 12 laps in 24 minutes, so each lap takes $\frac{24}{12} = 2$ minutes. Since $\frac{5}{2} = 2\frac{1}{2}$, she has improved her lap time by $\boxed{\frac{1}{2}}$ minute.

7.51   The northbound train travels 324 miles at 50 miles per hour, so its trip takes

$$\frac{324 \text{ miles}}{50 \frac{\text{miles}}{\text{hour}}} = \frac{324}{50} \text{ hours} = 6.48 \text{ hours}.$$

The southbound train travels 324 miles at 40 miles per hour, so its trip takes

$$\frac{324 \text{ miles}}{40 \frac{\text{miles}}{\text{hour}}} = \frac{324}{40} \text{ hours} = 8.1 \text{ hours}.$$

Combining these, the train covers $324 + 324 = 648$ miles in $6.48 + 8.1 = 14.58$ hours, for an average speed of

$$\frac{648 \text{ miles}}{14.58 \text{ hours}} = \frac{400}{9} \text{ mph} = 44\frac{4}{9} \text{ mph} \approx \boxed{44.4 \text{ miles per hour}}.$$

7.52   During the first half hour of its journey, the car from Boston goes

$$50 \frac{\text{miles}}{\text{hour}} \cdot \left(\frac{1}{2} \text{ hour}\right) = 25 \text{ miles}.$$

So, at 1:30 PM, the cars are $295 - 25 = 270$ miles apart. Thereafter, the distance between the two cars decreases by $50 + 40 = 90$ miles per hour. Since they are 270 miles apart, and the distance decreases by 90 miles per hour, it takes

$$\frac{270 \text{ miles}}{90 \frac{\text{miles}}{\text{hour}}} = 3 \text{ hours}$$

for the distance to decrease to 0. Therefore, the cars meet at $\boxed{\text{4:30 PM}}$.

7.53   In 30 days, the amount of water that evaporates is

$$(30 \text{ days}) \cdot \left(12.5 \frac{\text{gallons}}{\text{day}}\right) = 375 \text{ gallons}.$$

Therefore, the amount of water left after 30 days is $1000 - 375 = \boxed{625 \text{ gallons}}$.

7.54   Since candle B burns twice as fast as candle A, it takes half as long for candle B to burn out as candle A. Therefore, candle B burns for $\frac{1}{2} \cdot (72 \text{ minutes}) = 36$ minutes. Similarly, candle C burns for one-third as long as candle B, so candle C burns for $\frac{1}{3} \cdot (36 \text{ minutes}) = 12$ minutes. Therefore, if we burn one candle at a time, the candles can give $72 + 36 + 12 = \boxed{120 \text{ minutes}}$ of light.

7.55   Since Jamie can mow the lawn in 75 minutes, she can mow $\frac{1}{75}$ of her lawn each minute. Therefore, in 30 minutes, she mows $\frac{30}{75} = \frac{2}{5}$ of the lawn. This means that Bob mows the other $1 - \frac{2}{5} = \frac{3}{5}$ of the lawn in 30 minutes. There are several ways we can finish from here.

We could note that if Bob mows $\frac{3}{5}$ of the lawn in 30 minutes, then he can mow $\frac{1}{5}$ the lawn in $\frac{30}{3} = 10$ minutes, which means he can mow the lawn once in $\boxed{50 \text{ minutes}}$.

We could also use conversion factors to find his rate of lawn mowing. He mows $\frac{3}{5}$ of the lawn in 30 minutes, so his rate is

$$\frac{30 \text{ minutes}}{\frac{3}{5} \text{ lawn}} = \frac{30}{\frac{3}{5}} \frac{\text{minutes}}{\text{lawn}} = \boxed{50 \text{ minutes}} \text{ per lawn}.$$

7.56 Applying conversion factors for each of the exchanges, we have

$$2000 \text{ pounds} \cdot \frac{1 \text{ dollar}}{0.62 \text{ pounds}} \cdot \frac{12.1 \text{ pesos}}{1 \text{ dollar}} = \frac{2000 \cdot 12.1}{0.62} \text{ pesos} \approx \boxed{39{,}032 \text{ pesos}}.$$

7.57 We know that

4 short-order cooks can make 24 omelets in 10 minutes.

1 cook can make $\frac{1}{4}$ as many omelets as 4 cooks can, so

1 short-order cook can make 6 omelets in 10 minutes.

If the cook has 15 minutes instead of 10 minutes, then he has $\frac{15}{10} = \frac{3}{2}$ times as much time. So, he can make $\frac{3}{2}$ times as many omelets. Since $6 \cdot \frac{3}{2} = 9$, we have

1 short-order cook can make 9 omelets in 15 minutes.

Since we need 90 omelets in these 15 minutes, and each cook makes 9 omelets in 15 minutes, we need $\frac{90}{9} = \boxed{10 \text{ cooks}}$.

Challenge Problems

7.58 The common denominator of the fractions is 12, so we can convert all the terms in the ratio to integers by multiplying them by 12:

$$\frac{1}{6} : \frac{1}{3} : \frac{1}{4} = 12 \cdot \frac{1}{6} : 12 \cdot \frac{1}{3} : 12 \cdot \frac{1}{4} = 2 : 4 : 3.$$

The pile with the least number of dimes has $\frac{2}{2+4+3} = \frac{2}{9}$ of the dimes, so that pile has $\frac{2}{9} \cdot 45 = \boxed{10}$ dimes.

7.59 Because 5 workers can build a road in 20 days, each worker can build $\frac{1}{5}$ of a road in 20 days. Therefore, each worker can build 1 road working alone in $20 \cdot 5 = 100$ days. So, each worker builds $\frac{1}{100}$ road each day. Therefore, three workers together build $\frac{3}{100}$ of a road each day, so in 10 days they build $10 \cdot \frac{3}{100} = \frac{3}{10}$ of a road. At this point, they have $1 - \frac{3}{10} = \frac{7}{10}$ of the road to finish.

After the 11 extra workers join in, the 14 workers build $\frac{14}{100}$ of the road each day. So, to finish the remaining $\frac{7}{10}$ of a road, the workers must work for

$$\frac{\frac{7}{10} \text{ road}}{\frac{14}{100} \frac{\text{road}}{\text{day}}} = \frac{7}{10} \cdot \frac{100}{14} \text{ days} = 5 \text{ days}.$$

Therefore, the road is built in $10 + 5 = \boxed{15 \text{ days}}$.

7.60 When Paul finishes, Robert has run $60 - 10 = 50$ meters and Sam has run $60 - 20 = 40$ meters. Therefore, when Robert and Sam run for the same amount of time, Sam covers $\frac{40}{50} = \frac{4}{5}$ of the distance that Robert covers. So, while Robert runs the final 10 meters of the race, Sam runs $\frac{4}{5} \cdot 10 = 8$ meters. This means Robert's lead over Sam increases by 2 more meters, and he beats Sam by $10 + 2 = \boxed{12}$ meters.

7.61 Multiplying both parts of $4 : x^2$ by x gives $4 : x^2 = 4x : x^3$, and multiplying both parts of $x : 16$ by 4 gives $x : 16 = 4x : 64$. Substituting these into $4 : x^2 = x : 16$ gives $4x : x^3 = 4x : 64$, so $x^3 = 64$, which gives $x = \boxed{4}$.

We also could have written the ratios in the given equation as fractions to get $\frac{4}{x^2} = \frac{x}{16}$. Multiplying both sides by x^2 and by 16 gets rid of the denominators and gives $4 \cdot 16 = x \cdot x^2$, so $64 = x^3$ and $x = \boxed{4}$.

7.62 Let the cost of one of the smallest mirrors be x, so the cost of 25 of these mirrors is $25x$. The radius of the largest mirror is 5 times the radius of the smallest mirror. The cost of the mirror is proportional to the cube of the radius, so the cost of the largest mirror is $5^3 = 125$ times the cost of the smallest mirror. Therefore, the cost of the largest mirror is $125x$, and the desired ratio is $25x : 125x$. Dividing both parts of the ratio by $25x$ gives $\boxed{1 : 5}$.

7.63 The ratio of Kim's votes to Amy's votes is $3 : 2$, so Kim received $3x$ votes and Amy received $2x$ votes for some value of x. If 8 of the people had voted for Amy instead of Kim, then Kim would have $3x - 8$ votes and Amy would have $2x + 8$ votes. If this had happened, then Amy would have twice as many votes as Kim, so $2x + 8 = 2(3x - 8)$. Expanding the right-hand side gives $2x + 8 = 6x - 16$. Subtracting $2x$ from both sides and adding 16 to both sides gives $24 = 4x$, so $x = 6$. Since $3x + 2x = 5x$ people voted, the number of voters is $5 \cdot 6 = \boxed{30}$.

7.64 First, we determine how many joggers started the race. Since the ratio of runners to joggers was $2 : 19$ at the start of the race, the joggers are $\frac{19}{2+19} = \frac{19}{21}$ of the participants. There were 4200 participants at the start of the race, so $4200 \cdot \frac{19}{21} = 3800$ of them are joggers. The other $4200 - 3800 = 400$ are runners. Since 500 joggers dropped out, only $3800 - 500 = 3300$ finished. Therefore, the ratio of runners to joggers among those who finished is $400 : 3300 = \boxed{4 : 33}$.

7.65

(a) Bob spends 15 seconds more on each page, so he spends

$$760 \text{ pages} \cdot \frac{15 \text{ seconds}}{1 \text{ page}} = \boxed{11400 \text{ seconds}}$$

more than Chandra.

(b) In 90 seconds, Bob reads 2 pages and Chandra reads 3 pages. So, for any fixed amount of time, the ratio of the amount Bob reads to the amount Chandra reads is $2 : 3$. Therefore, Chandra must read $\frac{3}{2+3} = \frac{3}{5}$ of the book, and during that time Bob will read the other $\frac{2}{5}$. This means Chandra should read $\frac{3}{5} \cdot 760 = 456$ pages, which means she should stop reading after completing page $\boxed{456}$.

(c) In 90 seconds, Alice reads $\frac{90}{20} = 4.5$ pages. So, in 90 seconds, the three together read $4.5 + 2 + 3 = 9.5$ pages. Therefore, the total amount of time they need to read is

$$760 \text{ pages} \cdot \frac{90 \text{ seconds}}{9.5 \text{ pages}} = \boxed{7200 \text{ seconds}}.$$

7.66 We count the number of Dallas-bound buses that a single Houston-bound bus passes. There are two groups of Dallas-bound buses that we have to consider. First, there are the buses that were on the way to Dallas at the time the Houston-bound bus leaves Dallas. Second, there are the buses that leave Houston after the Houston-bound bus starts but before it arrives at Houston.

In order for a bus to be on the way to Dallas when the Houston-bound bus leaves, it must have left Houston less than 5 hours before the Houston-bound bus leaves. There is one such bus each hour, for a total of 5 buses.

In order for a Dallas-bound bus to leave Houston after the Houston-bound bus leaves but before the Houston-bound bus arrives, the Dallas-bound must leave Houston within 5 hours after the Houston-bound bus leaves. Again, there is one such bus each hour, for a total of 5 more buses that the Houston-bound bus passes.

Therefore, the Houston-bound bus passes $5 + 5 = \boxed{10}$ Dallas-bound buses.

CHAPTER 8

Percents

24® Cards

First Card $(8 \cdot 10 - 8)/3$.

Second Card $27/(1 + 8/64)$.

Third Card $8/(3 - 8/3)$.

Fourth Card $8 \cdot 10 - 8 \cdot 7$.

Exercises for Section 8.1

8.1.1

(a) $37\% = \boxed{\frac{37}{100}}$.

(b) $80\% = \frac{80}{100} = \boxed{\frac{4}{5}}$.

(c) $250\% = \frac{250}{100} = \boxed{\frac{5}{2}}$, which we can write as $\boxed{2\frac{1}{2}}$.

(d) $-25\% = \frac{-25}{100} = \boxed{-\frac{1}{4}}$.

(e) $-200\% = \frac{-200}{100} = \boxed{-2}$.

(f) $1810\% = \frac{1810}{100} = \boxed{\frac{181}{10}}$, which we can write as $\boxed{18\frac{1}{10}}$.

8.1.2

(a) $\frac{33}{50} = \frac{33 \cdot 2}{50 \cdot 2} = \frac{66}{100} = \boxed{66\%}$.

(b) $\frac{2}{5} = \frac{2 \cdot 20}{5 \cdot 20} = \frac{40}{100} = \boxed{40\%}$.

(c) $3\frac{1}{4} = \frac{13}{4} = \frac{13 \cdot 25}{4 \cdot 25} = \frac{325}{100} = \boxed{325\%}$. We might also have noticed that $3 = 300\%$ and $\frac{1}{4} = \frac{25}{100} = 25\%$, so $3\frac{1}{4} = 325\%$.

(d) We can't multiply 8 by an integer to get 100, but writing $\frac{3}{8}$ as a decimal makes converting to a percent easy. We have $\frac{3}{8} = 0.375$, so $-2\frac{3}{8} = -2.375 = \boxed{-237.5\%}$.

(e) $0 = \frac{0}{100} = \boxed{0\%}$.

(f) We move the decimal point two places to the right: $-192.5 = -192.50 = \boxed{-19250\%}$.

(g) We multiply the fraction by $100\% = 1$:

$$\frac{2}{7} = (100\%) \cdot \frac{2}{7} = \frac{200}{7}\% = \boxed{28\frac{4}{7}\%}.$$

(h) We move the decimal point two places to the right: $0.319 = \frac{31.9}{100} = \boxed{31.9\%}$.

8.1.3

(a) 30% of $200 = \frac{30}{100} \cdot 200 = \boxed{60}$.

(b) 55% of $120 = \frac{55}{100} \cdot 120 = \frac{11}{20} \cdot 120 = \frac{11 \cdot 120}{20} = 11 \cdot \frac{120}{20} = \boxed{66}$.

(c) 225% of $16 = 2\frac{1}{4} \cdot 16 = \frac{9}{4} \cdot 16 = 9 \cdot \frac{16}{4} = 9 \cdot 4 = \boxed{36}$.

(d) -80% of $35 = -\frac{80}{100} \cdot 35 = -\frac{4}{5} \cdot 35 = -4 \cdot \frac{35}{5} = -4 \cdot 7 = \boxed{-28}$.

(e) 15% of $380 = \frac{15}{100} \cdot 380 = \frac{3}{20} \cdot 380 = 3 \cdot \frac{380}{20} = 3 \cdot 19 = \boxed{57}$.

(f) -100% of $617 = -\frac{100}{100} \cdot 617 = \boxed{-617}$.

(g) 0% of $2{,}827{,}192 = \frac{0}{100} \cdot 2{,}827{,}192 = \boxed{0}$.

(h) $\frac{1}{5}\%$ of $2000 = \frac{1/5}{100} \cdot 2000 = (1/5) \cdot \frac{2000}{100} = \frac{1}{5} \cdot 20 = \boxed{4}$.

8.1.4

(a) We want the value of x such that $x\% = \frac{20}{80} = \frac{1}{4} = \frac{1 \cdot 25}{4 \cdot 25} = \frac{25}{100}$. So, $x = 25$ and the answer is $\boxed{25\%}$.

(b) We want the value of x such that $x\% = \frac{-60}{30} = -2 = \frac{-200}{100}$. So, $x = -200$ and the answer is $\boxed{-200\%}$.

(c) We want the value of x such that $x\% = \frac{51}{17} = 3 = \frac{300}{100}$. So, $x = 300$ and the answer is $\boxed{300\%}$.

(d) We want the value of x such that $x\% = \frac{(1/2)}{5} = \frac{1}{10} = \frac{10}{100}$. So, $x = 10$ and the answer is $\boxed{10\%}$.

(e) We want the value of x such that $x\% = \frac{(2/3)}{(5/6)} = \frac{2}{3} \cdot \frac{6}{5} = \frac{4}{5} = \frac{4 \cdot 20}{5 \cdot 20} = \frac{80}{100}$. So, $x = 80$ and the answer is $\boxed{80\%}$.

(f) We want the value of x such that $x\% = \frac{7}{-35} = -\frac{1}{5} = \frac{-20}{100}$. So, $x = -20$ and the answer is $\boxed{-20\%}$.

8.1.5

(a) We want the number x such that $11 = 20\% \cdot x$. Since $20\% = \frac{20}{100} = \frac{1}{5}$, we have $11 = \frac{1}{5} \cdot x$, so $x = \boxed{55}$.

(b) We want the number x such that $\frac{2}{3} = 30\% \cdot x$. Since $30\% = \frac{30}{100} = \frac{3}{10}$, we have $\frac{2}{3} = \frac{3}{10}x$. Multiplying both sides by $\frac{10}{3}$ gives $\frac{10}{3} \cdot \frac{2}{3} = x$, so $x = \boxed{\frac{20}{9}}$ or $\boxed{2\frac{2}{9}}$.

(c) We want the number x such that $3 = -40\% \cdot x$. Since $-40\% = -\frac{40}{100} = -\frac{2}{5}$, we have $3 = -\frac{2}{5}x$. Multiplying both sides by $-\frac{5}{2}$, we have $-\frac{5}{2} \cdot 3 = x$, so $x = \boxed{-\frac{15}{2}}$ or $\boxed{-7\frac{1}{2}}$.

(d) We want the number x such that $\frac{1}{7} = \frac{1}{2}\% \cdot x$. We have $\frac{1}{2}\% = 0.5\% = 0.005 = \frac{5}{1000} = \frac{1}{200}$, so our original equation is $\frac{1}{7} = \frac{1}{200} \cdot x$. Multiplying both sides by 200 gives $x = \boxed{\frac{200}{7}}$ or $\boxed{28\frac{4}{7}}$.

8.1.6 We have

$$\frac{7}{9} \text{ of } 180 = \frac{7}{9} \cdot 180 = 7 \cdot \frac{180}{9} = 140,$$

$$75\% \text{ of } 200 = \frac{75}{100} \cdot 200 = 75 \cdot \frac{200}{100} = 150,$$

so $\boxed{75\% \text{ of } 200}$ is greater.

8.1.7 $(60\% \text{ of } 75) + (75\% \text{ of } 60) = \frac{60}{100} \cdot 75 + \frac{75}{100} \cdot 60 = \frac{3}{5} \cdot 75 + \frac{3}{4} \cdot 60 = 45 + 45 = \boxed{90}$.

8.1.8 Since 70% of 10 is $70\% \cdot 10 = 0.7 \cdot 10 = 7$, we know that

$$40\% \text{ of } 70\% \text{ of } 10 = 40\% \text{ of } 7 = 0.40 \cdot 7 = \boxed{2.8}.$$

We can write 2.8 as a mixed number as $\boxed{2\frac{4}{5}}$, or as a fraction as $\boxed{\frac{14}{5}}$.

8.1.9 Let x be half the number. Since 2% of half the number is 5, we have 2% of $x = 5$. Since $2\% = \frac{2}{100} = \frac{1}{50}$, we have $\frac{1}{50} \cdot x = 5$, so $x = 250$. But x is just half the desired number, so the desired number is $2 \cdot 250 = \boxed{500}$.

Exercises for Section 8.2

8.2.1 There are $5 \cdot 5 = 25$ total 1×1 squares, and $2 \cdot 2 + 3 \cdot 3 = 4 + 9 = 13$ of these are shaded. So, $\frac{13}{25}$ of the small squares are shaded. Converting $\frac{13}{25}$ to a percent gives $\frac{13}{25} = \frac{13 \cdot 4}{25 \cdot 4} = \frac{52}{100} = \boxed{52\%}$ as the percent of the 5×5 square that is shaded.

8.2.2 There were $4 + 28 + 8 = 40$ people at the meeting, and 28 of them were students. So, the percent of people at the meeting who were students is $\frac{28}{40} = \frac{7}{10} = \frac{70}{100} = \boxed{70\%}$.

8.2.3 Since 35 of the 50 senators are female, the other $50 - 35 = 15$ are male. Therefore, the percent of senators who are male is $\frac{15}{50} = \frac{15 \cdot 2}{50 \cdot 2} = \frac{30}{100} = \boxed{30\%}$.

8.2.4

(a) Rhonda's commission is 6% of the sales price of \$270,000. So, her commission is

$$6\% \text{ of } \$270,000 = \frac{6}{100} \cdot \$270,000 = 6 \cdot \frac{\$270,000}{100} = 6 \cdot \$2,700 = \boxed{\$16,200}.$$

(b) Let x be the sales price of the condo in dollars. Since \$15,000 is 6% of the sales price, we seek the number x such that 6% of $\$x = \$15,000$. We have $6\% = \frac{6}{100} = \frac{3}{50}$, so we must have

$$\frac{3}{50} \cdot \$x = \$15,000.$$

Multiplying both sides by $\frac{50}{3}$ gives

$$\$x = \frac{50}{3} \cdot (\$15,000) = 50 \cdot \frac{\$15,000}{3} = 50(\$5,000) = \$250,000.$$

So, the sales price of the condo is $\boxed{\$250,000}$.

8.2.5 First, we determine how many questions Toni answered correctly. 70% of the arithmetic problems is $0.70 \cdot 10 = 7$ problems. 40% of the algebra problems is $0.40 \cdot 30 = 12$ problems. 60% of the geometry problems is $\frac{60}{100} \cdot 35 = \frac{3}{5} \cdot 35 = 21$ problems. This is a total of $7 + 12 + 21 = 40$ problems answered correctly.

Next, we find how many questions must be answered correctly to pass. There are $10 + 30 + 35 = 75$ questions on the test. She must answer 60% of these correctly to pass, which is a total of $60\% \cdot 75 = \frac{3}{5} \cdot 75 = 45$ questions.

Therefore, Tori needed to answer $45 - 40 = \boxed{5}$ more questions correctly to pass.

8.2.6 The new mixture has $30 + 5 = 35$ liters of paint. To determine the percentage of yellow tint, we find the number of liters of yellow tint in the mixture. The original 30 liters was 30% yellow tint, so the amount of yellow tint in the original mixture is

$$30\% \cdot (30 \text{ liters}) = 0.3 \cdot (30 \text{ liters}) = 9 \text{ liters}.$$

The new mixture has 5 more liters of yellow tint, for a total of 14 liters. Therefore, the percentage of yellow tint in the new mixture is $\frac{14}{35} = \frac{2}{5} = \frac{40}{100} = \boxed{40\%}$.

8.2.7 Since the car has 3 gallons when it is 80% empty, these 3 gallons fill $100\% - 80\% = 20\%$ of the tank. This means that the 3 gallons are $20\% = \frac{20}{100} = \frac{1}{5}$ of the whole tank. So, the whole tank holds $3 \cdot 5 = 15$ gallons. Therefore, when the tank is 80% full, it has $80\% \cdot 15 = \frac{80}{100} \cdot 15 = \frac{4}{5} \cdot 15 = \boxed{12}$ gallons of gas. (We also might have noticed that when it was 80% empty, the tank held 3 gallons, so 80% of the tank would hold $15 - 3 = 12$ gallons.)

8.2.8 Since Emily won 65% of the games, Peter won the other $100\% - 65\% = 35\%$. Therefore, the ratio of the number of Emily's wins to the number of Peter's wins is $65 : 35$. Dividing both parts of this ratio by 5 gives $13 : 7$. Since Peter won 7 games, we conclude that Emily won $\boxed{13}$ games.

8.2.9 The season has $50 + 40 = 90$ games, so 70% of the team's games is a total of $70\% \cdot 90 = 0.7 \cdot 90 = 63$ games. The Fighting Tomatoes have already won 30 games, so they need to win $63 - 30 = \boxed{33}$ more games.

8.2.10 The two schools combined have 300 students. The table tells us that 30% of the 100 students in West Parkville are 6th graders. Therefore, West Parkville has $30\% \cdot 100 = 30$ 6th graders. The table also tells us that 45% of East Parkville's 200 students are 6th graders. So, there are $45\% \cdot 200 = \frac{45}{100} \cdot 200 = 90$ students in 6th grade at East Parkville.

The two schools together have $30 + 90 = 120$ students in grade 6 out of the 300 total students in the two schools. So, the percentage of students in the two schools combined who are in grade 6 is $\frac{120}{300} = \frac{4}{10} = \boxed{40\%}$.

8.2.11 If only 10% of the students had failed the exam, then $100\% - 10\% = 90\%$ would have passed the exam. This is an additional $90\% - 75\% = 15\%$ more than the 75% who did actually pass the exam. Since there are 500 students total, the additional 15% who passed the exam consist of $15\% \cdot 500 = \frac{15}{100} \cdot 500 = 15 \cdot 5 = \boxed{75}$ students.

Exercises for Section 8.3

8.3.1

(a) 20% of 15 is $\frac{20}{100} \cdot 15 = \frac{1}{5} \cdot 15 = 3$, so 20% more than 15 is $15 + 3 = \boxed{18}$.

(b) 30% of 40 is $0.30 \cdot 40 = 12$, so 30% less than 40 is $40 - 12 = \boxed{28}$.

(c) 150% of $\frac{2}{3}$ is $\frac{150}{100} \cdot \frac{2}{3} = \frac{3}{2} \cdot \frac{2}{3} = 1$, so 150% more than $\frac{2}{3}$ is $\frac{2}{3} + 1 = \boxed{\frac{5}{3}}$.

(d) 50% of $\frac{1}{7}$ is $\frac{1}{2} \cdot \frac{1}{7} = \frac{1}{14}$, so 50% more than $\frac{1}{7}$ is $\frac{1}{7} + \frac{1}{14} = \boxed{\frac{3}{14}}$.

(e) 80% of $\frac{3}{10}$ is $\frac{80}{100} \cdot \frac{3}{10} = \frac{4}{5} \cdot \frac{3}{10} = \frac{6}{25}$, so 80% less than $\frac{3}{10}$ is $\frac{3}{10} - \frac{6}{25} = \frac{15}{50} - \frac{12}{50} = \boxed{\frac{3}{50}}$.

Alternatively, we might have noted that if we take 80% of a number away from a number, we are left with 20% of the number. So, 80% less than $\frac{3}{10}$ equals 20% of $\frac{3}{10}$, which is $\frac{1}{5} \cdot \frac{3}{10} = \frac{3}{50}$, as before.

(f) 60% of 4.8 is $0.6 \cdot 4.8 = 2.88$, so 60% more than 4.8 is $4.8 + 2.88 = \boxed{7.68}$.

8.3.2 Most of the increases involve doubling the dollar value, which is an increase of 100%. Only 3 steps involve a different percent increase.

In going from question 2 to question 3, the prize increases by \$100 from \$200. This is an increase of $\frac{100}{200} = \frac{50}{100} = 50\%$.

In going from question 3 to question 4, the prize increases by \$200 from \$300. Since a 50% increase from \$300 would be an increase of $0.5 \cdot \$300 = \$150$, an increase of \$200 from \$300 is more than a 50% increase.

In going from question 11 to question 12, the prize increases by \$61,000 from \$64,000. A 50% increase from \$64,000 would be just \$32,000, so this increase is also more than 50%.

All of the other steps from one question to the next involve a 100% increase, so the smallest increase occurs between $\boxed{\text{questions 2 and 3}}$.

8.3.3 Last week, the candy bars cost $\frac{\$5}{4} = \$1.25$ each. This week, they cost $\frac{\$4}{5} = \$0.80$ each. So, the price decreased by $\$1.25 - \$0.80 = \$0.45$ from last week's \$1.25 price per bar. This represents a $\frac{0.45}{1.25} = \frac{45}{125} = \frac{9}{25} = \frac{9 \cdot 4}{25 \cdot 4} = \boxed{36\% \text{ decrease}}$.

8.3.4 20% off of the \$2.50 price is $\frac{1}{5} \cdot \$2.50 = \$0.50$ savings on each folder. Since Karl bought 5 folders, he could have saved $5 \cdot \$0.50 = \boxed{\$2.50}$.

8.3.5 Ana's raise was an increase of $0.2 \cdot \$2000 = \$400$, so her new salary was $\$2000 + \$400 = \$2400$. Then, her pay was decreased by 20% of this \$2400, which is $0.2 \cdot \$2400 = \$480$. So, her monthly pay after both changes was $\$2400 - \$480 = \boxed{\$1920}$.

Alternatively, we could reason that Ana's June salary is 120% of her May salary, and her July salary is 80% of her June salary. So her July salary equals

$$80\% \text{ of } 120\% \text{ of } \$2000 = 80\% \cdot 120\% \cdot \$2000 = (0.8)(1.2)(\$2000) = \boxed{\$1920}.$$

8.3.6 The sale reduced the price by $25\% \cdot \$80 = \frac{1}{4} \cdot \$80 = \$20$, making the new price $\$80 - \$20 = \$60$. The tax is 10% of this sale price, or $0.1 \cdot \$60 = \$6$. Therefore, the total cost is $\$60 + \$6 = \boxed{\$66}$.

8.3.7 First, we compute the final value of AA. The increase of 20% from \$100 makes the price \$120. Then, the price decreases by 20%, which is a decrease of $0.2 \cdot \$120 = \$24$ to $\$120 - \$24 = \$96$.

Next, we compute the final value of BB. The increase of 25% from \$100 makes the price \$125. Then, the price decreases by 25%, which is a decrease of $\frac{1}{4} \cdot \$125 = \$31.25$ to $\$125 - \$31.25 = \$93.75$.

The value of CC stays unchanged at \$100 throughout. So, from low to high, the stocks are

$$\boxed{\text{BB at } \$93.75, \text{ AA at } \$96, \text{ CC at } \$100}.$$

8.3.8 The original total price of the pets was $\$80 + \$40 = \$120$. Chris saved $0.4 \cdot \$80 = \$32$ on the iguana and $0.55 \cdot \$40 = \frac{11}{20} \cdot \$40 = \$22$ on the parakeet, for a total savings of $\$32 + \$22 = \$54$. Therefore, the amount he saved was $\frac{54}{120} = \frac{9}{20} = \frac{45}{100} = \boxed{45\%}$ of the total of the original prices.

8.3.9 Let the November price of a doll be x dollars. An increase of 25% is an increase of $\frac{1}{4} \cdot x = \frac{x}{4}$ dollars, so the December price is $x + \frac{x}{4} = \frac{4x}{4} + \frac{x}{4} = \frac{5x}{4}$ dollars. Then, the price is decreased from $\frac{5x}{4}$ dollars by 20%. This is a decrease of $20\% \cdot \frac{5x}{4} = \frac{1}{5} \cdot \frac{5x}{4} = \frac{x}{4}$ dollars, so the sale price is $\frac{5x}{4} - \frac{x}{4} = x$ dollars. The sale price is $\boxed{\text{the same}}$ as the November price, no matter what the original price was!

8.3.10 Each time Dave wins, he increases his total money by 50%. If he increases his money by 50%, then his new amount is 150% of his old amount. This means that his new amount is 1.5 times his old amount.

Each time Dave loses, he decreases his total money by 50%, so his new amount of money is 0.5 times his old amount of money.

Therefore, each time Dave wins, he multiplies his money by 1.5, and each time he loses, he multiplies his money by 0.5. So, when he wins three times and loses two times, he multiplies his money by 1.5 three times and multiplies his money by 0.5 two times. Since multiplication is commutative, he will end up with the same amount of money no matter what order these multiplications occur. His final amount of money is the result of multiplying \$1,000 by 1.5 three times and by 0.5 two times, which leaves him with

$$\$1{,}000 \cdot 1.5^3 \cdot 0.5^2 = \boxed{\$843.75}.$$

This is a little surprising: even though he won 3 hands but lost only 2, he still lost money overall! Think about why this is true.

8.3.11

(a) An increase from 100 of $p\%$ is an increase of $p\% \cdot 100 = \frac{p}{100} \cdot 100 = p$. So our quantity, after the increase, is $100 + p$. A decrease of $p\%$ from this amount is a decrease of

$$p\% \cdot (100 + p) = \frac{p}{100} \cdot (100 + p) = \frac{p(100 + p)}{100} = \frac{100p + p^2}{100}.$$

Decreasing $100 + p$ by this amount leaves

$$
\begin{aligned}
100 + p - \frac{100p + p^2}{100} &= \frac{100^2 + 100p}{100} - \frac{100p + p^2}{100} \\
&= \frac{100^2 + 100p - (100p + p^2)}{100} \\
&= \frac{100^2 + 100p - 100p - p^2}{100} \\
&= \frac{100^2 - p^2}{100} = \boxed{100 - \frac{p^2}{100}}.
\end{aligned}
$$

Since p^2 must be positive, this quantity must be $\boxed{\text{smaller than } 100}$.

(b) We follow essentially the same steps as in the previous part. A decrease from 100 of $q\%$ is a decrease of q, to $100 - q$. A increase of $q\%$ from this amount is an increase of

$$
q\% \cdot (100 - q) = \frac{q}{100} \cdot (100 - q) = \frac{q(100 - q)}{100} = \frac{100q - q^2}{100}.
$$

Increasing $100 - q$ by this amount gives

$$
\begin{aligned}
100 - q + \frac{100q - q^2}{100} &= \frac{100^2 - 100q}{100} + \frac{100q - q^2}{100} \\
&= \frac{100^2 - 100q + 100q - q^2}{100} \\
&= \frac{100^2 - q^2}{100} = \boxed{100 - \frac{q^2}{100}}.
\end{aligned}
$$

Since q^2 must be positive, this quantity must be $\boxed{\text{smaller than } 100}$. Moreover, notice that if we let $q = p$, we get the same expression as in part (a). What does that mean?

Review Problems

8.21 We have

$$
\begin{aligned}
40\% \text{ of } 20\% \text{ of } 10\% \text{ of } 80{,}000 &= 40\% \text{ of } 20\% \text{ of } (10\% \text{ of } 80{,}000) \\
&= 40\% \text{ of } 20\% \text{ of } (0.10 \cdot 80{,}000) \\
&= 40\% \text{ of } 20\% \text{ of } 8000 \\
&= 40\% \text{ of } (0.2 \cdot 8000) \\
&= 40\% \text{ of } 1600 \\
&= 0.4 \cdot 1600 \\
&= \boxed{640}.
\end{aligned}
$$

8.22 18% of 50 is $\frac{18}{100} \cdot 50 = \frac{9}{50} \cdot 50 = 9$. So, we want to know what percent of 24 is 9. Since $\frac{9}{24} = \frac{3}{8} = 0.375 = 37.5\%$, we know that 9 is $\boxed{37.5\%}$ of 24.

8.23 Since there are 240 boys, there are $960 - 240 = 720$ girls. Therefore, the percentage of students who are girls is $\frac{720}{960} = \frac{3}{4} = 0.75 = \boxed{75\%}$.

8.24 There are 15 scores in the list. There are 6 scores in the list that have a grade of C, so the percentage of students who receive a C is $\frac{6}{15} = \frac{2}{5} = \boxed{40\%}$.

8.25 45 is 36% of the number of cups in the coffee maker. So, we must find the number x such that 36% of x is 45. Since $36\% = \frac{36}{100} = \frac{9}{25}$, we must have $\frac{9}{25} \cdot x = 45$. Multiplying both sides by $\frac{25}{9}$ gives $x = \frac{25}{9} \cdot 45 = 25 \cdot 5 = 125$. So, the coffee maker makes $\boxed{125}$ cups of coffee.

8.26 Judy had $1 + 1 + 5 = 7$ hits that were not singles, so the remaining $35 - 7 = 28$ hits were singles. Therefore, the percent of hits that were singles was $\frac{28}{35} = \frac{4}{5} = \frac{80}{100} = \boxed{80\%}$.

8.27 Larry's 50 state quarters have a face value of $50 \cdot \$0.25 = \$12.50$. 600% of a number is 6 times the number, so Larry receives $6 \cdot \$12.50 = \boxed{\$75}$ for his whole collection.

8.28 Katie pays 3% less by driving to the neighboring state. So, she saves $3\% \cdot \$300 = 0.03 \cdot \$300 = \boxed{\$9}$.

8.29 Since Sally made 55% of her first 20 shots, she made $\frac{55}{100} \cdot 20 = \frac{11}{20} \cdot 20 = 11$ of her first 20 shots. Since she made 56% of her first 25 shots, she made $\frac{56}{100} \cdot 25 = \frac{14}{25} \cdot 25 = 14$ of her first 25 shots. Therefore, she made $14 - 11 = \boxed{3}$ of her last 5 shots.

8.30 From 2010 to 2011, enrollment increases by 20% from 200, which is an increase of $0.2 \cdot 200 = 40$ students. So, enrollment is 240 students in 2011. From 2011 to 2012, enrollment increases by 20% from 240, which is an increase of $0.2 \cdot 240 = 48$ students. So, enrollment is $240 + 48 = \boxed{288}$ in 2012.

8.31 A 10% increase of $2200 is an increase of $\frac{1}{10} \cdot \$2200 = \$220$, to $\$2200 + \$220 = \$2420$. A decrease of 15% from $2420 is a decrease of $0.15 \cdot \$2420 = \$363$, which makes the price $\$2420 - \$363 = \boxed{\$2057}$.

8.32 On the 10-problem test, she answers $0.7 \cdot 10 = 7$ questions correctly. On the 20-problem test, she answers $0.8 \cdot 20 = 16$ questions correctly. On the 30-problem test, she answers $0.9 \cdot 30 = 27$ questions correctly. So, she answers $7 + 16 + 27 = 50$ of the 60 total questions correctly, which means she answers $\frac{50}{60} = \frac{5}{6}$ of the questions correctly. Since $\frac{5}{6} = 0.8\overline{3}$, her percentage to the nearest whole percent is $\boxed{83\%}$.

8.33 After one such raise, Paul's new salary is 106% of his old salary, which means that his new salary is 1.06 times his old salary. Similarly, each of his raises multiplies his salary by 1.06. So, four such raises multiplies his salary by $1.06^4 \approx 1.262$, which is an increase of 0.262 times his original salary. Therefore, to the nearest percent, his salary has increased by $\boxed{26\%}$.

8.34 Since $25\% = \frac{1}{4}$, Jack sold $\frac{1}{4} \cdot 128 = 32$ apples to Jill, leaving him with $128 - 32 = 96$ apples. He then sold $\frac{1}{4} \cdot 96 = 24$ to June, leaving him with $96 - 24 = 72$. After giving his teacher the most shiny one of these, he was left with $\boxed{71}$ apples.

8.35 The 20% discount from $100 reduces the price by $20 to $80. The discount coupon reduces the price to $\$80 - \$5 = \$75$. The tax of 8% of this sales price equals $\frac{8}{100} \cdot \$75 = \frac{2}{25} \cdot \$75 = \$6$, so the total amount the shopper pays is $\$75 + \$6 = \boxed{\$81}$.

8.36 Polly's 15% loss in the first year reduced her $250 by $\frac{15}{100} \cdot \$250 = \frac{3}{20} \cdot \$250 = 3 \cdot \frac{\$250}{20} = 3 \cdot \$12.50 = \$37.50$, to $212.50. Then, in the second year her $212.50 is increased by 20%, which is an increase of

$0.2 \cdot \$212.50 = \$42.50$. So, at the end of her second year, her investment is $\$212.50 + \$42.50 = \$255$. Therefore, her original investment of $250 increased in value by $5 total. Since a 1% increase in value is $2.50, her $5 increase represents a $\boxed{2\% \text{ gain}}$.

Alternatively, the loss in the first year multiplies Polly's investment by 0.85, and the gain in the second year multiplies the investment by 1.2. So over the two-year period, the investment is multiplied by $(0.85)(1.2) = 1.02$, which is a 2% gain.

8.37 A sale price of 50% off halves the price from $180 to $90. An additional discount of 20% off of this new price is a discount of $0.2 \cdot \$90 = \$18$. This makes the Saturday price $\$90 - \$18 = \boxed{\$72}$.

8.38 Since a bork equals 100 borklets, the 5 bork ice cream cone costs 500 borklets.

First, we compute the final price if we use Jack's strategy of applying the coupon first. The coupon gives a discount of $\frac{1}{10} \cdot 500 = 50$ borklets, making the price 450 borklets. Then, the 20% tax applied to this price increases the total by $\frac{1}{5} \cdot 450 = 90$ borklets, to 540 borklets.

Next, we compute the price with Jill's strategy of applying the tax first. A 20% tax on 500 borklets is an increase of $\frac{1}{5} \cdot 500 = 100$ borklets, to 600 borklets. A 10% discount from 600 borklets is a discount of $\frac{1}{10} \cdot 600 = 60$ borklets, to $600 - 60 = 540$ borklets.

So, $\boxed{\text{the two strategies result in the same final cost}}$.

Is this just a coincidence? Try changing the original numbers in the problem, and see if the two strategies still produce the same final cost.

Challenge Problems

8.39 Since $125\% = 5 \cdot 25\%$, we know that 125% of a number is 5 times 25% of that same number. Since 25% of n is 18, we know that 125% of n is $5 \cdot 18 = \boxed{90}$.

Alternatively, we could have found n. Since 25% of n is 18, we know that $\frac{1}{4}$ of n is 18. This means that $n = 18 \cdot 4 = 72$. We then compute that 125% of n is $\frac{125}{100} \cdot 72 = \frac{5}{4} \cdot 72 = \boxed{90}$, as before.

8.40 The ratio of 30% of a number to 20% of that same number is $30 : 20 = 3 : 2$. So, 30% of a number is $\frac{3}{2}$ times 20% the same number. Since 20% of the number in the problem is 12, we know that 30% of the same number is $\frac{3}{2} \cdot 12 = \boxed{18}$.

Alternatively, we could have found the number. We are given that $20\% = \frac{1}{5}$ of the number is 12, so the number is $12 \cdot 5 = 60$. Therefore, 30% of the same number is $0.3 \cdot 60 = \boxed{18}$.

8.41 20% of x equals $0.2x$ and 12% of $x + 20$ is

$$0.12(x + 20) = 0.12x + 0.12 \cdot 20 = 0.12x + 2.4.$$

Therefore, we must have $0.2x = 0.12x + 2.4$. Multiplying both sides by 100 gets rid of the decimals and leaves $20x = 12x + 240$. Subtracting $12x$ from both sides gives $8x = 240$, so $x = \boxed{30}$.

8.42 The numbers of yellow beans in the bags are $\frac{1}{2} \cdot 26 = 13$, $\frac{1}{4} \cdot 28 = 7$, and $\frac{1}{5} \cdot 30 = 6$, respectively. So, the total number of yellow beans is $13 + 7 + 6 = 26$. The total number of beans is $26 + 28 + 30 = 84$, so, to the nearest whole percent, the percentage of beans that are yellow is $\frac{26}{84} \approx 0.3095 \approx \boxed{31\%}$.

8.43 Since there are 12 inches in a foot, Cody's height is $12 \cdot 5 + 6 = 66$ inches. Cody grew half as much as Tyler, and they started from the same height. So, while Tyler grew by 20%, Cody grew by 10%. Therefore, Cody's height is 66 inches after he grew by 10%. Since Cody is now 10% taller than he was, his current height is $100\% + 10\% = 110\%$ of his original height. So, if we let Cody's original height be x inches, we must have $1.1 \cdot x = 66$. Dividing both sides by 1.1 gives $x = 60$, so Cody was originally 60 inches tall.

Tyler was the same height as Cody, 60 inches, before growing by 20%. While growing, Tyler's height increased by $20\% \cdot 60 = 0.2 \cdot 60 = 12$ inches. So, he is now $\boxed{72 \text{ inches}}$, which is $\boxed{6 \text{ feet}}$.

8.44 Since Miki gets 8 ounces of pear juice from 3 pears, she gets $\frac{8}{3}$ ounces of juice from 1 pear. Similarly, Miki gets 8 ounces of orange juice from 2 oranges, so she gets $\frac{8}{2} = 4$ ounces of juice from 1 orange. When Miki combines 1 pear and 1 orange, she gets $\frac{8}{3} + 4 = \frac{20}{3}$ ounces total. Of these $\frac{20}{3}$ ounces, $\frac{8}{3}$ ounces are pear juice. So, the pear juice is $\frac{8/3}{20/3} = \frac{8}{3} \cdot \frac{3}{20} = \frac{8}{20} = \frac{2}{5} = \boxed{40\%}$ of the total.

8.45 Let x be the number of games the Unicorns played during the season. The Unicorns won 60% of the first $x - 10$ games, which is $60\% \cdot (x - 10)$ games. Combining this with the 8 games they won in the last 10 games gives a total of

$$60\% \cdot (x - 10) + 8 = 0.6(x - 10) + 8 = 0.6x - 6 + 8 = 0.6x + 2$$

wins. We also know that the Unicorns won 65% of their x games, which is a total of $65\% \cdot x = 0.65x$ wins. Setting this equal to our other expression for the total number of wins gives $0.65x = 0.6x + 2$, so $0.05x = 2$. Writing $0.05 = \frac{5}{100} = \frac{1}{20}$, we have $\frac{1}{20}x = 2$, so $x = \boxed{40}$.

8.46 A decrease of $a\%$ is a decrease of 100 to $100 - a$. A decrease of this by $b\%$ is a decrease of

$$\frac{b}{100} \cdot (100 - a) = \frac{b}{100} \cdot 100 - \frac{b}{100} \cdot a = b - \frac{ab}{100}.$$

Therefore, the result of the $b\%$ decrease is

$$100 - a - \left(b - \frac{ab}{100}\right) = 100 - a - b + \frac{ab}{100} = 100 - \left(a + b - \frac{ab}{100}\right).$$

So, the total decrease from 100 is $a + b - \frac{ab}{100}$. Since we started with 100, this is a $\boxed{a + b - \frac{ab}{100}}$ percent decrease.

8.47 Our original 500 grams of grapes consists of $(80\%) \cdot 500 = \frac{4}{5} \cdot 500 = 400$ grams of water and $500 - 400 = 100$ grams of pulp. After removing enough water to leave raisins, we still have 100 grams of pulp, which should be 80% of the weight of the raisins. Thus, the weight of the raisins is $100/80\% = 100/\frac{4}{5} = 100 \cdot \frac{5}{4} = \boxed{125 \text{ grams}}$.

CHAPTER 9

Square Roots

24® Cards

First Card $9 + 6 \cdot 5/2$.

Second Card $7 \cdot 5 - 9 - 2$.

Third Card $9 \cdot 2 \cdot 8/6$.

Fourth Card $6 + 9 \cdot (7 - 5)$.

Exercises for Section 9.1

9.1.1

(a) $\sqrt{196} = \sqrt{14^2} = \boxed{14}$. If we didn't recognize that 196 is 14^2, we might instead reason as follows: $\sqrt{196} = \sqrt{4 \cdot 49} = \sqrt{(2 \cdot 7)^2} = 2 \cdot 7 = 14$.

(b) $\sqrt{441} = \sqrt{21^2} = \boxed{21}$. If we didn't recognize that 441 is 21^2, we might instead reason as follows: $\sqrt{441} = \sqrt{9 \cdot 49} = \sqrt{(3 \cdot 7)^2} = 3 \cdot 7 = 21$.

(c) By the definition of square root, we have $\sqrt{37^2} = \boxed{37}$.

(d) $\sqrt{2^{12}} = \sqrt{2^{6 \cdot 2}} = \sqrt{(2^6)^2} = 2^6 = \boxed{64}$.

(e) $\sqrt{3600 \cdot 25} = \sqrt{60^2 \cdot 5^2} = \sqrt{(60 \cdot 5)^2} = 60 \cdot 5 = \boxed{300}$.

(f) $\sqrt{8 \cdot 6 \cdot 147} = \sqrt{(2^3) \cdot (2 \cdot 3) \cdot (3 \cdot 7^2)} = \sqrt{2^4 \cdot 3^2 \cdot 7^2} = \sqrt{(2^2 \cdot 3 \cdot 7)^2} = 2^2 \cdot 3 \cdot 7 = \boxed{84}$.

9.1.2 We start by finding the prime factorization of 1368900. We have $1368900 = 2^2 \cdot 3^4 \cdot 5^2 \cdot 13^2$, so

$$\sqrt{1368900} = \sqrt{2^2 \cdot 3^4 \cdot 5^2 \cdot 13^2} = \sqrt{(2 \cdot 3^2 \cdot 5 \cdot 13)^2} = 2 \cdot 3^2 \cdot 5 \cdot 13 = \boxed{1170}.$$

9.1.3 We have

$$\sqrt{(-7)^4} = \sqrt{((-1) \cdot 7)^4} = \sqrt{(-1)^4 \cdot 7^4} = \sqrt{7^4} = \sqrt{(7^2)^2} = 7^2 = \boxed{49}.$$

9.1.4 The quantity $\sqrt{n^2}$ equals the *nonnegative* number whose square is n^2. We know that squaring n results in n^2, but n is negative, so $\sqrt{n^2}$ cannot equal n. However, the square of $-n$ is also n^2, and $-n$ is positive when n is negative. So, if n is negative, then $\sqrt{n^2} = \boxed{-n}$.

9.1.5 We have $\sqrt{3^2 + 4^2} = \sqrt{9 + 16} = \sqrt{25} = \boxed{5}$. Note that the answer is NOT $3 + 4$.

9.1.6 If we square a number that ends in 1, the result ends in 1. So, it is impossible for 331 to be the square root of 110889. (For the record, $\sqrt{110889}$ is 333.)

9.1.7 Since 9 is the square root of 81, we know that the square of $\boxed{-9}$ is also 81.

9.1.8 Since $4^2 = 16$, we have $\sqrt{16} = 4$, which means $n = 16$. Since $n = 16$, we have $n^2 = \boxed{256}$.

9.1.9 We know that $2x + 1$ must be the number whose square root is 13. Since 13^2 is the number whose square root is 13, we must have $2x + 1 = 13^2$. Therefore, $2x = 13^2 - 1 = 168$, so $x = \frac{168}{2} = \boxed{84}$.

9.1.10 We know that $t^2 - 15$ must be the number whose square root is 7. Since 7^2 is the number whose square root is 7, we must have $t^2 - 15 = 49$. Adding 15 to both sides gives $t^2 = 64$. There are two numbers whose square is 64: $\boxed{8 \text{ and } -8}$. Both are solutions to the equation.

Exercises for Section 9.2

9.2.1

(a) We have $8^2 = 64$ and $9^2 = 81$, so $\sqrt{78}$ is between 8 and 9. Since $8.5^2 = 72.25$, we know that $8.5 < \sqrt{78} < 9$, so the nearest integer to $\sqrt{78}$ is $\boxed{9}$.

(b) We have $14^2 = 196$ and $15^2 = 225$, so $14 < \sqrt{200} < 15$, and we expect that $\sqrt{200}$ is closer to 14 than to 15. Since $14.5^2 = 210.25$, we have $14 < \sqrt{200} < 14.5$, so the closest integer to $\sqrt{200}$ is $\boxed{14}$.

(c) Since $60^2 = 3600$, we know that $\sqrt{4004}$ is greater than 60. Computing squares of numbers slightly larger than 60, we find that $63^2 = 3969$ and $64^2 = 4096$, so $63 < \sqrt{4004} < 64$. Since $63.5^2 = 4032.25$, we have $63 < \sqrt{4004} < 63.5$, and the integer closest to $\sqrt{4004}$ is $\boxed{63}$.

9.2.2 We have $2^2 < 7 < 3^2$, so $2 < \sqrt{7} < 3$. We also have $14^2 < 220 < 15^2$, so $14 < \sqrt{220} < 15$. Therefore, the integers between $\sqrt{7}$ and $\sqrt{220}$ are $3, 4, 5, 6, \ldots, 14$. There are $\boxed{12}$ numbers in this list.

9.2.3 Since $9^2 = 81$, we know that $\sqrt{83}$ is just a little bit more than 9. Since $6^2 = 36$, we know that $\sqrt{35}$ is just a little bit less than 6. When we subtract a number that is a little less than 6 from a number that is a little more than 9, we get a number that is a bit more than $\boxed{3}$. To make sure that $\sqrt{83} - \sqrt{35}$ isn't greater than 4, we can estimate both $\sqrt{83}$ and $\sqrt{35}$ more closely. We find that $9.1 < \sqrt{83} < 9.2$ and $5.9 < \sqrt{35} < 6$. Since $\sqrt{83}$ is less than 9.2 and $\sqrt{35}$ is greater than 5.9, their difference is less than $9.2 - 5.9 = 3.3$.

9.2.4 $\left(\sqrt{14}\right)^4 = \left(\sqrt{14}\right)^{2 \cdot 2} = \left(\left(\sqrt{14}\right)^2\right)^2 = (14)^2 = \boxed{196}$.

9.2.5 *Solution 1: Approximate $\sqrt{5}$.* We approximate $\sqrt{5}$ and then multiply the result by 4. We have $2.2^2 = 4.84$ and $2.3^2 = 5.29$, so $2.2 < \sqrt{5} < 2.3$. Multiplying by 4, we have $8.8 < 4\sqrt{5} < 9.2$. Since $4\sqrt{5}$ is between 8.8 and 9.2, the integer closest to $4\sqrt{5}$ is $\boxed{9}$.

Solution 2: Square $4\sqrt{5}$. We have $(4\sqrt{5})^2 = 4^2(\sqrt{5})^2 = 16(5) = 80$. Since $8^2 < (4\sqrt{5})^2 < 9^2$, we know that $4\sqrt{5}$ is between 8 and 9. Furthermore, we have $8.5^2 = 72.25$, so $4\sqrt{5}$ is between 8.5 and 9. Therefore, the integer closest to $4\sqrt{5}$ is $\boxed{9}$.

9.2.6 We have $10{,}000^2 = 100{,}000{,}000$ and $100{,}000^2 = 10{,}000{,}000{,}000$, so the square root of 108,868,356 is between 10,000 and 100,000. Therefore, it has $\boxed{5}$ digits. (For the record, $\sqrt{108{,}868{,}356} = 10{,}434$.)

9.2.7 Since $5^2 < 30 < 6^2$, we have $5 < \sqrt{30} < 6$. Since $7^2 < 50 < 8^2$, we have $7 < \sqrt{50} < 8$. Combining these, we know that $12 < \sqrt{30} + \sqrt{50} < 14$. But is $\sqrt{30} + \sqrt{50}$ less than or greater than 13? Since $\sqrt{50}$ is very close to 7 (because $7^2 = 49$) and $\sqrt{30}$ is between 5 and 6, we expect that $\sqrt{30} + \sqrt{50}$ is less than 13. To make sure, we notice that $5.5^2 = 30.25$, so $\sqrt{30} < 5.5$, and $7.1^2 = 50.41$, so $\sqrt{50} < 7.1$. Therefore, $\sqrt{30} + \sqrt{50} < 5.5 + 7.1 = 12.6$. So, $\sqrt{30} + \sqrt{50}$ is between $\boxed{12 \text{ and } 13}$.

9.2.8 We compare the three by writing each as a decimal, estimating where necessary. Since $\frac{68}{4}$ equals 17, we know that 50% of $\frac{68}{4}$ is 8.5. Next, dividing 9 into 75 gives $8.\overline{3}$. (We might also note $\frac{75}{9} = \frac{25}{3} = 8\frac{1}{3}$.) Finally, we note that $8.5^2 = 72.25$, so $8.5 < \sqrt{75}$. Therefore, $\boxed{\sqrt{75}}$ is the largest of the three numbers.

Exercises for Section 9.3

9.3.1

(a) $\sqrt{2} \cdot \sqrt{18} = \sqrt{2 \cdot 18} = \sqrt{36} = \boxed{6}$.

(b) $\sqrt{8} \cdot \sqrt{50} = \sqrt{8 \cdot 50} = \sqrt{400} = \boxed{20}$.

(c) $\sqrt{120} \cdot \sqrt{30} = \sqrt{120 \cdot 30} = \sqrt{3600} = \sqrt{36} \cdot \sqrt{100} = 6 \cdot 10 = \boxed{60}$.

(d) $\sqrt{6} \cdot \sqrt{15} \cdot \sqrt{10} = \sqrt{6 \cdot 15 \cdot 10} = \sqrt{900} = \boxed{30}$.

(e) We know that multiplying 50 by 2 will give us 100, which is a perfect square, so we regroup the numbers in the product accordingly:

$$\sqrt{50} \cdot \sqrt{6} \cdot \sqrt{27} = \sqrt{50 \cdot 6 \cdot 27} = \sqrt{300 \cdot 27} = \sqrt{100 \cdot 3 \cdot 27} = \sqrt{100} \cdot \sqrt{81} = 10 \cdot 9 = \boxed{90}.$$

(f) $2^3 + \sqrt{32} \cdot 2\sqrt{2} \div 8 = 8 + \frac{2\sqrt{32 \cdot 2}}{8} = 8 + \frac{2\sqrt{64}}{8} = 8 + \frac{2(8)}{8} = 8 + 2 = \boxed{10}$.

(g) $\sqrt{\frac{1}{9} + \frac{1}{16}} = \sqrt{\frac{16}{9 \cdot 16} + \frac{9}{9 \cdot 16}} = \sqrt{\frac{16+9}{9 \cdot 16}} = \sqrt{\frac{25}{144}} = \frac{\sqrt{25}}{\sqrt{144}} = \boxed{\frac{5}{12}}$.

(h) $\sqrt{2\frac{1}{4}} + \sqrt{1\frac{7}{9}} = \sqrt{\frac{9}{4}} + \sqrt{\frac{16}{9}} = \frac{\sqrt{9}}{\sqrt{4}} + \frac{\sqrt{16}}{\sqrt{9}} = \frac{3}{2} + \frac{4}{3} = \frac{9}{6} + \frac{8}{6} = \frac{17}{6} = \boxed{2\frac{5}{6}}$.

(i) We have $9^2 = 81$, and squaring a number with three digits after the decimal point gives a number with 6 digits after the decimal point. So, we have $0.009^2 = 0.000081$, which means $\sqrt{0.000081} = \boxed{0.009}$.

9.3.2 $\sqrt{64t^{64}} = \sqrt{64} \cdot \sqrt{t^{64}} = 8\sqrt{t^{32 \cdot 2}} = 8\sqrt{(t^{32})^2} = \boxed{8t^{32}}$. (Note that t^{32} must be nonnegative, so we can write $\sqrt{(t^{32})^2} = t^{32}$.)

The hardest of these steps is realizing that $\sqrt{t^{64}}$ is t^{32} instead of t^8. If you're not convinced, try applying exponent laws to both $(t^{32})^2$ and $(t^8)^2$ to see which equals t^{64}.

9.3.3 We have $\sqrt{250} = \sqrt{25 \cdot 10} = \sqrt{25} \cdot \sqrt{10} = 5\sqrt{10}$. Since $\sqrt{10}$ is approximately 3.16, we expect $5\sqrt{10}$ to be approximately $5(3.16) = 15.8$. Let's make sure we have the nearest tenth, and that we haven't made a rounding error. Since $\sqrt{10}$ to the nearest hundredth is 3.16, we know that $3.155 \le \sqrt{10} < 3.165$. Multiplying by 5, we see that $15.775 \le 5\sqrt{10} < 15.825$. So, $5\sqrt{10}$ approximated to the nearest tenth is $\boxed{15.8}$.

9.3.4 $\sqrt{x} \cdot \sqrt{z} = \sqrt{xz} = \sqrt{\frac{5}{27} \cdot \frac{5}{3}} = \sqrt{\frac{25}{81}} = \frac{\sqrt{25}}{\sqrt{81}} = \boxed{\frac{5}{9}}$.

9.3.5 We have

$$A = \sqrt{1.44} = \sqrt{(1.2)^2} = 1.2,$$
$$B = \frac{13}{11} = 1\frac{2}{11} = 1.\overline{18},$$
$$C = \sqrt{8} - 2\sqrt{2} = \sqrt{4 \cdot 2} - 2\sqrt{2} = 2\sqrt{2} - 2\sqrt{2} = 0,$$
$$D = \frac{3}{5} + \frac{3}{4} = \frac{12}{20} + \frac{15}{20} = \frac{27}{20} = 1.35.$$

So, from least to greatest, we have $\boxed{C, B, A, D}$.

9.3.6

(a) $\sqrt{363} = \sqrt{121 \cdot 3} = \sqrt{121} \cdot \sqrt{3} = \boxed{11\sqrt{3}}$.

(b) $\sqrt{525} = \sqrt{25 \cdot 21} = \sqrt{25} \cdot \sqrt{21} = \boxed{5\sqrt{21}}$.

(c) We start with the prime factorization of 3168. We have

$$\sqrt{3168} = \sqrt{2^5 \cdot 3^2 \cdot 11} = \sqrt{2^4} \cdot \sqrt{3^2} \cdot \sqrt{2 \cdot 11} = 4 \cdot 3 \cdot \sqrt{22} = \boxed{12\sqrt{22}}.$$

9.3.7 We have $\sqrt{75} = \sqrt{25 \cdot 3} = 5\sqrt{3}$ and $\sqrt{27} = \sqrt{9 \cdot 3} = 3\sqrt{3}$, so

$$3\sqrt{75} + 2\sqrt{27} = 3(5\sqrt{3}) + 2(3\sqrt{3}) = 15\sqrt{3} + 6\sqrt{3} = \boxed{21\sqrt{3}}.$$

9.3.8 First, we group the constants and we group the x terms:

$$\sqrt{75x} \cdot \sqrt{2x} \cdot \sqrt{14x} = \sqrt{(75x)(2x)(14x)} = \sqrt{(75 \cdot 2 \cdot 14)(x \cdot x \cdot x)} = \sqrt{75 \cdot 2 \cdot 14} \cdot \sqrt{x^3}.$$

We have

$$\sqrt{75 \cdot 2 \cdot 14} = \sqrt{(3 \cdot 5^2)(2)(2 \cdot 7)} = \sqrt{2^2 \cdot 3 \cdot 5^2 \cdot 7} = \sqrt{2^2} \cdot \sqrt{5^2} \cdot \sqrt{3 \cdot 7} = (2)(5)\sqrt{21} = 10\sqrt{21},$$

and

$$\sqrt{x^3} = \sqrt{x^2} \cdot \sqrt{x}.$$

Since we are given that x is nonnegative, we have $\sqrt{x^2} = x$, so $\sqrt{x^3} = x\sqrt{x}$. Therefore, we have

$$\sqrt{75 \cdot 2 \cdot 14} \cdot \sqrt{x^3} = (10\sqrt{21})(x\sqrt{x}) = \boxed{10x\sqrt{21x}}.$$

9.3.9 We have

$$\frac{\sqrt{375} + \sqrt{60}}{\sqrt{5}} = \frac{\sqrt{375}}{\sqrt{5}} + \frac{\sqrt{60}}{\sqrt{5}} = \sqrt{\frac{375}{5}} + \sqrt{\frac{60}{5}} = \sqrt{75} + \sqrt{12} = \sqrt{25 \cdot 3} + \sqrt{4 \cdot 3} = 5\sqrt{3} + 2\sqrt{3} = \boxed{7\sqrt{3}}.$$

Review Problems

9.23

(a) $\sqrt{(27)(12)} = \sqrt{(3^3)(3 \cdot 4)} = \sqrt{3^4} \cdot \sqrt{4} = \sqrt{81} \cdot 2 = 9 \cdot 2 = \boxed{18}$.

(b) $\sqrt{2 \cdot 18 \cdot 40 \cdot 10} = \sqrt{2 \cdot 18} \cdot \sqrt{40 \cdot 10} = \sqrt{36} \cdot \sqrt{400} = 6 \cdot 20 = \boxed{120}$.

(c) $\sqrt{7 \cdot 2} \cdot \sqrt{2^3 \cdot 7^3} = \sqrt{7 \cdot 2 \cdot 2^3 \cdot 7^3} = \sqrt{2^4 \cdot 7^4} = \sqrt{(2^2 \cdot 7^2)^2} = 2^2 \cdot 7^2 = \boxed{196}$.

(d) $\sqrt{24} \cdot 2\sqrt{54} = \sqrt{4 \cdot 6} \cdot 2\sqrt{9 \cdot 6} = 2\sqrt{6} \cdot 2 \cdot 3\sqrt{6} = (2 \cdot 2 \cdot 3)\left(\sqrt{6} \cdot \sqrt{6}\right) = 12(6) = \boxed{72}$.

(e) $\sqrt{3} \cdot \sqrt{5} \cdot \sqrt{15} = \sqrt{15} \cdot \sqrt{15} = \boxed{15}$.

(f) $\sqrt{24} \cdot \sqrt{18} \cdot \sqrt{12} = \sqrt{4 \cdot 6} \cdot \sqrt{9 \cdot 2} \cdot \sqrt{4 \cdot 3} = 2\sqrt{6} \cdot 3\sqrt{2} \cdot 2\sqrt{3} = (2 \cdot 3 \cdot 2)\left(\sqrt{6} \cdot \sqrt{2} \cdot \sqrt{3}\right) = 12\sqrt{6 \cdot 2 \cdot 3} = 12\sqrt{6^2} = \boxed{72}$.

We also could have combined the three original numbers under a single radical in our first step:

$$\sqrt{24} \cdot \sqrt{18} \cdot \sqrt{12} = \sqrt{24 \cdot 18 \cdot 12} = \sqrt{4 \cdot 6 \cdot 6 \cdot 3 \cdot 3 \cdot 4} = \sqrt{4^2 \cdot 6^2 \cdot 3^2} = \sqrt{(4 \cdot 6 \cdot 3)^2} = 4 \cdot 6 \cdot 3 = \boxed{72}.$$

Furthermore, we could have noticed that $24 = 2 \cdot 12$ to make the computation easier:

$$\sqrt{24} \cdot \sqrt{18} \cdot \sqrt{12} = \left(\sqrt{2} \cdot \sqrt{12}\right) \cdot \sqrt{18} \cdot \sqrt{12} = \left(\sqrt{2} \cdot \sqrt{18}\right) \cdot \left(\sqrt{12} \cdot \sqrt{12}\right) = \sqrt{36} \cdot 12 = \boxed{72}.$$

(g) $\sqrt{5\frac{4}{9}} = \sqrt{\frac{49}{9}} = \frac{\sqrt{49}}{\sqrt{9}} = \frac{7}{3} = \boxed{2\frac{1}{3}}$.

(h) $\sqrt{12\frac{1}{4}} = \sqrt{\frac{49}{4}} = \frac{\sqrt{49}}{\sqrt{4}} = \frac{7}{2} = \boxed{3\frac{1}{2}}$.

(i) $\sqrt{2.89} = \sqrt{\frac{289}{100}} = \frac{\sqrt{289}}{\sqrt{100}} = \frac{17}{10} = \boxed{1.7}$.

(j) $\frac{\sqrt{24}}{\sqrt{30}} \div \frac{\sqrt{20}}{3\sqrt{25}} = \frac{\sqrt{24}}{\sqrt{30}} \cdot \frac{3\sqrt{25}}{\sqrt{20}} = \sqrt{\frac{24}{30}} \cdot 3 \cdot \sqrt{\frac{25}{20}} = \sqrt{\frac{4}{5}} \cdot 3 \cdot \sqrt{\frac{5}{4}} = 3 \cdot \sqrt{\frac{4}{5} \cdot \frac{5}{4}} = 3\sqrt{1} = \boxed{3}$.

(k) $\sqrt{3^5 + 3^5 + 3^5} = \sqrt{3^5(1 + 1 + 1)} = \sqrt{3^5(3)} = \sqrt{3^6} = \sqrt{(3^3)^2} = 3^3 = \boxed{27}$.

(l) $\sqrt{5^5 + 5^5 + 5^5 + 5^5 + 5^5} = \sqrt{5^5(1 + 1 + 1 + 1 + 1)} = \sqrt{5^5(5)} = \sqrt{5^6} = \sqrt{(5^3)^2} = 5^3 = \boxed{125}$.

9.24 $\sqrt{5^3 - 2^2} = \sqrt{125 - 4} = \sqrt{121} = \boxed{11}$.

9.25 First, we evaluate $\sqrt{1296}$ by finding the prime factorization of 1296. We find that $1296 = 2^4 \cdot 3^4$, so $\sqrt{1296} = \sqrt{2^4 \cdot 3^4} = \sqrt{(2^2 \cdot 3^2)^2} = 2^2 \cdot 3^2 = 36$. So, we have $\sqrt{28 + \sqrt{1296}} = \sqrt{28 + 36} = \sqrt{64} = \boxed{8}$.

9.26 The nonnegative perfect cubes less than 100 are $0^3 = 0$, $1^3 = 1$, $2^3 = 8$, $3^3 = 27$, and $4^3 = 64$. (The next largest, 5^3, has three digits.) We don't have to worry about the negative cubes, because negative numbers do not have integer square roots. Of these perfect cubes, only $\boxed{0, 1, \text{ and } 64}$ have integer square roots. These three numbers are perfect sixth powers. Is that a coincidence?

9.27 Both 4 and -4 have squares equal to 16, so the sum of the possible values of x is $4 + (-4) = \boxed{0}$.

9.28 We have $\sqrt{n} = \sqrt{81} - \sqrt{16} = 9 - 4 = 5$. Since the square root of n is 5, we have $n = 5^2 = \boxed{25}$.

9.29

(a) Since $\sqrt{9 + 4y} = 11$, the value of $9 + 4y$ is the number whose square root is 11, which is $11^2 = 121$. Therefore, we have $9 + 4y = 121$. Subtracting 9 from both sides gives $4y = 112$, and dividing by 4 gives $y = \boxed{28}$.

(b) Subtracting 6 from both sides gives $-\sqrt{z + 1} = 3$. Multiplying both sides by -1 gives $\sqrt{z + 1} = -3$. By the definition of square root, the expression $\sqrt{z + 1}$ must be nonnegative. So, the equation $\sqrt{z + 1} = -3$ has no solutions, which means there are $\boxed{\text{no solutions}}$ to the original equation.

9.30 We have $4^2 = 16$ and $5^2 = 25$, so $4 < \sqrt{23} < 5$. To make sure that $\sqrt{23}$ is closer to 5 than to 4, we note that $4.5^2 = 20.25$, so $4.5 < \sqrt{23} < 5$. But the question asks about $-\sqrt{23}$. Since $\sqrt{23}$ is between 4.5 and 5, we know that $-\sqrt{23}$ is between -4.5 and -5. This means that the integer closest to $-\sqrt{23}$ is $\boxed{-5}$.

9.31

(a) We have $10^2 = 100$, which is less than 101. So 10 is less than $\sqrt{101}$, which means that $10 - \sqrt{101}$ is $\boxed{\text{negative}}$.

(b) We have $10^2 = 100$ and $\left(3\sqrt{11}\right)^2 = 3^2\left(\sqrt{11}\right)^2 = 9(11) = 99$, so $10 > 3\sqrt{11}$, which means that $10 - 3\sqrt{11}$ is $\boxed{\text{positive}}$.

(c) To determine which is larger, $4\sqrt{33}$ or $5\sqrt{21}$, we square both. We have

$$\left(4\sqrt{33}\right)^2 = 4^2\left(\sqrt{33}\right)^2 = 16(33) = 528,$$
$$\left(5\sqrt{21}\right)^2 = 5^2\left(\sqrt{21}\right)^2 = 25(21) = 525,$$

so $4\sqrt{33}$ is greater than $5\sqrt{21}$, which means that $4\sqrt{33} - 5\sqrt{21}$ is $\boxed{\text{positive}}$.

9.32 Since $6^2 = 36$ and $7^2 = 49$, we have $6 < \sqrt{37} < 7$. To figure out what two consecutive integers $5\sqrt{11}$ is between, we start by squaring $5\sqrt{11}$. We have $\left(5\sqrt{11}\right)^2 = 5^2\left(\sqrt{11}\right)^2 = 25(11) = 275$. Since $16^2 = 256$ and $17^2 = 289$, we have $16 < 5\sqrt{11} < 17$. So, the integers between $\sqrt{37}$ and $5\sqrt{11}$ are $7, 8, 9, \ldots, 16$. There are $\boxed{10}$ numbers in this list.

9.33 At the top of the Empire State Building, we have $h = 1250$. Putting this value in the formula $d = \sqrt{1.5h}$, we have $d = \sqrt{1.5(1250)} = \sqrt{1875}$. We now must approximate $\sqrt{1875}$ to the nearest integer. We have $43^2 = 1849$ and $44^2 = 1936$, so $43 < \sqrt{1875} < 44$ and we expect that $\sqrt{1875}$ is closer to 43 than to 44. To make sure, we compute $43.5^2 = 1892.25$, which means that $43 < \sqrt{1875} < 43.5$. Therefore, to the nearest mile, you can see $\boxed{43}$ miles to the horizon.

9.34 We have $8^2 = 64$ and $9^2 = 81$, so $8 < \sqrt{80} < 9$. We also have $10^2 = 100$ and $11^2 = 121$, so $10 < \sqrt{120} < 11$. Since $\sqrt{80}$ is between 8 and 9, and $\sqrt{120}$ is between 10 and 11, the sum $\sqrt{80} + \sqrt{120}$ is between $8 + 10$ and $9 + 11$. This tells us that $18 < \sqrt{80} + \sqrt{120} < 20$. But is $\sqrt{80} + \sqrt{120}$ greater than or less than 19?

Since $9^2 = 81$ and $11^2 = 121$, we know that $\sqrt{80}$ is a tiny bit less than 9 and $\sqrt{120}$ is a tiny bit less than 11. So, the sum $\sqrt{80} + \sqrt{120}$ is a tiny bit less than $9 + 11$, which is 20. To make sure that $\sqrt{80} + \sqrt{120}$ is greater than 19, we can compute $8.5^2 = 72.25$ and $10.5^2 = 110.25$. So, $\sqrt{80}$ is greater than 8.5 and $\sqrt{120}$ is greater than 10.5. This means that $\sqrt{80} + \sqrt{120}$ is between 19 and 20. Therefore, the greatest integer that is less than $\sqrt{80} + \sqrt{120}$ is $\boxed{19}$.

9.35 We order the numbers by ordering their squares:

$$15^2 = 225,$$
$$\left(4\sqrt{14}\right)^2 = 4^2\left(\sqrt{14}\right)^2 = 16(14) = 224,$$
$$\left(3\sqrt{26}\right)^2 = 3^2\left(\sqrt{26}\right)^2 = 9(26) = 234,$$
$$\left(6\sqrt{6}\right)^2 = 6^2\left(\sqrt{6}\right)^2 = 36(6) = 216.$$

So, in order from least to greatest, we have $\boxed{6\sqrt{6}, 4\sqrt{14}, 15, 3\sqrt{26}}$.

9.36 Since $6^2 = 36$ and $7^2 = 49$, we have $6 < \sqrt{42.3} < 7$. Since 42.3 is closer to 6^2 than to 7^2, we might expect that $\sqrt{42.3}$ is closer to 6 than to 7. But to make sure, we compute $6.5^2 = 42.25$, which is less than 42.3! So, we have $6.5 < \sqrt{42.3} < 7$, which means that $\sqrt{42.3}$ is closer to $\boxed{7}$ than to 6, even though 42.3 is closer to 6^2 than to 7^2.

9.37 We have

$$\frac{3\sqrt{3}}{12\sqrt{12}} = \frac{3}{12} \cdot \frac{\sqrt{3}}{\sqrt{12}} = \frac{1}{4} \cdot \sqrt{\frac{3}{12}} = \frac{1}{4} \cdot \sqrt{\frac{1}{4}} = \frac{1}{4} \cdot \frac{\sqrt{1}}{\sqrt{4}} = \frac{1}{4} \cdot \frac{1}{2} = \frac{1}{8}.$$

So, $3\sqrt{3}$ is $\frac{1}{8}$ of $12\sqrt{12}$. To convert $\frac{1}{8}$ to a percent, we solve $\frac{x}{100} = \frac{1}{8}$. Multiplying both sides by 100 gives $x = \frac{1}{8} \cdot 100 = \frac{100}{8} = 12.5$. Therefore, $3\sqrt{3}$ is $\boxed{12.5\%}$ of $12\sqrt{12}$.

9.38

(a) $\sqrt{360} = \sqrt{36} \cdot \sqrt{10} = \boxed{6\sqrt{10}}$.

(b) $\sqrt{936} = \sqrt{9} \cdot \sqrt{104} = 3\sqrt{4}\sqrt{26} = 3 \cdot 2\sqrt{26} = \boxed{6\sqrt{26}}$.

(c) We start by finding the prime factorization of 10164. We have $10164 = 2^2 \cdot 3 \cdot 7 \cdot 11^2$, so

$$\sqrt{10164} = \sqrt{2^2 \cdot 3 \cdot 7 \cdot 11^2} = \sqrt{2^2} \cdot \sqrt{11^2} \cdot \sqrt{3 \cdot 7} = 2 \cdot 11 \cdot \sqrt{21} = \boxed{22\sqrt{21}}.$$

9.39 We start by simplifying both radicals. We have $\sqrt{98} = \sqrt{49} \cdot \sqrt{2} = 7\sqrt{2}$ and $\sqrt{50} = \sqrt{25} \cdot \sqrt{2} = 5\sqrt{2}$. Therefore, we have $\sqrt{98} - \sqrt{50} = 7\sqrt{2} - 5\sqrt{2} = 2\sqrt{2}$. In the text, we found that $\sqrt{2}$ to the nearest tenth is 1.4, so $2\sqrt{2}$ is approximately 2.8. This means the closest integer to $\sqrt{98} - \sqrt{50}$ is $\boxed{3}$. (We might also have noticed that $2\sqrt{2} = \sqrt{8}$, and $\sqrt{8}$ is closer to 3 than to 2.)

9.40 The value of $k\sqrt{5}$ is less than 10 if $\left(k\sqrt{5}\right)^2$ is less than 10^2. We have $\left(k\sqrt{5}\right)^2 = k^2\left(\sqrt{5}\right)^2 = 5k^2$. We have $5k^2 < 100$ if $k^2 < 20$. The only positive integers whose squares are less than 20 are 1, 2, 3, and 4. Therefore, there are $\boxed{4}$ positive integer values of k such that $k\sqrt{5}$ is less than 10.

9.41 We have $\sqrt{60} = \sqrt{4} \cdot \sqrt{15} = 2\sqrt{15}$ and $\sqrt{135} = \sqrt{9} \cdot \sqrt{15} = 3\sqrt{15}$. Therefore, we have

$$4\sqrt{60} - 2\sqrt{135} = 4(2\sqrt{15}) - 2(3\sqrt{15}) = 8\sqrt{15} - 6\sqrt{15} = \boxed{2\sqrt{15}}.$$

9.42 We could use the distributive property, but that looks pretty scary! Instead, let's try simplifying $\sqrt{27}$ and $\sqrt{75}$ and hope something convenient happens. We have $\sqrt{27} = \sqrt{9} \cdot \sqrt{3} = 3\sqrt{3}$ and $\sqrt{75} = \sqrt{25} \cdot \sqrt{3} = 5\sqrt{3}$, so

$$\left(\sqrt{3} - \sqrt{27} + \sqrt{75}\right)^2 = \left(\sqrt{3} - 3\sqrt{3} + 5\sqrt{3}\right)^2 = \left(3\sqrt{3}\right)^2 = 3^2\left(\sqrt{3}\right)^2 = 9(3) = \boxed{27}.$$

Challenge Problems

9.43

(a) Since t squared is 9^6 and t is positive, we know that t is the square root of 9^6. We then have

$$t = \sqrt{9^6} = \sqrt{(9^3)^2} = 9^3 = \boxed{729}.$$

(b) As in the first part, t must be the square root of 9^5. At first, it doesn't look like 9^5 is a perfect square. However, 9 is a perfect square, and we have

$$t = \sqrt{9^5} = \sqrt{(3^2)^5} = \sqrt{3^{10}} = \sqrt{(3^5)^2} = 3^5 = \boxed{243}.$$

9.44 Since $\frac{1}{\sqrt{z}} = 5$, we know that $\sqrt{z}$ and 5 are reciprocals. The reciprocal of 5 is $\frac{1}{5}$, so we must have $\sqrt{z} = \frac{1}{5}$. Since $\left(\frac{1}{5}\right)^2 = \frac{1}{25}$, the number whose square root is $\frac{1}{5}$ is $z = \boxed{\frac{1}{25}}$.

9.45 The value of $\sqrt{r + 200}$ is a positive integer if and only if $r + 200$ is a positive perfect square. Since r is negative, we know that $r + 200$ is less than 200. Therefore, in order for r to be negative and $\sqrt{r + 200}$ to be a positive integer, the value of $r + 200$ must be a positive perfect square less than 200. The largest square less than 200 is $14^2 = 196$. Letting $r = -4$ gives us $\sqrt{r + 200} = \sqrt{196} = 14$. Similarly, for each perfect square less than 200, there is a negative value of r such that $r + 200$ equals that perfect square, so $\sqrt{r + 200}$ is an integer. Since 14^2 is the largest perfect square less than 200, there are $\boxed{14}$ positive perfect squares less than 200, and hence 14 negative values of r such that $\sqrt{r + 200}$ is a positive integer.

9.46

(a) The geometric mean of 24 and 150 is the square root of their product, $\sqrt{24 \cdot 150}$. Since

$$\sqrt{24 \cdot 150} = \sqrt{4 \cdot 6 \cdot 25 \cdot 6} = \sqrt{4} \cdot \sqrt{25} \cdot \sqrt{6 \cdot 6} = 2 \cdot 5 \cdot 6 = 60,$$

the geometric mean of 24 and 150 is $\boxed{60}$.

(b) $\boxed{\text{Yes}}$. For example, the geometric mean of $\sqrt{3}$ and $\sqrt{27}$ is

$$\sqrt{\sqrt{3} \cdot \sqrt{27}} = \sqrt{\sqrt{81}} = \sqrt{9} = 3.$$

As another example, the geometric mean of $\frac{3}{2}$ and $\frac{2}{3}$ is $\sqrt{\frac{3}{2} \cdot \frac{2}{3}} = \sqrt{1} = 1$.

9.47 If the square root of x^3 is 27, then x^3 must equal 27^2. Since $27 = 3^3$, we have $(27)^2 = (3^3)^2 = 3^{3 \cdot 2} = 3^{2 \cdot 3} = (3^2)^3 = 9^3$. Therefore, if $x^3 = 27^2$, then $x^3 = 9^3$, which means $x = \boxed{9}$.

9.48 The numerator of the first fraction cancels with the denominator of the third fraction, the numerator of the second fraction cancels with the denominator of the fourth fraction, and so on. The only terms that don't cancel are the first two denominators and the last two numerators, leaving

$$\sqrt{\frac{(n+1)(n+2)}{(1)(2)}}.$$

This square root is an integer if and only if $\frac{(n+1)(n+2)}{2}$ is a perfect square, so now we want the smallest positive integer n for which $\frac{(n+1)(n+2)}{2}$ is a perfect square. The numbers 1 through 6 do not produce a perfect square, but $n = 7$ does. Therefore, $n = \boxed{7}$ is the first positive integer for which $\frac{(n+1)(n+2)}{2}$ is a perfect square.

9.49 We present two solutions.

Solution 1: Notice that 3 divides 9 evenly. Seeing that 3 divides evenly into 9, we write 9 as $\sqrt{81}$ and use the properties of square roots:

$$\frac{9}{2\sqrt{3}} = \frac{\sqrt{81}}{2\sqrt{3}} = \frac{1}{2} \cdot \frac{\sqrt{81}}{\sqrt{3}} = \frac{1}{2}\sqrt{\frac{81}{3}} = \frac{1}{2}\sqrt{27} = \frac{1}{2}\sqrt{9 \cdot 3} = \boxed{\frac{3\sqrt{3}}{2}}.$$

Solution 2: Strategically multiply by 1. Since $\sqrt{3} \cdot \sqrt{3} = 3$, we can get rid of the square root in the denominator by multiplying the fraction by $\frac{\sqrt{3}}{\sqrt{3}}$, which equals 1. Multiplying a number by 1 doesn't change its value, so we have

$$\frac{9}{2\sqrt{3}} = \frac{9}{2\sqrt{3}} \cdot \frac{\sqrt{3}}{\sqrt{3}} = \frac{9\sqrt{3}}{2\sqrt{3} \cdot \sqrt{3}} = \frac{9\sqrt{3}}{2 \cdot 3} = \frac{9\sqrt{3}}{6} = \boxed{\frac{3\sqrt{3}}{2}}.$$

9.50 We can use the second strategy from the previous problem. First, we have

$$\sqrt{\frac{1}{5}} = \frac{\sqrt{1}}{\sqrt{5}} = \frac{1}{\sqrt{5}}.$$

Next, we multiply by $\frac{\sqrt{5}}{\sqrt{5}}$, and we have

$$\frac{1}{\sqrt{5}} \cdot \frac{\sqrt{5}}{\sqrt{5}} = \frac{\sqrt{5}}{5}.$$

Since $\sqrt{5}$ is approximately 2.236, we divide 2.236 by 5 to see that $\frac{\sqrt{5}}{5}$ is approximately $\boxed{0.45}$ to the nearest hundredth.

9.51

(a) $4^x \cdot 4^x = 4^{x+x} = \boxed{4^{2x}}$.

(b) If $4^x = 2$, then $4^x \cdot 4^x = 2 \cdot 2 = \boxed{4}$.

(c) In the first two parts, we found that $4^x \cdot 4^x$ equals both 4^{2x} and 4. Therefore, we must have $4^{2x} = 4$, which means $4^{2x} = 4^1$. This equation holds if $2x = 1$, or $x = \boxed{\frac{1}{2}}$.

In general, $n^{1/2}$ is another way of writing $\sqrt{n}$. You'll learn more about fractional exponents in Art of Problem Solving's *Introduction to Algebra*.

9.52

(a) Since $2^3 = 8$, we have $\sqrt[3]{8} = \boxed{2}$.

(b) The prime factorization of 216 is $2^3 \cdot 3^3$, so we see that $216 = 6^3$, which means that $\sqrt[3]{216} = \boxed{6}$.

(c) Yes. Since $(-10)^3 = -1000$, we have $\sqrt[3]{-1000} = \boxed{-10}$. Since the cube of a negative number is negative, we can define cube roots of negative numbers.

(d) We have $0^3 = 0$ and $1^3 = 1$, so $\sqrt[3]{0} = 0$ and $\sqrt[3]{1} = 1$. Every number greater than 1 has a cube that is greater than itself. We also have to check negative numbers. We have $(-1)^3 = -1$, but every number less than -1 has a cube that is less than itself. Therefore, the only integers that equal their own cube roots are $\boxed{-1, 0, 1}$.

(e) Since $3^4 = 81$, we have $\sqrt[4]{81} = \boxed{3}$.

(f) The prime factorization of 256 is 2^8. So, we see that $256 = 2^8 = (2^2)^4$, which means $\sqrt[4]{256} = 2^2 = \boxed{4}$.

9.53

(a) Applying the distributive property gives

$$\left(\sqrt{11} - \sqrt{7}\right)\left(\sqrt{11} + \sqrt{7}\right) = \sqrt{11}\left(\sqrt{11} + \sqrt{7}\right) - \sqrt{7}\left(\sqrt{11} + \sqrt{7}\right)$$
$$= \sqrt{11}\sqrt{11} + \sqrt{11}\sqrt{7} - \sqrt{7}\sqrt{11} - \sqrt{7}\sqrt{7}$$
$$= 11 + \sqrt{77} - \sqrt{77} - 7$$
$$= \boxed{4}.$$

Notice that there are no radicals in our final answer.

(b) In the previous part, when we multiplied an expression of the form $\sqrt{x} - \sqrt{y}$ by $\sqrt{x} + \sqrt{y}$, the square root terms canceled out. With this in mind, we expect that multiplying $\sqrt{5} - \sqrt{2}$ by $\sqrt{5} + \sqrt{2}$ will eliminate the square roots. Indeed, we find

$$\left(\sqrt{5} - \sqrt{2}\right)\left(\sqrt{5} + \sqrt{2}\right) = \sqrt{5}\left(\sqrt{5} + \sqrt{2}\right) - \sqrt{2}\left(\sqrt{5} + \sqrt{2}\right)$$
$$= \sqrt{5}\sqrt{5} + \sqrt{5}\sqrt{2} - \sqrt{2}\sqrt{5} - \sqrt{2}\sqrt{2}$$
$$= 5 + \sqrt{10} - \sqrt{10} - 2$$
$$= 3.$$

So, to write $\frac{1}{\sqrt{5}-\sqrt{2}}$ without any square roots in the denominator, we multiply by $\frac{\sqrt{5}+\sqrt{2}}{\sqrt{5}+\sqrt{2}}$. This fraction equals 1, so it doesn't change the value of our original number when we multiply by it:

$$\frac{1}{\sqrt{5} - \sqrt{2}} = \frac{1}{\sqrt{5} - \sqrt{2}} \cdot \frac{\sqrt{5} + \sqrt{2}}{\sqrt{5} + \sqrt{2}} = \frac{\sqrt{5} + \sqrt{2}}{\left(\sqrt{5} - \sqrt{2}\right)\left(\sqrt{5} + \sqrt{2}\right)} = \boxed{\frac{\sqrt{5} + \sqrt{2}}{3}}.$$

9.54 We could simply plug in higher and higher numbers for k and hope we get lucky, but that could take a long time. Instead, we note that $84k$ must be a perfect square in order for $\sqrt{84k}$ to be an integer. So, we find the prime factorization of 84, and then think about what prime factors k needs in order to make $84k$ a perfect square. We have $84 = 2^2 \cdot 3 \cdot 7$, so $84k = 2^2 \cdot 3 \cdot 7 \cdot k$. Therefore, k needs a factor of 3 and a factor of 7 in order to make $84k$ a perfect square. Letting $k = 3 \cdot 7 = \boxed{21}$ gives $\sqrt{84k} = 2\sqrt{21k} = 2\sqrt{21 \cdot 21} = 2 \cdot 21 = 42$.

9.55 First, we subtract 7^2 from both sides, and we have

$$\frac{1}{x^2} = 25^2 - 7^2 = 625 - 49 = 576.$$

Since $\frac{1}{x^2} = 576$, we know that 576 is the reciprocal of x^2. Therefore, x^2 is the reciprocal of 576, which means $x^2 = \frac{1}{576}$. Since $576 = 24^2$, we have $x^2 = \frac{1}{24^2} = \left(\frac{1}{24}\right)^2$. The two values of x that satisfy this equation are $\boxed{\frac{1}{24} \text{ and } -\frac{1}{24}}$.

9.56 First, let's make the giant fraction on the left side simpler:

$$\frac{\left(\frac{1}{2}\right)^2 + \left(\frac{1}{3}\right)^2}{\left(\frac{1}{4}\right)^2 + \left(\frac{1}{5}\right)^2} = \frac{\frac{1}{4} + \frac{1}{9}}{\frac{1}{16} + \frac{1}{25}} = \frac{\frac{9}{36} + \frac{4}{36}}{\frac{25}{400} + \frac{16}{400}} = \frac{\frac{13}{36}}{\frac{41}{400}} = \frac{13}{36} \cdot \frac{400}{41} = \frac{(13)(400)}{(36)(41)}.$$

So, now the equation is simply $\dfrac{13x}{41y} = \dfrac{(13)(400)}{(41)(36)}$, or

$$\frac{13}{41} \cdot \frac{x}{y} = \frac{13}{41} \cdot \frac{400}{36}.$$

Multiplying both sides by $\frac{41}{13}$ eliminates the $\frac{13}{41}$ on both sides and leaves

$$\frac{x}{y} = \frac{400}{36} = \frac{100}{9}.$$

We seek $\sqrt{x} \div \sqrt{y}$. Since $\sqrt{x} \div \sqrt{y} = \frac{\sqrt{x}}{\sqrt{y}} = \sqrt{\frac{x}{y}}$, and $\frac{x}{y} = \frac{100}{9}$, we have

$$\sqrt{\frac{x}{y}} = \sqrt{\frac{100}{9}} = \frac{\sqrt{100}}{\sqrt{9}} = \boxed{\frac{10}{3}}.$$

9.57 The expression is not defined if $\frac{x+1}{x-1}$ is negative, or if the denominator is 0. The denominator is 0 only when $x = 1$. Since $x + 1$ is always 2 greater than $x - 1$, the expression $\frac{x+1}{x-1}$ can only be negative when $x + 1$ is positive and $x - 1$ is negative. We have $x + 1 > 0$ when $x > -1$, and we have $x - 1 < 0$ when $x < 1$. So we have $x + 1$ positive and $x - 1$ negative when $-1 < x < 1$ (that is, when x is between -1 and 1). Combining this with when the denominator is 0, we see that the original expression is not defined when $\boxed{-1 < x \le 1}$.

9.58 The number whose square root is 9 is $9^2 = 81$. So, we must have $(r - 3)^2 = 81$. Here, we must be careful. Both 9 and -9 have a square equal to 81, so there are two possibilities: $r - 3 = 9$ or $r - 3 = -9$. If $r - 3 = 9$, then $r = 12$. If $r - 3 = -9$, then $r = -6$. The sum of these two possibilities is $12 + (-6) = \boxed{6}$.

9.59 If we multiply both sides by h to get rid of the fraction on the left side, we have

$$3\sqrt{27} = \frac{h^2}{27\sqrt{3}}.$$

Next, we multiply both sides by $27\sqrt{3}$ to get rid of the fraction on the right side. This gives us $(3\sqrt{27})(27\sqrt{3}) = h^2$. Simplifying the left side gives

$$(3\sqrt{27})(27\sqrt{3}) = (3 \cdot 27)(\sqrt{27} \cdot \sqrt{3}) = 81\sqrt{27 \cdot 3} = 81\sqrt{81} = 81(9) = 729.$$

So, we have $h^2 = 729$. There are two values that satisfy this equation, $h = \sqrt{729}$ and $h = -\sqrt{729}$. Since $\sqrt{729} = \sqrt{81 \cdot 9} = \sqrt{81} \cdot \sqrt{9} = 9 \cdot 3 = 27$, the two values of h that satisfy the equation are $\boxed{27 \text{ and } -27}$.

9.60 We multiply both sides by $\sqrt{5 - 2x}$ to get the $\sqrt{5 - 2x}$ out of the denominator of the fraction. This gives us

$$\sqrt{5 - 2x} \cdot \sqrt{5 - 2x} = \frac{10}{\sqrt{5 - 2x}} \cdot \sqrt{5 - 2x}.$$

After canceling the $\sqrt{5 - 2x}$ terms on the right, and noting that the two square roots on the left are the same, we have

$$\left(\sqrt{5 - 2x}\right)^2 = 10.$$

Therefore, we must have $5 - 2x = 10$. Subtracting 5 from both sides gives $-2x = 5$. Dividing by -2 gives $x = \boxed{-5/2}$.

9.61

(a) Since $\sqrt{a}$ is positive, we can multiply both sides of $\sqrt{a} > \sqrt{b}$ by $\sqrt{a}$ to give $\left(\sqrt{a}\right)^2 > \sqrt{ab}$, so $a > \sqrt{ab}$. Similarly, multiplying both sides of $\sqrt{a} > \sqrt{b}$ by $\sqrt{b}$ gives $\sqrt{ab} > b$. Combining $a > \sqrt{ab}$ and $\sqrt{ab} > b$ gives $a > \sqrt{ab} > b$, so $a > b$.

(b) We can follow the same steps as the first part, replacing ">" everywhere with "=".

(c) We can follow the same steps as the first part, replacing ">" everywhere with "<".

(d) In the first three parts, we proved the following:

$$\text{If } \sqrt{a} > \sqrt{b}, \text{ then } a > b.$$

$$\text{If } \sqrt{a} = \sqrt{b}, \text{ then } a = b.$$

$$\text{If } \sqrt{a} < \sqrt{b}, \text{ then } a < b.$$

For any nonnegative a and b, exactly one of $\sqrt{a} > \sqrt{b}$, $\sqrt{a} = \sqrt{b}$, and $\sqrt{a} < \sqrt{b}$ must be true. Only one of them, $\sqrt{a} > \sqrt{b}$, leads to $a > b$. So, if we know that $a > b$, then we know that $\sqrt{a} > \sqrt{b}$.

9.62 By definition, if x is positive, the square of $\sqrt{x}$ is x. Multiplying two numbers that are at least 1 gives a product that is at least 1, so we cannot have $\sqrt{x} \geq 1$ and $x < 1$. Therefore, $\sqrt{x}$ must also be between 0 and 1.

We now know that $\sqrt{x} < 1$ and that $\sqrt{x}$ is positive. Since $\sqrt{x}$ is positive, we can multiply both sides of $\sqrt{x} < 1$ by $\sqrt{x}$ to give $\left(\sqrt{x}\right)^2 < \sqrt{x}$. Since $\left(\sqrt{x}\right)^2 = x$, we have $x < \sqrt{x}$.

9.63 The square root of $6n$ is an integer if and only if $6n$ is a perfect square. In order for $6n$ to be a perfect square, all the primes in its prime factorization must have even powers. Since $6 = 2 \cdot 3$, we therefore know that n must have at least one factor of 2 and at least one factor of 3. So, we know that n is 6 times some integer. Let m be this integer, so $n = 6m$. Then, we can write $6n$ as $6(6m)$, or $36m$. Now, we have $\sqrt{6n} = \sqrt{36m} = \sqrt{36} \cdot \sqrt{m} = 6\sqrt{m}$.

Since m must be an integer, the expression $6\sqrt{m}$ equals an integer if and only if m is a perfect square. This means that $6n$ is a perfect square if and only if m is a perfect square. Since n must have two digits and it must be 6 times a perfect square, the possibilities for n are $6 \cdot 4 = 24$, $6 \cdot 9 = 54$, and $6 \cdot 16 = 96$. Therefore, there are $\boxed{3}$ two-digit values of n for which $\sqrt{6n}$ is an integer.

9.64

(a) If $\frac{p}{q} = \sqrt{2}$, then $\frac{p}{q}$ is the number we square to get 2. So, we have $\left(\frac{p}{q}\right)^2 = 2$, which means $\frac{p^2}{q^2} = \boxed{2}$.

(b) Since $\frac{p^2}{q^2} = 2$, we know that $p^2 = 2q^2$. Since q is an integer, $2q^2$ must be an even integer, which means that p^2 is even. Since p is an integer whose square is even, p must also be even.

(c) Substituting $p = 2r$ into the equation $p^2 = 2q^2$ gives $(2r)^2 = 2q^2$. Expanding the left side gives $4r^2 = 2q^2$, and dividing by 2 gives $2r^2 = q^2$. Since r is an integer, $2r^2$ is an even integer. Because q^2 is an even integer, we know that q is even also.

(d) In part (a), we started by supposing that there is some fraction $\frac{p}{q}$ in simplest form that equals $\sqrt{2}$. A fraction is only in simplest form if its numerator and denominator have no common factors greater than 1. But in parts (b) and (c), we showed that if the fraction $\frac{p}{q}$ equals $\sqrt{2}$, then both the numerator and denominator must be even. So, we started with "there is a fraction in simplest form that equals $\sqrt{2}$," and reached the conclusion "the fraction is not in simplest form." But a fraction cannot be both in simplest form and not in simplest form! Since our starting point, "there is a fraction in simplest form that equals $\sqrt{2}$," leads to the impossible situation that a fraction both is and is not in simplest form, we know that our starting point itself is impossible. Therefore, there is no quotient of integers that equals $\sqrt{2}$.

CHAPTER **10**

_____Angles

24® Cards

First Card $(2 + 4/10) \cdot 10$.

Second Card $1 \cdot 3 \cdot 10 - 6$.

Third Card $16/(4 - 10/3)$.

Fourth Card $(10 + 5) \cdot 3 - 21$.

Exercises for Section 10.1

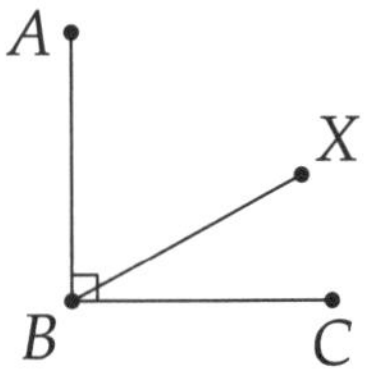

10.1.1   Since $\angle ABX + \angle XBC = \angle ABC$, we have $\angle ABX = \angle ABC - \angle XBC = 90° - 28° = \boxed{62°}$.

10.1.2   Since the rays are equally spaced, the 5 acute angles have the same measure. The angles around a point add to $360°$, so each of these 5 acute angles measures $\frac{360°}{5} = \boxed{72°}$.

10.1.3   Use the straight side of your protractor to extend the sides of the angle until the angle is large enough to measure with your protractor! The measure of the angle is $\boxed{42°}$.

10.1.4   Since the two angles together form a straight angle, the angles are supplementary. Therefore, we have $(x° + 10°) + x° = 180°$. Simplifying the left side gives $2x° + 10° = 180°$. Subtracting $10°$ from both sides gives $2x° = 170°$. Dividing by 2 gives $x° = 85°$, so $x = \boxed{85}$.

10.1.5   Let x be the measure of the smaller angle. Since the ratio of the two angle measures is $1 : 5$, the measure of the larger angle is $5x$. Since the two angles together form a right angle, we have $x + 5x = 90°$, so $6x = 90°$. Dividing by 6 gives $x = \boxed{15°}$.

10.1.6   Let x be the measure of the angle's supplement. The information in the problem tells us that the angle has a measure of $2x + 15°$. Since these angles are supplementary, we have $x + (2x + 15°) = 180°$. Simplifying the left side gives $3x + 15° = 180°$. Subtracting $15°$ from both sides gives $3x = 165°$, and dividing by 3 gives $x = 55°$. Therefore, the desired angle has measure $2x + 15° = \boxed{125°}$.

10.1.7 A full circle is 360°, so the "long way around" angle has measure $360° - 18° = \boxed{342°}$.

10.1.8 A right angle cuts off one-quarter of a circle centered at the vertex of the angle. Therefore, the measure of a right angle is $\frac{1}{4}(500) = \boxed{125}$ clerts.

10.1.9 Let x be the measure of $\angle COD$. Since $\overline{OD}$ divides $\angle COE$ into two equal angles, we have $\angle DOE = \angle COD = x$. Since $\angle BOF$ and $\angle COE$ are vertical angles, we have $\angle BOF = \angle COE = 2x$. Since we have $\angle COB : \angle BOF = 7 : 2$ and $\angle BOF = 2x$, we have $\angle COB = 7x$. Together, $\angle COB$ and $\angle COE$ form a straight angle, so $7x + 2x = 180°$. Simplifying gives $9x = 180°$, so $x = \boxed{20°}$.

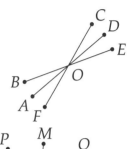

10.1.10 First we show that $\angle MBP = \angle MBQ$. Since $\angle MBA = \angle MBC$ and $\angle PBA = \angle QBC$, we know that

$$\angle MBP = \angle MBA - \angle PBA = \angle MBC - \angle QBC = \angle MBQ,$$

so $\angle MBP = \angle MBQ$. Now we can find the measures of all of the angles in the diagram. We are given that $\angle MBQ = 28°$, so $\angle MBP = 28°$ and $\angle PBQ = \angle MBQ + \angle MBP = 56°$. Since $\angle PBQ = \angle QBC$, we have $\angle CBP = 2\angle PBQ = \boxed{112°}$.

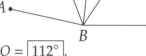

Exercises for Section 10.2

10.2.1 Since $\overline{AD} \parallel \overline{BC}$, we know that $\angle A + \angle B = 180°$, so $\angle B = 180° - \angle A = 107°$. Similarly, since $\overline{AB} \parallel \overline{CD}$, we have $\angle B + \angle C = 180°$. Therefore, $\angle C = 180° - \angle B = \boxed{73°}$. Notice that $\angle A = \angle C$. Is that a coincidence?

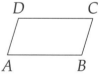

10.2.2 Because $j \parallel k$, the acute angles in the diagram are supplementary to the obtuse angles in the diagram. Specifically, this means that the labeled angle measures must add to 180°:

$$x + x + 30° = 180°.$$

Therefore, we have $2x + 30° = 180°$, so $2x = 150°$ and $x = \boxed{75°}$.

10.2.3 Since $\overline{AD} \parallel \overline{FG}$, we have $\angle CED = \angle CFG$, so $\angle CED = 2t$. The three angles $\angle AEB$, $\angle BEC$, and $\angle CED$ together form a straight angle, so they sum to 180°. Therefore, we have $t + 3t + 2t = 180°$. Simplifying gives $6t = 180°$, and dividing by 6 gives $t = 30°$, which means $\angle CED = 2t = \boxed{60°}$.

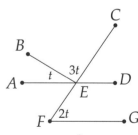

10.2.4 $\boxed{\text{Yes}}$. Since j and k are parallel, line ℓ makes the same angles with k that it makes with j. So, because ℓ makes a 90° angle with j, it also makes a 90° angle with k. This means that lines ℓ and k are perpendicular.

10.2.5 There are 2 regions that only have one of the lines as a border and there are 7 regions between a pair of consecutive parallel lines, for a total of $2 + 7 = \boxed{9}$.

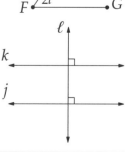

10.2.6 Since $c \parallel d$ and a is perpendicular to d, we know that a is perpendicular to c as well. We mark the angle between a and c as a right angle in our diagram on the right.

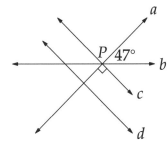

We next consider the three angles at point P, which is the intersection of lines a, b, and c. The measures of the three angles with vertex P that together make line a must have sum $180°$. So, the acute angle between b and c must have measure $180° - 90° - 47° = 43°$. Since c and d are parallel, the acute angle between b and d has the same measure as the acute angle between b and c, which is $\boxed{43°}$.

10.2.7 $\boxed{\text{No}}$. The other angle between m and k is $180° - 89.99° = 90.01°$, which is not equal to the corresponding angle that m makes with j. If j and k were parallel, these two angles would have to be equal. So, lines j and k are not parallel.

10.2.8 We draw line ℓ through A parallel to m and n. Since $\ell \parallel n$, we have $\angle BAD = 46°$. Since $\ell \parallel m$, we have $\angle CAD = 31°$. Therefore, we have $\angle BAC = \angle BAD - \angle CAD = \boxed{15°}$.

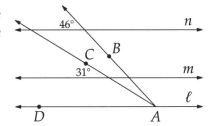

Exercises for Section 10.3

10.3.1 From $\triangle ABC$, we have $\angle ABC = 180° - \angle A - \angle C = 180° - 47° - 72° = 61°$, so $\angle ABD = 180° - \angle ABC = \boxed{119°}$.

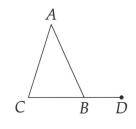

10.3.2 Let x be the measure of the smallest angle. Since the angles are in the ratio $1:4:5$, their measures are x, $4x$, and $5x$ for some value of x. The angles of a triangle sum to $180°$, so $x + 4x + 5x = 180°$, which means $10x = 180°$. Dividing by 10 gives $x = 18°$, so the smallest angle of the triangle has measure $\boxed{18°}$.

10.3.3 The sum of the measures of the angles of a polygon with n sides is $(n - 2)(180°)$, so the sum of the measures of the angles of a polygon with 20 sides is $(18)(180°)$. Since the angles of a regular polygon are congruent, each angle of a regular polygon with 20 sides has measure

$$\frac{(18)(180°)}{20} = 18 \cdot \frac{180°}{20} = 18 \cdot 9° = \boxed{162°}.$$

10.3.4 The angles of a quadrilateral add to $360°$, so each angle of a regular quadrilateral has measure $360°/4 = 90°$. (Of course, a "regular quadrilateral" is a square!) The angles of a hexagon add to $(6 - 2)(180°) = 720°$, so each angle of a regular hexagon has measure $720°/6 = 120°$. The three angles at point Q must sum to $360°$, so we have $\angle RQS = 360° - 120° - 90° = \boxed{150°}$.

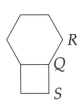

10.3.5 There are lots of ways to solve this problem. The acute angles of a right triangle sum to 90°, so considering right triangle ABC gives $\angle B = 90° - \angle A = 49°$. Then, right triangle BFG gives $\angle FGB = 90° - \angle B = 41°$. From straight angle CGB, we have $\angle CGF = 180° - \angle BGF = \boxed{139°}$.

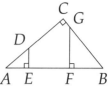

We also could have noted that $\angle ADE = 90° - \angle A = 49°$ from right triangle ADE. Then, we have $\angle CDE = 180° - \angle ADE = 131°$. We now know four of the angle measures in pentagon $CDEFG$. The sum of the angles in a pentagon is $(5-2)(180°) = 540°$, so we find $\angle CGF$ by subtracting the measures of the other four angles from 540°. We have $\angle CGF = 540° - 3(90°) - 131° = \boxed{139°}$.

10.3.6 We find x by finding $\angle BCD$ of quadrilateral $ABCD$. We first find the measures of the other three angles of $ABCD$. We have $\angle ADC = 65°$ because vertical angles are congruent. The 72° angle at A and $\angle DAB$ together make a straight angle, so they are supplementary. Therefore, we have $\angle DAB = 180° - 72° = 108°$. Similarly, we have $\angle ABC = 180° - 51° = 129°$. We now know the measures of three of the angles of $ABCD$, so we can find the fourth angle by using the fact that the angles of a quadrilateral sum to 360°. We have

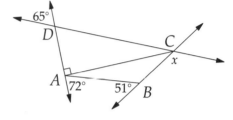

$$\angle BCD + 65° + 108° + 129° = 360°,$$

so $\angle BCD = 58°$. Finally, the angle with measure x is supplementary to $\angle BCD$, so $x = 180° - \angle BCD = \boxed{122°}$.

10.3.7 We have $\angle ADE = 72°$ because vertical angles are congruent. We also have $\angle BCA = 180° - 108° = 72°$. Since $\overleftrightarrow{BC} \parallel \overleftrightarrow{DE}$, we have $\angle ABC = \angle ADE = 72°$ and $\angle DEA = \angle BCA = 72°$. From the vertical angles at E, we then have $x = \angle DEA = \boxed{72°}$. From $\triangle ABC$, we have $\angle BAC = 180° - \angle ABC - \angle ACB = 36°$, so $y = 180° - \angle BAC = \boxed{144°}$.

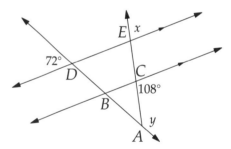

10.3.8 $\boxed{\text{No}}$. A reflex angle has measure greater than 180°, so the sum of the measures of three reflex angles is greater than $3(180°)$, which is 540°. But the sum of all five angles in a pentagon is $(5-2)(180°) = 540°$, so it is impossible to have three reflex angles among the interior angles of a pentagon.

10.3.9 Each interior-exterior angle pair sums to 180° because together the angles make a straight angle. In a polygon with n sides, there are n such pairs, and the sum of all of these pairs is $180n$ degrees. The sum of just the interior angles is $180(n-2)$ degrees. Expanding $180(n-2)$ gives $180n - 360$ degrees as the sum of all the interior angles. We know that adding the exterior angles together with all the interior angles gives a total of $180n$ degrees, so the exterior angles must sum to 360 degrees.

See if you can also use the "walk around the perimeter" method from the textbook to explain why the sum of the exterior angles of a convex polygon is 360°. Does this method work for concave polygons, too?

Review Problems

10.21   Intersecting lines form two pairs of congruent angles. Let x be the measure of each smaller angle, so $5x$ is the measure of each larger angle. Each smaller angle can be combined with a larger angle to form a straight angle, so we must have $x + 5x = 180°$. This gives us $6x = 180°$, so $x = 30°$. Therefore, the angles formed by the lines have measures $\boxed{30° \text{ and } 150°}$.

10.22   The angles around a point must sum to $360°$, so we have $\angle BAE = 360° - 65° - 132° - 71° = \boxed{92°}$.

10.23   Let $x = \angle XOY$, so $\angle WOZ = 5x$. Since $\overrightarrow{OW} \perp \overrightarrow{OY}$, we have

$$\angle WOX = 90° - \angle XOY = 90° - x.$$

Since we also have $\overrightarrow{OX} \perp \overrightarrow{OZ}$, we have $\angle XOZ = 90°$, and

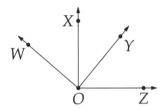

$$\angle WOZ = \angle WOX + \angle XOZ = (90° - x) + 90° = 180° - x.$$

We now have both $\angle WOZ = 5x$ and $\angle WOZ = 180° - x$, so $5x = 180° - x$. Adding x to both sides gives $6x = 180°$, so $x = \boxed{30°}$.

10.24   At 8:30, the hour hand points exactly between the 8 and the 9 on the face of the clock, as shown in the diagram on the right. There are $360/12 = 30$ degrees between adjacent numbers on the clock, so at 8:30 the hour hand is 15 degrees past the 8. The 8 is $2 \cdot 30 = 60$ degrees past the 6, which is where the minute hand is pointing. So, the smaller angle between the hour and minute hand at 8:30 is $15° + 60° = \boxed{75°}$.

10.25   In the 21 minutes it takes to eat dessert, the restaurant goes through $\frac{21}{56}$ of a rotation. Since a full rotation is $360°$, the restaurant revolves during dessert by

$$\frac{21}{56} \cdot 360° = \frac{3}{8} \cdot 360° = 3 \cdot \frac{360°}{8} = 3 \cdot 45° = \boxed{135°}.$$

10.26   Since $\angle ACB = 180°$, we have $\angle ACX + \angle XCB = 180°$. Since $\angle ACX$ is 50% greater than $\angle XCB$, we have $\angle ACX = 1.5(\angle XCB)$, so $1.5\angle XCB + \angle XCB = 180°$. Therefore, we have $2.5\angle XCB = 180°$, so $\angle XCB = \frac{180°}{2.5} = 72°$. We have $\angle BCY = 180° - \angle ACY = 90°$, so $\angle XCY = \angle BCY - \angle XCB = 90° - 72° = \boxed{18°}$.

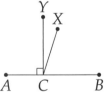

10.27   Because $l \parallel m$, the angles labeled in the diagram must add to $180°$. Therefore, we have the equation $3x + 4° + 131° = 180°$. Simplifying the left side gives $3x + 135° = 180°$, so $3x = 45°$. Dividing by 3 gives $x = \boxed{15°}$.

10.28 Yes. A diagram for this problem is shown at the left below. If a is perpendicular to both b and c, then the corresponding angles indicated in the diagram are indeed equal. So b must be parallel to c.

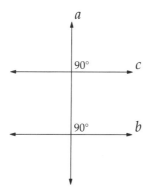

Figure 10.1: Diagram for Problem 10.28   Figure 10.2: Diagram for Problem 10.29

10.29 The diagram for this problem is at the right above.

(a) The angles at Q together make a straight angle, so $2x + (x + 12°) = 180°$. Simplifying the left side gives $3x + 12° = 180°$, so $3x = 168°$ and $x = \boxed{56°}$.

(b) We have $x = 56°$ from part (a), so $\angle P = 56°$ and $\angle PQR = 2x = 112°$. Now that we have the measures of two angles of $\triangle PQR$, we can find the third:

$$\angle R = 180° - \angle P - \angle PQR = \boxed{12°}.$$

Extra challenge: Could we have determined that $\angle R = 12°$ without ever finding x?

10.30

(a) Because $\overline{EF} \parallel \overline{DH}$, we must have $\angle D + \angle E = 180°$. The box at $\angle D$ tells us that $\angle D = 90°$, so $\angle E = 180° - \angle D = \boxed{90°}$.

(b) Because $\overline{EF} \parallel \overline{DH}$, we have $\angle EFH + \angle FHD = 180°$. Therefore, we have $\angle EFH = 180° - \angle FHD = 180° - 75° = \boxed{105°}$.

(c) Because $\overline{EG} \parallel \overline{DH}$, we have $\angle HFG = \angle FHD = 75°$. In $\triangle FGH$, we have

$$\angle HFG + \angle G + \angle FHG = 180°,$$

so $75° + 58° + \angle FHG = 180°$, which gives $\angle FHG = \boxed{47°}$.

10.31 Let the first angle mentioned have measure x, so the second angle has measure $x - 20°$ and the third angle has measure $2x$. Since these three angles are the angles of a triangle, we must have $x + (x - 20°) + 2x = 180°$. Simplifying the left side gives $4x - 20° = 180°$. Adding $20°$ to both sides gives $4x = 200°$, and dividing by 4 gives $x = 50°$. Therefore, the first angle has measure $\boxed{50°}$, the second has measure $x - 20° = \boxed{30°}$, and the third has measure $2x = \boxed{100°}$.

10.32 In triangle ABC, we must have $\angle A + \angle C + \angle ABC = 180°$, so $\angle A + \angle C = 180° - \angle ABC$. We also have $\angle ABC + \angle ABD = 180°$, so $\angle ABD = 180° - \angle ABC$. Since $\angle A + \angle C$ and $\angle ABD$ both equal $180° - \angle ABC$, we have $\angle ABD = \angle A + \angle C$.

10.33 Let x be the measure of the smallest angle, so the angles have measures x, $2x$, $3x$, and $4x$. The sum of the angles in a quadrilateral is $360°$, so we have $x + 2x + 3x + 4x = 360°$, which means that $10x = 360°$. Dividing by 10 gives $x = \boxed{36°}$.

10.34 Since $\overline{EH} \parallel \overline{ID}$, we have $y = \angle HDI = \boxed{36°}$. Since we have $\angle EFD + \angle GFD = 180°$, we find that

$$\angle GFD = 180° - 95° = 85°.$$

From $\triangle GFD$, we have

$$\angle FGD = 180° - \angle GFD - \angle FDG = 180° - 85° - 36° = 59°.$$

Since $\overline{HI} \parallel \overline{EG}$, we have $x + \angle EGD = 180°$, so

$$x = 180° - \angle EGD = 180° - 59° = \boxed{121°}.$$

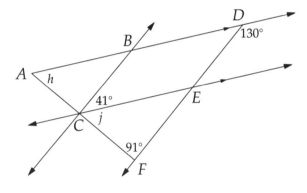

10.35 First, we have $\angle ADF = 180° - 130° = 50°$. From $\triangle ADF$, we find

$$h = 180° - \angle ADF - \angle AFD = 180° - 50° - 91° = \boxed{39°}.$$

Since $\overleftrightarrow{CE} \parallel \overleftrightarrow{AD}$, we have $\angle FCE = \angle FAD$, so $j = h = \boxed{39°}$.

10.36 A nonagon has 9 sides. The sum of the interior angles in a nonagon is $(9 - 2)(180°) = 7(180°)$. Each angle in a regular nonagon has the same measure, so each measures $\frac{7(180°)}{9} = 7\left(\frac{180°}{9}\right) = 7(20°) = \boxed{140°}$.

Another way to do this problem is to use the fact that the exterior angles of any polygon add to $360°$. (We showed this earlier in Exercise 10.3.9.) So, each exterior angle of a regular nonagon has measure $\frac{360°}{9} = 40°$. Therefore, each interior angle has measure $180° - 40° = \boxed{140°}$.

Challenge Problems

10.37 Let x be the measure of the angle. Its complement then has measure $90° - x$ and its supplement has measure $180° - x$. Since 25% of the supplement equals the complement, we have $\frac{1}{4}(180° - x) = 90° - x$. Multiplying both sides by 4 gives $180° - x = 4(90° - x)$. Expanding the right side gives $180° - x = 360° - 4x$. Subtracting $180°$ from both sides, and adding $4x$ to both sides, gives $3x = 180°$, so $x = \boxed{60°}$.

10.38 If we can find $\angle BFD$, we can use $\triangle BFD$ to find $\angle B + \angle D$. We also see that $\angle BFD + \angle AFG = 180°$, so if we can find $\angle AFG$, we can solve the problem. In $\triangle AGF$, we have $\angle A + \angle AFG + \angle AGF = 180°$, so $\angle AFG + \angle AGF = 180° - \angle A = 160°$. We also know that $\angle AFG = \angle AGF$. Combining this with $\angle AFG + \angle AGF = 160°$, we have $\angle AFG = \angle AGF = 80°$. Now, we have $\angle BFD = 180° - \angle AFG = 180° - 80° = 100°$. From triangle BFD, we have $\angle B + \angle D + \angle BFD = 180°$, so $\angle B + \angle D = 180° - \angle BFD = 180° - 100° = \boxed{80°}$.

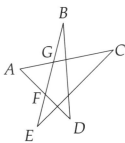

10.39 The point will be back in its original position when it has rotated by a multiple of $360°$. We could list multiples of $150°$ until we hit a multiple of $360°$, but instead, we compute what portion of a circle is $150°$. We have $\frac{150°}{360°} = \frac{5 \cdot 30°}{12 \cdot 30°} = \frac{5}{12}$. So, each day, the disk is rotated $\frac{5}{12}$ of a circle. After n days, it has been rotated $\frac{5n}{12}$ of a circle. The smallest positive integer n for which $\frac{5n}{12}$ is an integer is $n = 12$. We have to be a bit careful—the disk is rotated on the first Saturday, so 12 spins of the disk is 11 days after Saturday, which brings us to $\boxed{\text{Wednesday}}$.

10.40 We have $\angle JLH = 180° - 141° = 39°$. Since $\overleftrightarrow{IK} \parallel \overleftrightarrow{JL}$, we have $\angle HJL = \angle HIK = 116°$. Next, we see that

$$\angle KJL = \angle HJL - \angle HJK = 116° - 79° = 37°.$$

From $\triangle JKL$, we have

$$q = 180° - \angle KJL - \angle JLK = 180° - 37° - 39° = \boxed{104°}.$$

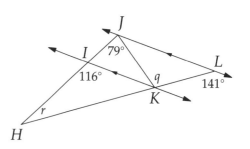

From triangle HJL, we have $r + \angle HJL + \angle JLH = 180°$, so $r + 116° + 39° = 180°$. This gives us $r + 155° = 180°$, so $r = \boxed{25°}$.

10.41 Since vertical angles are congruent, we have $\angle TUS = 42°$. We also have

$$\angle TSU = 180° - \angle TSQ = 180° - 151° = 29°,$$

so $\angle PSQ = \angle TSU = 29°$. The sum of the interior angles of $RPSU$ must be $360°$, even though one of the angles is a reflex angle. The angle of $RPSU$ at vertex S has measure $\angle PSQ + \angle QST + \angle TSU = 209°$, so

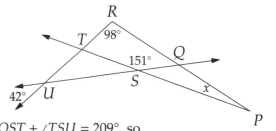

$$x = 360° - 209° - 42° - 98° = \boxed{11°}.$$

10.42 From triangle AFD, we have

$$\angle AFD = 180° - \angle FAD - \angle FDA = 180° - 2x - 2y = 180° - 2(x + y).$$

So, if we can find $x + y$, we can find $\angle AFD$. From quadrilateral $ABCD$, we have $\angle DAB + \angle B + \angle C + \angle CDA = 360°$, so $3x + 110° + 100° + 3y = 360°$. Simplifying the left side gives $3x + 3y + 210° = 360°$. Subtracting $210°$ from both sides gives $3x + 3y = 150°$. Dividing by 3 gives $x + y = 50°$. Substituting this into our expression for $\angle AFD$ gives $\angle AFD = 180° - 2(50°) = 180° - 100° = \boxed{80°}$.

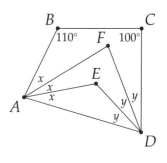

10.43   There is a pentagon in the middle of the star. If we extend the sides of any of the pentagon's interior angles, we reach two of the points of the star. In this way, we can build a triangle for each vertex of the pentagon. One of these triangles is shown in bold on the right. There are five such triangles, one for each vertex of the pentagon. If we add the measures of the angles in all five triangles (counting an angle twice if it appears in two triangles), we get $5 \cdot 180° = 900°$, because the angles of each triangle  sum to $180°$. When adding up all these angles, we include each angle of the pentagon once and each of the points of the star twice. The angles of the pentagon sum to $540°$, leaving $900° - 540° = 360°$. This must equal twice the sum of the angles at the points of the star (since these angles were included twice in the sum), so the angles at the points of the star must sum to $360°/2 = 180°$. (There are many ways to solve this problem; see if you can find others.)

10.44   A decagon has 10 sides, so the sum of its interior angles is $(10 - 2)(180°) = 1440°$. Suppose the decagon has n acute angles. The sum of these n angles is less than $90n$ degrees. The remaining $10 - n$ angles must each be less than $360°$ (the angles may be reflex angles). So, the sum of all of the angle measures must be less than

$$90n + 360(10 - n)$$

degrees. Expanding the product gives $90n + 3600 - 360n$ degrees, and simplifying gives $3600 - 270n$ degrees.

We now know that if there are n acute angles in the decagon, then the sum of the angles in the decagon must be less than $3600 - 270n$ degrees. We also know that the sum of all the interior angles must be 1440 degrees. So, it is only possible to make a decagon with n acute angles if $3600 - 270n > 1440$. Adding $270n$ to both sides of the inequality and subtracting 1440 from both sides gives $2160 > 270n$. Dividing by 270 gives $\frac{2160}{270} > n$. Simplifying the fraction gives $8 > n$. This tells us that it is impossible to have 8 or more acute angles. The diagram at the right shows that we can make a decagon with 7 acute angles.

CHAPTER 11

Perimeter and Area

24® Cards

First Card $(11 \cdot 11 - 1)/5$.

Second Card $(2 + 2/11) \cdot 11$.

Third Card $11 \cdot 13 - 7 \cdot 17$.

Fourth Card $(13 \cdot 19 + 17)/11$.

Exercises for Section 11.1

11.1.1 Starting from the upper left corner and going clockwise, we add the side lengths and find a perimeter of
$$3 + 1 + 2 + 4 + 1 + 2 + 6 + 4 + 2 + 2 + 1 + 2 + 5 + 7 = \boxed{42}.$$

11.1.2 The diagram at the right shows the effect of including a margin. The dashed rectangle is the remaining area in which I can draw. Since the margin is 1 inch on all sides, the length and width of the dashed rectangle are 2 inches shorter than the length and width of the solid rectangle. This leaves a 14 inch by 10 inch region in which I can still draw. The perimeter of this region is $2(14 + 10) = 2(24) = \boxed{48 \text{ inches}}$.

11.1.3 Since $AX : AY = 1 : 4$, we have $AY = 4AX$, which means $XY = 3AX$, and X is closer to A than Y is. Since $BY = AX$, we have
$$AB = AX + XY + BY = AX + 3AX + AX = 5AX.$$
Therefore, we have $5AX = 12$, so $AX = \frac{12}{5}$. We then have
$$BX = XY + BY = 3AX + AX = 4AX = 4\left(\frac{12}{5}\right) = \boxed{\frac{48}{5} \text{ inches}}.$$

11.1.4 The Triangle Inequality tells us that $6.5 + s$ must be greater than 10. So, s is at least $\boxed{4}$.

11.1.5 Let x be the length of the first side of the triangle, so $3x$ is the length of the second side. Since the triangle is isosceles, the third side must have length x or $3x$. But the third side cannot have length x, since the lengths x, x, $3x$ do not satisfy the Triangle Inequality. Therefore, the sides of the triangle have lengths x, $3x$, and $3x$. The perimeter of the triangle is $x + 3x + 3x = 7x$, so we have $7x = 140$, which means $x = 20$. The base of the triangle thus has length $\boxed{20}$.

11.1.6 Each solid side of a square has length equal to one of the sides of the triangle. There are three solid sides equal to each side of the triangle, so the perimeter of the nine-sided figure is three times the perimeter of the triangle. Therefore, the desired perimeter is $3(17) = \boxed{51}$.

11.1.7 Let the length of the pool be l and the width be w. The perimeter of the pool is $2l + 2w$. Since 18 times the perimeter equals 60 lengths, we have $18(2l + 2w) = 60l$. Expanding the product on the left gives $18(2l) + 18(2w) = 60l$, so $36l + 36w = 60l$. Subtracting $36l$ from both sides gives $36w = 24l$. Dividing both sides by l gives $\frac{36w}{l} = 24$, and dividing both sides by 36 gives $\frac{w}{l} = \frac{24}{36} = \frac{2}{3}$. Therefore, the ratio of the width to the length is $\boxed{2:3}$.

11.1.8 Let the base have length b and each leg have length a. Since the perimeter is 25, we must have $2a + b = 25$, so $b = 25 - 2a$. Since a and b must be positive, a is at least 1 and a cannot be greater than 12. (If a were greater than 12, then b would be negative.) However, we can't forget about the Triangle Inequality! We must also have $a + a > b$, which means $2a > b$. Substituting $b = 25 - 2a$ into $2a > b$ gives us $2a > 25 - 2a$. Adding $2a$ to both sides gives $4a > 25$, and dividing by 4 gives $a > \frac{25}{4}$, or $a > 6\frac{1}{4}$. So, a must be at least 7 and at most 12, which means the possible values of a are $\boxed{7, 8, 9, 10, 11, 12}$.

Exercises for Section 11.2

11.2.1 Let l be the length of the original rectangle, and let w be the rectangle's width. Then, the area of the original rectangle is lw. If the length is increased by 1 inch, then the area of the new rectangle is $(l + 1)(w) = lw + w$. Since this area is 12 greater than the original area, which was lw, we have $lw + w = lw + 12$, so $w = 12$. If instead the width is increased by 2, then the area of the new rectangle is $l(w + 2) = lw + 2l$. Since this is 42 greater than the original area, which was lw, we have $lw + 2l = lw + 42$. This means that $2l = 42$, so $l = 21$. Therefore, the area of the original rectangle is $(12)(21) = \boxed{252}$.

11.2.2 The area of the rectangle is $(AB)(BC)$ and the area of the triangle is $(BE)(BC)/2$. Since these areas are equal, we must have $AB = BE/2$, so $BE = 2AB$. Therefore, we have $AE = AB + BE = 3AB$, so $AB/AE = AB/(3AB) = \boxed{1/3}$.

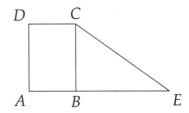

11.2.3 Each tile has area $(2)(3) = 6$ square inches. Each side of the square region is 24 inches, so the area of the square region is $(24)(24) = 576$ square inches. So, we need at least $576/6 = 96$ tiles to cover the square region. We still have to make sure we can fit the 96 tiles snugly within the square region. We can do so because the side length of the square is a multiple of each side length of the rectangle. We can place the tiles in 12 rows of 8 tiles, where each row is 2 inches wide. Therefore, the least number of tiles needed to completely cover the region is $\boxed{96}$.

11.2.4 The area of the pentagon is the difference between the area of the rectangle and the right triangle. The area of the rectangle is $(AB)(AD) = (27)(11) = 297$. To find the area of the right triangle, we find the lengths of its legs. Since $BC = AD = 11$, we have $CE = BC - BE = 5$. Since $DC = AB = 27$, we have $FC = DC - DF = 12$, so $[ECF] = (CE)(FC)/2 = 30$. Therefore, the area of the pentagon is $297 - 30 = \boxed{267}$.

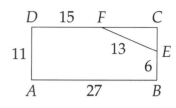

11.2.5 We can find the area of $\triangle STU$ by finding the length of the altitude from U to $\overline{ST}$. Drawing this altitude (dashed in the diagram at the right) completes a rectangle, as shown. So, the length of this altitude equals the length of $\overline{SV}$, which means that the area of $\triangle STU$ is $(12)(6)/2 = \boxed{36}$.

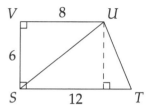

11.2.6 Suppose the original length is l and the original width is w, so the original area is lw. If we increase the length by 20%, then the new length is $1.2l$. If we decrease the width by 10%, then the new width is $0.9w$. So, the new area is $(1.2l)(0.9w) = 1.08lw$. Since the original area was lw, the area has been increased by $\boxed{8\%}$.

11.2.7 As explained in the text, a median of a triangle divides the triangle into two triangles with equal area. Therefore, median $\overline{BX}$ divides $\triangle ABC$ into two triangles with area $26/2 = 13$ cm^2. Since Y is the midpoint of $\overline{BX}$, we know that $\overline{AY}$ is a median of $\triangle ABX$. This means that $\overline{AY}$ divides $\triangle ABX$ into two triangles with equal area. Therefore, $[AYB] = [ABX]/2 = \boxed{13/2 \text{ cm}^2}$.

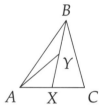

11.2.8 We have $[ABE] = (AE)(AB)/2 = 16$ and $[ABC] = (6)(4)/2 = 12$. We also have $[ADE] = [ABE] - [ABD]$ and $[BDC] = [ABC] - [ABD]$. Subtracting the expression for $[BDC]$ from the expression for $[ADE]$ gives

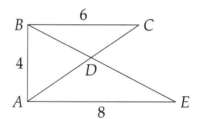

$$[ADE] - [BDC] = ([ABE] - [ABD]) - ([ABC] - [ABD]) = [ABE] - [ABC] = \boxed{4}.$$

Another way to think of this is that the overlap of $\triangle ABE$ and $\triangle ABC$ is $\triangle ABD$. So, when we subtract $[ABC]$ from $[ABE]$, the area of $\triangle ABD$ "cancels out" and we are left with $[ADE] - [BDC]$.

Exercises for Section 11.3

11.3.1 The diameter of the track is 100 yards, so the circumference of the track is 100π yards. Since $\pi \approx 3.14$, the circumference of the track is approximately 314 yards. Since there are 1760 yards in a mile, there are $4 \cdot 1760 = 7040$ yards in four miles. Jane covers approximately 314 yards in each lap, so she covers four miles in $7040/314$ laps. Since $7040/314 \approx 22.4$, Jane must run at least $\boxed{23}$ complete laps to cover at least 4 miles.

11.3.2 If we triple the diameter of a circle, we also triple the radius of the circle. Since the area of a circle equals π times the square of the radius, when we triple the radius, we multiply the area by 9. So, we'll need 9 times as many seeds, or $\boxed{27{,}000}$ seeds.

11.3.3 The three lines together intersect each other in a total of at most three points (the vertices of a triangle determined by the three lines). Each line can intersect the circle in at most two points, so the largest possible number of intersections between the lines and the circle is $3 \cdot 2 = 6$. So, the total largest possible number of intersections is $3 + 6 = \boxed{9}$. An example of an arrangement with 9 intersections is shown at the right.

11.3.4 The circumference of a circle is π times the diameter. So, the ratio of the diameters of the two circles is also $3 : 5$, which means that the ratio of the radii of the two circles is also $3 : 5$. Since the area of a circle is π times the square of the radius, the ratio of the areas of the circles equals the ratio of the squares of the radii, which is $\boxed{9 : 25}$.

11.3.5 The shaded regions are what remains when right triangle ABC is removed from the semicircle. So, the area of the shaded region is the difference between the area of the semicircle and the area of $\triangle ABC$. The semicircle has radius $20/2 = 10$, so its area is $\frac{1}{2}(\pi \cdot 10^2) = 50\pi$. Since $\triangle ABC$ is a right triangle with legs 12 and 16, its area is $\frac{12 \cdot 16}{2} = 96$. Therefore, the total shaded area is $\boxed{50\pi - 96}$.

11.3.6 The larger shaded portion results from removing a semicircle with diameter $3 \cdot 4$ cm $= 12$ cm from a semicircle with diameter $4 \cdot 4$ cm $= 16$ cm. The semicircle with diameter 16 cm has radius 8 cm. A circle with radius 8 cm has area 64π cm^2, so the semicircle with radius 8 cm has area 32π cm^2. Similarly, the semicircle with diameter 12 cm has radius 6 cm and area $\frac{1}{2}\left(6^2\pi\right) = 18\pi$ cm^2. Therefore, the area of the larger shaded portion is $32\pi - 18\pi = 14\pi$ cm^2.

The smaller shaded portion results from removing a semicircle with diameter 4 cm from a semicircle with diameter 8 cm. The semicircle with diameter 8 cm has radius 4 cm and area $\frac{1}{2}\left(4^2\pi\right) = 8\pi$. The semicircle with diameter 4 cm has radius 2 cm and area $\frac{1}{2}\left(2^2\pi\right) = 2\pi$. So, the smaller shaded portion has area $8\pi - 2\pi = 6\pi$ cm^2. Combining this with the larger shaded portion's area, the total shaded area is $14\pi + 6\pi = \boxed{20\pi \text{ cm}^2}$.

Review Problems

11.17 First, we start with a diagram:

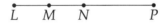

M is the midpoint of $\overline{LN}$, so MN is $\frac{1}{2}$ of LN. Similarly, N is the midpoint of $\overline{LP}$, so LN is $\frac{1}{2}$ of LP. Combining these, we see that MN is $\frac{1}{2} \cdot \frac{1}{2} = \frac{1}{4}$ of LP. Therefore, $MN : LP = \boxed{1 : 4}$.

11.18 Let the original length be l and the original width be w. After increasing both by 10%, the length is $1.1l$ and the width is $1.1w$.

(a) The original perimeter is $2(l + w)$. The perimeter of the new rectangle is

$$2(1.1l + 1.1w) = 2(1.1)(l + w) = 1.1 \cdot 2(l + w).$$

Therefore, the new perimeter is 1.1 times the old perimeter, which means it is $\boxed{10\%}$ greater than the old perimeter.

(b) The original area is lw. The area of the new rectangle is

$$(1.1l)(1.1w) = 1.21lw.$$

The new area is 1.21 times the old area, so it is $\boxed{21\%}$ greater than the old area.

11.19 Let the smallest of the five sides in the ratio $1:2:3:4:5$ have length x, so the five sides have lengths $x, 2x, 3x, 4x,$ and $5x$. Then, the perimeter of the hexagon is $x + 2x + 3x + 4x + 5x + 45$. Simplifying this gives a perimeter of $15x + 45$. We are given that the perimeter is 300, so $15x + 45 = 300$. Subtracting 45 from both sides gives $15x = 255$, and dividing by 15 gives $x = 17$. So, the sides have lengths 17, 34, 51, 68, 85, and 45. Therefore, the positive difference between the lengths of the longest and shortest sides is $85 - 17 = \boxed{68 \text{ units}}$.

11.20 The ratio of the length of the drawing to the width of the drawing is $12:8$. Simplifying this ratio, we have

$$\text{longer dimension : shorter dimension} = 3:2.$$

Since the shorter dimension of the room is 20 feet, we have

$$\text{longer dimension : 20 feet} = 3:2.$$

Multiplying both parts of the ratio on the right by 10 gives

$$\text{longer dimension : 20 feet} = 30:20,$$

so the longer dimension is 30 feet. (We also might have used the $3:2$ ratio to note that the longer dimension is $\frac{3}{2}$ times the shorter dimension. This means that the longer dimension of the room is $\frac{3}{2}(20) = 30$ feet.)

The perimeter of the room is $2(30 + 20) = \boxed{100 \text{ feet}}$.

11.21 We have to be careful about the corners of the room. There are four corner tiles. Along one width of the room, the corner tiles cover $2 \cdot 9 = 18$ inches already. The width of the room is $15 \cdot 12 = 180$ inches, so we must tile 162 inches more. This requires $\frac{162}{9} = 18$ more tiles. The other width of the room also requires 18 more tiles. As with each width, 18 inches of each length of the room are already covered by corner tiles. The length of the room is $18 \cdot 12 = 216$ inches, so $216 - 18 = 198$ inches remain to tiled. This requires $\frac{198}{9} = 22$ tiles for each length. Including all 4 corners, both widths and both lengths, we need $4 + 18 + 18 + 22 + 22 = \boxed{84}$ tiles.

11.22 Suppose we start in the lower left corner of the figure and walk once clockwise around the figure. We travel upward along vertical segments on the left side of the figure a total distance of $6 + 2 + 2 = 10$ cm. Because all the angles are right angles, we travel back downward along vertical segments on the right side of the figure the same total distance. Similarly, we travel leftward across the bottom of the figure a total distance of $8 + 6 = 14$ cm, and travel rightward the same total distance across the top of the figure. This gives us a total perimeter of $2(10 + 14) = \boxed{48 \text{ cm}}$.

11.23   The resulting figure is shown at the right, with the removed triangles dashed. We remove two segments of length 3 inches along each side of the original triangle, leaving $9 - 3 - 3 = 3$ inches remaining. The resulting figure also has three sides that were not sides of the original triangle. Each of these sides was a side of one of the removed triangles, so each has length 3 inches as well. Therefore, the resulting figure has 6 sides of length 3 inches, which means its perimeter is $\boxed{18 \text{ in}}$.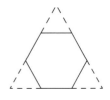

11.24   *BUVWXYZACMNO* has 12 vertices, so it has 12 sides. Since the polygons are regular, and the triangle shares a side with each of the other two polygons, all 12 sides of *BUVWXYZACMNO* have the same length. Moreover, the 3 sides of $\triangle ABC$ have the same length as the sides of *BUVWXYZACMNO*. Since *BUVWXYZACMNO* has 4 times as many sides as *ABC*, and all sides of both polygons have the same length, the perimeter of *ABC* is $\frac{1}{4}$ the perimeter of *BUVWXYZACMNO*, or $\frac{1}{4}(160) = \boxed{40}$.

11.25   In a pentagon train, each end contributes 4 sides to the perimeter, and each pentagon in the interior contributes 3 sides to the perimeter. So, a pentagon train with 85 pentagons has $4 \cdot 2 + 3 \cdot 83 = 8 + 249 = 257$ sides. Since each side has length 1 inch, the perimeter of the pentagon train is $\boxed{257 \text{ inches}}$.

11.26   Suppose the third side has length x. From the Triangle Inequality, we must have $7 + x > 19$, which means x must be greater than 12. We must also have $7 + 19 > x$, which means x must be less than 26. (We must also have $19 + x > 7$, but that is always true if x is positive.) Since x must be greater than 12 and less than 26, the possible values of x are $13, 14, \ldots, 25$. There are $\boxed{13}$ such numbers. (To see why, try subtracting 12 from each number in the list.)

11.27   The perimeter of a square is 4 times the side length of the square, so the ratio of the side lengths of the squares is also $2 : 7$. The area of a square is the square of its side length. So, the ratio of the areas of the squares is $2^2 : 7^2$, or $\boxed{4 : 49}$.

11.28   Since the area of the square is 36 square centimeters, each side of the square has length $\sqrt{36} = 6$ centimeters. Therefore, the perimeter of the square is $4(6) = 24$ centimeters. Let w be the width of the rectangle, so the length is $2w$ and the perimeter is $2(2w + w) = 2(3w) = 6w$. Since the perimeters of the rectangle and the square are the same, we have $6w = 24$, so $w = 4$. Therefore, the length of the rectangle is $2w = 8$ centimeters and the area of the rectangle is $4(8) = \boxed{32 \text{ cm}^2}$.

11.29   Drawing $\overline{CD}$ completes rectangle *ABCD*. The area of unshaded triangle *CED* is half the area of the rectangle *ABCD*, so the combined area of the shaded triangles is also half the area of the rectangle. We have $[ABCD] = (AB)(BC) = 96 \text{ cm}^2$, so the total area of the shaded regions is $\frac{96}{2} = \boxed{48 \text{ cm}^2}$.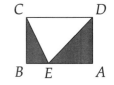

We also could have solved this problem by rearranging the triangles to form a single triangle. We do so by sliding one of the shaded triangles triangle so that $\overline{AD}$ and $\overline{BC}$ coincide. This forms a triangle with base 12 and height 8. So, the total shaded area is $(8)(12)/2 = \boxed{48 \text{ cm}^2}$.

11.30 In order to use as few tiles as possible, we need to use as many of the large tiles as we can. The area of the floor is $9 \cdot 11 = 99$ square feet, and the area of each rectangular tile is $2 \cdot 3 = 6$ square feet. Since 99 divided by 6 is $16\frac{1}{2}$, we can use at most 16 of the larger tiles. This would leave $99 - 16 \cdot 6 = 3$ squares uncovered, so we need 3 of the square tiles as well. But can we fit 16 rectangular tiles on the floor? Yes, as shown in the diagram at the right. So, the least total number of tiles is $16 + 3 = \boxed{19}$.

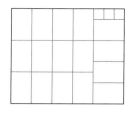

11.31 The area of $\triangle ABC$ is $\frac{1}{2} \cdot 5 \cdot 8 = \frac{1}{2} \cdot 40 = 20$. Let h be the desired length of the altitude from B to $\overline{AC}$. The area of $\triangle ABC$ can also be expressed as $\frac{1}{2}(AC)(h) = \frac{1}{2}(6)(h) = 3h$. Therefore, we must have $3h = 20$, so $h = \boxed{20/3}$.

11.32 We draw $\overline{OT}$ from O perpendicular to $\overline{MP}$ as shown, thereby splitting $MNOP$ into rectangle $MNOT$ and right triangle OPT. Since $MNOT$ is a rectangle, we have $OT = MN = 6$ and $MT = NO = 4$, so $TP = MP - MT = 5$. Therefore, we have

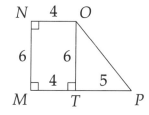

$$[MNOP] = [MNOT] + [OPT] = (MN)(NO) + \frac{(OT)(TP)}{2}$$

$$= (6)(4) + \frac{(6)(5)}{2} = \boxed{39}.$$

11.33 The area of a circle is π times the square of the radius. So, the ratio of the areas of the circles is $5^2 : 2^2$, which is $\boxed{25 : 4}$.

11.34 A line intersects a circle in at most two points. So, a hexagon intersects a circle in at most $6 \cdot 2 = \boxed{12}$ points. At the right is a configuration in which a hexagon intersects a circle in 12 points.

11.35 Circle Y has a diameter of 8, so its radius is 4. Circle Z has a radius of $\sqrt{9} = 3$. Since $3 < \pi < 4$, the desired order is $\boxed{Z, X, Y}$.

11.36 The lot and the largest circular region the sprinkler can reach are shown at the right. The diameter of the circle equals the side length of the lot, so the radius of the circle is 100 feet. Therefore, the area of the circle is $100^2 \pi \approx (10000)(3.1416) = 31416$ ft^2. The area of the square is $200^2 = 40000$ ft^2, so the desired percentage is $\frac{31416}{40000} \approx \boxed{79\%}$.

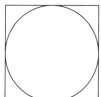

11.37 Since the radius of each quarter-circle is 6 inches, the length and width of the rectangle are 12 inches and 6 inches, respectively. So, the area of the rectangle is $(12)(6) = 72$ square inches. Each quarter-circle has area $\frac{1}{4}(\pi \cdot 6^2) = 9\pi$ square inches, so the quarter-circles together have area 18π square inches. Subtracting the area of the quarter-circles from the area of the rectangle leaves the area of the shaded region, which is $\boxed{72 - 18\pi \text{ square inches}}$.

11.38 There are 12 inches in a foot, so there are $12/2 = 6$ semicircles on the top half of the pattern and 6 semicircles on the bottom. (Because 2 inches divides into 12 inches evenly, the semicircle that is cut in half at the beginning will match up perfectly with the piece on the other end to form a full semicircle.) Twelve semicircles together make 6 full circles. Each circle has radius 1 inch, and therefore area π square inches, so the total shaded area in a 1-foot length of the pattern is $\boxed{6\pi \text{ in}^2}$.

Challenge Problems

11.39 Since $AB : BD = 5 : 7$, we know that there is some x such that $AB = 5x$ and $BD = 7x$. Similarly, since $AC : CD = 13 : 11$, there is some y such that $AC = 13y$ and $CD = 11y$. So, we have both $AD = AB + BD = 12x$ 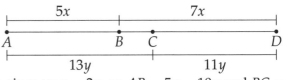 and $AD = AC + CD = 24y$, which means $12x = 24y$. This gives us $x = 2y$, so $AB = 5x = 10y$ and $BC = AD - AB - CD = 24y - 10y - 11y = 3y$. Therefore, we have $AB : BC : CD = 10y : 3y : 11y = \boxed{10 : 3 : 11}$.

Another approach is to choose a length for $\overline{AD}$. A convenient choice is 24, since this will allow us to avoid fractions as segment lengths. Because $AC : CD = 13 : 11$ and $AC + CD = 24$, we have $AC = 13$ and $CD = 11$. Because $AB : BD = 5 : 7$ and $AB + BD = 24$, we have $AB = 10$ and $BD = 14$. We then have $BC = BD - CD = 14 - 11 = 3$, so $AB : BC : CD = \boxed{10 : 3 : 11}$, as before.

11.40 The area of the rectangle is $(AD)(DC)$. Since $\overline{YD}$ is the altitude from vertex Y of $\triangle AXY$, we have $[AXY] = (AX)(DY)/2$. Since X is the midpoint of $\overline{AD}$, we have $AX = AD/2$. Since Y is the midpoint of $\overline{DC}$, we have $DY = DC/2$. So, we have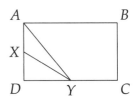

$$[AXY] = \frac{(AX)(DY)}{2} = \frac{(AD/2)(DC/2)}{2} = \frac{(AD)(DC)}{8},$$

which means that the area of $\triangle AXY$ is $\boxed{\frac{1}{8}}$ the area of the rectangle.

11.41 Let h be the length of the altitude from P to $\overline{QR}$. This is the height from P for triangles PQT, PQR, and PTR. So, $[PQT] = (QT)(h)/2$ and $[PQR] = (QR)(h)/2$. When we divide the first equation by the second, we have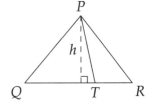

$$\frac{[PQT]}{[PQR]} = \frac{(QT)(h)/2}{(QR)(h)/2} = \frac{(QT)(h)}{2} \cdot \frac{2}{(QR)(h)} = \frac{QT}{QR}.$$

In other words, if two triangles share an altitude, then the ratio of the triangles' areas equals the ratio of the lengths of the bases to which the altitude is drawn. Here, we have $[PQT] = 75$ and $[PQR] = [PQT] + [PTR] = 75 + 40 = 115$, so we have

$$\frac{QT}{QR} = \frac{[PQT]}{[PQR]} = \frac{75}{115} = \boxed{\frac{15}{23}}.$$

11.42 In each of the first three parts, we are given XY and asked for $[WXYZ]$. We have $[WXYZ] = (XY)(ZY)$, so we only need ZY to find $[WXYZ]$. We are also given $[ZXA]$. Since $\overline{XY}$ is the altitude from X in $\triangle ZXA$, we have $[ZXA] = (XY)(ZA)/2$. So, we can use the given value of $[ZXA]$ together with the value of XY in each part to find ZA. Then, we use ZA to find ZY. Since $ZA = 3AY$, we have $AY = \frac{ZA}{3}$, so $ZY = ZA + AY = ZA + \frac{ZA}{3} = \frac{4ZA}{3}$.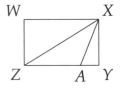

(a) Since $[ZXA] = (XY)(ZA)/2$, we have $36 = 12(ZA)/2$, so $36 = 6ZA$. Therefore, we have $ZA = 6$, so $ZY = \frac{4ZA}{3} = \frac{4 \cdot 6}{3} = 8$, and $[WXYZ] = (ZY)(XY) = \boxed{96}$.

(b) Since $[ZXA] = (XY)(ZA)/2$, we have $36 = 9(ZA)/2$, so $72 = 9ZA$. Therefore, we have $ZA = 8$, so $ZY = \frac{4ZA}{3} = \frac{4 \cdot 8}{3} = \frac{32}{3}$, and $[WXYZ] = (ZY)(XY) = \frac{32}{3} \cdot 9 = \boxed{96}$.

(c) Since $[ZXA] = (XY)(ZA)/2$, we have $36 = 6(ZA)/2$, so $36 = 3ZA$. Therefore, we have $ZA = 12$, so $ZY = \frac{4ZA}{3} = \frac{4 \cdot 12}{3} = 16$, and $[WXYZ] = (ZY)(XY) = \boxed{96}$.

(d) We found the same result for $[WXYZ]$ in all three parts. It's unlikely that this is a coincidence. Let's investigate. The area of $WXYZ$ is twice the area of $\triangle ZXY$. Triangles ZXY and ZXA have the same height from vertex X, and base $\overline{ZY}$ of $\triangle ZXY$ is $\frac{4}{3}$ times as long as base $\overline{ZA}$ of $\triangle ZXA$, so $[ZXY] = \frac{4}{3}[ZXA] = \frac{4}{3}(36) = 48$. Finally, we have $[WXYZ] = 2[ZXY] = 96$, no matter what XY is.

11.43 The sum of the angles in a polygon with n sides is $(n-2)(180°)$, so each angle in a regular polygon has measure $\frac{(n-2)(180°)}{n}$. Since this must equal $168°$, we have

$$\frac{(n-2)(180°)}{n} = 168°.$$

Multiplying both sides by n gives $(n-2)(180°) = (n)(168°)$. Expanding the product on the left gives $(n)(180°) - 360° = (n)(168°)$. Adding $360°$ to both sides, and subtracting $(n)(168°)$ from both sides, gives $(n)(12°) = 360°$. Dividing both sides by $12°$ gives $n = 30$. Since the polygon has 30 congruent sides and perimeter 120 cm, each side has length $120/30 = \boxed{4 \text{ cm}}$.

A much faster way to tackle this problem is to use the fact that the exterior angles of a polygon add to $360°$. Since each interior angle is $168°$, each exterior angle is $180° - 168° = 12°$. Since all the exterior angles add to $360°$ and each measures $12°$, there must be $\frac{360°}{12°} = 30$ of them. We then use the perimeter to determine that each side has length $\boxed{4 \text{ cm}}$, as before.

11.44 Since $AB : AD = 2 : 5$, we have $AB = 2x$ and $AD = 5x$ for some value of x. Since AB and AD must be integers, and the greatest common divisor of 2 and 5 is 1, the value of x must be an integer. Similarly, since $AD : CD = 3 : 4$, there is some integer y for which $AD = 3y$ and $CD = 4y$. Combining our expressions for AD, we have $5x = 3y$. The smallest positive integers x and y for which this equation holds are $x = 3$ and $y = 5$. So, we have $AB = 2x = 6$, $AD = 5x = 15$, $CD = 4y = 20$, and $BC = AD = 15$. Finally, the perimeter of $ABCD$ is $6 + 15 + 20 + 15 = \boxed{56}$.

11.45 Drawing all three long diagonals of a regular hexagon splits the regular hexagon into 6 equilateral triangles, as shown on the right. $\overline{AD}$ consists of two sides of these triangles, so each equilateral triangle has side length $16/2 = 8$. The sides of the regular hexagon consist of six sides of these triangles, so the perimeter of the hexagon is $6 \cdot 8 = \boxed{48}$.

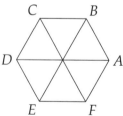

11.46 At each of her turns her path has a $180° - 20° = 160°$ angle. Suppose she forms a regular polygon with her walk. (We don't know for sure that the angles will work out; we are just checking if it is possible that they do so.) Then, the $20°$ by which she turns is an exterior angle of the polygon. Since the exterior angles of the polygon sum to $360°$ and each has measure $20°$, the polygon must have $\frac{360°}{20°} = 18$ angles. Checking, we see that each interior angle of a regular polygon with 18 sides is indeed $\frac{(18-2)(180°)}{18} = 160°$. Therefore, Rebecca's walk forms a regular polygon with 18 sides. Each side has length 100 feet, so she walks $18 \cdot 100 = \boxed{1800 \text{ feet}}$. (See the solution to Problem 11.43 to see how to find the number of sides of the polygon without considering the exterior angles.)

For a considerably more challenging problem, try to figure out how far Rebecca would walk if she turned 165° clockwise every 100 feet!

11.47 We break the upper semicircle into two quarter-circles and slide those two quarter-circular regions into the locations shown in the diagram at right. Now, the shaded region is a rectangle whose width is the radius of each semicircle and whose length is the diameter of each semicircle. Since the radius of each semicircle is 1, the area of the rectangle is $(1)(2) = \boxed{2}$.

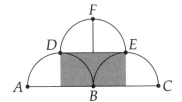

11.48 The only way Teri cannot make a triangle is if the sum of the lengths of the two shortest sticks is not greater than the length of the longest stick. If she cuts x off each stick, then the sticks have lengths $9 - x$, $12 - x$, and $14 - x$. The sum of the two shortest sticks then is $(9 - x) + (12 - x)$, which equals $21 - 2x$. If she can't make a triangle, then this sum must be less than or equal to the length of the longest stick, so $21 - 2x \le 14 - x$. Adding $2x$ to both sides and subtracting 14 from both sides gives $7 \le x$. Therefore, the smallest amount Teri could have cut off each stick is $\boxed{7 \text{ inches}}$. If she cuts 7 inches off of each stick, she'll have sticks with lengths 2 inches, 5 inches, and 7 inches. Since $2 + 5$ is not greater than 7, she cannot make a triangle with these sticks.

11.49 We apply the Triangle Inequality separately to $\triangle ABC$ and $\triangle ADC$. Applying it to $\triangle ABC$, we see that AC must be greater than 2 and less than 18. Applying it to $\triangle ADC$, we see that AC must be greater than 4 and less than 28. Combining these, we see that AC must be greater than 4 and less than 18, so the possible lengths are $5, 6, 7, \ldots, 17$. There are $\boxed{13}$ numbers in this list.

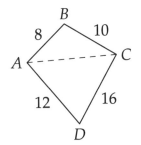

11.50 We do a little "outside the box" thinking, as shown at the right. We extend the pattern beyond the given initial square past each side, as shown. We see that each of the triangle pieces inside the original square fits together with a quadrilateral piece inside the square to form a square that is the same size as the black square. Therefore, inside the original square, the 8 white pieces that are quadrilaterals or triangles fit together to make 4 squares that are the same size as the black square. This means that the black square's area is $\frac{1}{5}$ the area of $ABCD$. The area of $ABCD$ is $(20)(20) = 400$ square units, so the area of the black square is $\boxed{80 \text{ square units}}$.

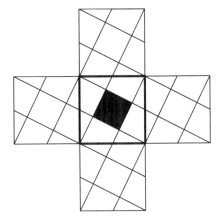

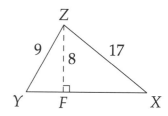

CHAPTER 12

_____ **Right Triangles and Quadrilaterals**

$24^®$ Cards

First Card $12 + 12 + 12 - 12.$

Second Card $12/(18/12 - 1).$

Third Card $12/(3 - 5/2).$

Fourth Card $(11 - 7) \cdot 12 - 24.$

Exercises for Section 12.1

12.1.1 Applying the Pythagorean Theorem to $\triangle PQR$ gives $PQ^2 + QR^2 = PR^2$, so $PQ^2 = PR^2 - QR^2 = 12^2 - 9^2 = 144 - 81 = 63$. Taking the square root gives $PQ = \sqrt{63} = \sqrt{9}\sqrt{7} = \boxed{3\sqrt{7}}$.

We could use the Pythagorean Theorem on $\triangle TUV$ to find TV, but we can simplify our work with our knowledge of Pythagorean triples. The legs have lengths $3 \cdot 5$ and $3 \cdot 12$, and the legs in the Pythagorean triple $\{5, 12, 13\}$ have lengths 5 and 12, so we know that the hypotenuse of $\triangle TUV$ has length $3 \cdot 13$. Therefore, $TV = \boxed{39}$.

Let F be the unlabeled endpoint of the dashed segment, so $\triangle ZYF$ and $\triangle ZXF$ are right triangles. Applying the Pythagorean Theorem to $\triangle ZYF$ gives $YF^2 + ZF^2 = ZY^2$, so $YF^2 = ZY^2 - ZF^2 = 81 - 64 = 17$. Therefore, $YF = \sqrt{17}$. Applying the Pythagorean Theorem to $\triangle ZXF$ gives $FZ^2 + FX^2 = ZX^2$, so $FX^2 = ZX^2 - FZ^2 = 289 - 64 = 225$. Therefore, $FX = \sqrt{225} = 15$. (We might also have recognized the $\{8, 15, 17\}$ Pythagorean triple.) Combining these results gives $XY = FX + YF = \boxed{15 + \sqrt{17}}$.

12.1.2 Bill's path is shown in bold at the right. He starts at A, goes south to B, then east to C, then south to D. In total, he goes south 1 mile and east $\frac{3}{4}$ mile. In other words, he could have gotten from A to D by going 1 mile south to point E, then $\frac{3}{4}$ mile east to point D. From right triangle ADE, we can find the distance between A and D. We have $AE = 1$ and $DE = \frac{3}{4}$, so

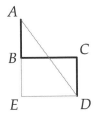

$$AD^2 = AE^2 + DE^2 = 1 + \frac{9}{16} = \frac{25}{16}.$$

Taking the square root, we find $AD = \boxed{\frac{5}{4} \text{ miles}}$.

We also could have noted that $ED = 3 \cdot \frac{1}{4}$ and $AE = 4 \cdot \frac{1}{4}$, so the ratio of the legs of $\triangle ADE$ matches the ratio of the smallest two numbers in the Pythagorean triple $\{3, 4, 5\}$. Therefore, the hypotenuse of $\triangle ADE$ is $5 \cdot \frac{1}{4} = \boxed{\frac{5}{4} \text{ miles}}$, as before.

12.1.3 As shown in the diagram at the right, the diagonal of a rectangle is the hypotenuse of a right triangle whose legs are a length and a width of the rectangle. Letting d be the length of the diagonal and letting l and w be the length and width of the rectangle, we have $d^2 = l^2 + w^2$. Taking the square root gives us our formula:

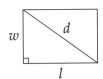

$d = \boxed{\sqrt{l^2 + w^2}}$. Note that the other diagonal of the rectangle is also the hypotenuse of a right triangle with legs of lengths l and w, so it also has length $\sqrt{l^2 + w^2}$.

12.1.4 The poles are in bold in the diagram at the right. We build a right triangle and a rectangle by drawing a segment ($\overline{DE}$ in the diagram) from the top of the short pole so that the segment is perpendicular to the tall pole. The rope is the hypotenuse $\overline{AD}$ of $\triangle ADE$. Since $BCDE$ is a rectangle, we have $DE = BC = 45$ and $EB = CD = 15$. Therefore, we have $AE = AB - EB = 39 - 15 = 24$, so right triangle ADE has legs with lengths 24 and 45. We could use the Pythagorean Theorem to find AD, but we can use Pythagorean triples to find AD even faster.

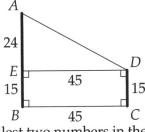

Since $24 = 8 \cdot 3$ and $45 = 15 \cdot 3$, the legs of $\triangle ADE$ are in the same ratio as the smallest two numbers in the $\{8, 15, 17\}$ Pythagorean triple. So, the hypotenuse of $\triangle ADE$ is $17 \cdot 3 = \boxed{51 \text{ feet}}$.

12.1.5 The rectangle has area $(12)(16) = 192$ square units and the square has area $12^2 = 144$ square units. One leg of the right triangle is a side of the square and the other is a longer side of the rectangle, so the legs of the triangle have lengths 12 and 16. This means the area of the triangle is $(12)(16)/2 = 96$ square units. The diameter of the semicircle is the hypotenuse of the right triangle. We can use the Pythagorean Theorem to find the hypotenuse. Or, we can note that the legs of the right triangle have lengths $3 \cdot 4$ and $4 \cdot 4$. Using the Pythagorean triple $\{3, 4, 5\}$, we see that the hypotenuse has length $5 \cdot 4 = 20$. Since the diameter of the semicircle is 20, the radius of the semicircle is 10. The area of a circle with radius 10 is $\pi(10)^2$, which equals 100π, so the area of the semicircle is half of 100π, or 50π square units. Adding the areas of all four pieces gives a total area of $192 + 144 + 96 + 50\pi = \boxed{432 + 50\pi}$ square units.

12.1.6 The peculiar form of the two numbers reveals a common factor. We can write the numbers as $49 \cdot 100001$ and $63 \cdot 100001$. Moreover, 49 and 63 have 7 as a common factor, so we can write the numbers as $7 \cdot 7 \cdot 100001$ and $9 \cdot 7 \cdot 100001$, or $7 \cdot 700007$ and $9 \cdot 700007$. Therefore, if we find the hypotenuse of a right triangle with legs 7 and 9, then we can multiply that by 700007 to get the desired hypotenuse. If c is the hypotenuse of a right triangle with legs 7 and 9, then $c^2 = 7^2 + 9^2 = 49 + 81 = 130$, so $c = \sqrt{130}$.

Therefore, the desired hypotenuse has length $\boxed{700007\sqrt{130}}$.

Exercises for Section 12.2

12.2.1 Triangle XYZ is an isosceles right triangle, so its hypotenuse is $\sqrt{2}$ times the length of each leg. Therefore, $XY = \boxed{8}$ and $XZ = \boxed{8}$.

For $\triangle TUV$, we first note that because $\angle U = \angle V$, we have $TU = TV = \boxed{20}$. Next, we must find UV. Let F be the unlabeled endpoint of the dashed segment, so $TF = 16$. Applying the Pythagorean Theorem to $\triangle TFV$ gives $TF^2 + FV^2 = TV^2$, so $FV^2 = TV^2 - TF^2 = 20^2 - 16^2 = 400 - 256 = 144$. Therefore, we have $FV = \sqrt{144} = 12$. (We could also have used the Pythagorean triple $\{12, 16, 20\}$.) Since $\triangle TUV$ is isosceles with $\angle U = \angle V$, the altitude $\overline{TF}$ divides base $\overline{UV}$ into two congruent segments. Therefore, we have $UV = 2(FV) = \boxed{24}$.

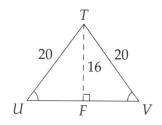

We immediately have $QR = QP = \boxed{14}$. We split $PQRS$ into two triangles by drawing $\overline{PR}$, as shown at the right. Since $QP = QR$, we know that $\triangle QPR$ is isosceles with $\angle QPR = \angle QRP$. Since $\angle Q = 60°$, these two base angles must sum to $180° - 60° = 120°$. Therefore, each measures $60°$, which means $\triangle QPR$ is equilateral. Therefore, $PR = 14$, $\angle RPS = 120° - \angle QPR = 60°$, and $\angle PRS = 90° - 60° = 30°$.

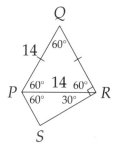

Turning to $\triangle PRS$, we have $\angle S = 180° - \angle RPS - \angle PRS = 90°$, so triangle PRS is a 30-60-90 triangle with hypotenuse $PR = 14$. The leg opposite the $30°$ equals half the hypotenuse, so $PS = \boxed{7}$. The longer leg is $\sqrt{3}$ times the shorter leg, so $SR = \boxed{7\sqrt{3}}$.

12.2.2 Since $AB = BC$, triangle ABC is isosceles. The angles opposite the congruent sides are congruent, so $\angle A = \angle C$. Since $\angle B = 68°$, we know that the other two angles of the triangle sum to $180° - 68° = 112°$. These two angles have the same measure, so each equals $112°/2 = \boxed{56°}$.

12.2.3 Since $3\angle K = 90°$, we have $\angle K = 30°$. We also have $\angle J = 90°$, so $\triangle JKL$ is a 30-60-90 triangle with hypotenuse $\overline{KL}$, short leg $\overline{JL}$ (opposite the $30°$ angle), and long leg $\overline{JK}$. The hypotenuse of a 30-60-90 triangle is twice the length of the short leg of the triangle, so $KL = 2JL = \boxed{8}$. The long leg of a 30-60-90 triangle is $\sqrt{3}$ times the short leg, so $JK = \boxed{4\sqrt{3}}$.

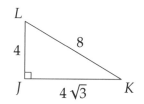

12.2.4 Let h be the length of the altitude to the base with length 24. Then, since the triangle's area is 60, we must have $24h/2 = 60$. Therefore, $12h = 60$, which means $h = 5$. As discussed in the text, the altitude to the base of an isosceles triangle divides the base into two equal segments. So, in the diagram on the right, we have $BH = HC = 24/2 = 12$. Therefore, the legs of right triangle ABH have lengths 5 and 12. Applying the Pythagorean Theorem, or recalling the Pythagorean triple $\{5, 12, 13\}$, tells us that the hypotenuse has length $\boxed{13}$.

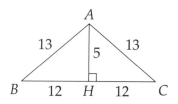

12.2.5

(a) Since $\overline{OB}$ and $\overline{OA}$ are radii of the circle, they both have length 12. Therefore, $\triangle OAB$ is isosceles, and $\angle A = \angle B$. Since $\angle O = 60°$, we know that $\angle A + \angle B = 180° - \angle O = 120°$. Combining this with $\angle A = \angle B$, we have $\angle A = \angle B = 60°$, so all three angles of $\triangle OAB$ are congruent. This means that $\triangle OAB$ is equilateral, which means $AB = OA = OB = \boxed{12}$.

(b) Since $\angle O$ is $\frac{60°}{360°} = \frac{1}{6}$ of a circle, the shorter arc from A to B is $\frac{1}{6}$ of the circumference of the circle. The diameter of the circle is $2 \cdot 12 = 24$, so the circumference of the circle is 24π. Therefore, the length of the shorter arc from A to B is $\frac{1}{6}(24\pi) = \boxed{4\pi}$.

(c) Just as the shorter arc cut off by $\angle AOB$ is $\frac{1}{6}$ the circumference of the circle, the sector cut off by $\angle AOB$ has area equal to $\frac{1}{6}$ the area of the circle. The area of the circle is $12^2\pi = 144\pi$, so the area of the sector is $\frac{1}{6}(144\pi) = \boxed{24\pi}$.

12.2.6

(a) Let $\triangle XYZ$ be an equilateral triangle with side length s, as shown in the diagram at the right. We draw altitude $\overline{ZD}$ and form right triangle XZD. The altitude splits base $\overline{XY}$ in half, so $XD = \frac{s}{2}$. Since $\triangle XYZ$ is equilateral, we have $\angle X = 60°$, which means $\triangle XZD$ is a 30-60-90 triangle. Therefore, $ZD = XD\sqrt{3} = \frac{s\sqrt{3}}{2}$. So, we have

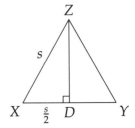

$$[XYZ] = \frac{(XY)(ZD)}{2} = \frac{(s)\left(\frac{\sqrt{3}}{2}\cdot s\right)}{2} = \left(s^2 \cdot \frac{\sqrt{3}}{2}\right)\frac{1}{2} = \boxed{\frac{s^2\sqrt{3}}{4}}.$$

(b) The region between $\overline{AB}$ and the shorter arc from A to B is what's left after removing $\triangle OAB$ from the sector in part (c) of the previous problem. So, the area of the desired region equals the area of the sector minus the area of $\triangle OAB$. Since $\triangle OAB$ is equilateral, we can use the formula we found in part (a) of this problem, which gives us

$$[OAB] = \frac{12^2\sqrt{3}}{4} = \frac{144\sqrt{3}}{4} = 36\sqrt{3}.$$

From part (c) of the previous problem, the area of the sector is 24π, so the desired area is $\boxed{24\pi - 36\sqrt{3}}$.

Exercises for Section 12.3

12.3.1 The perimeter of the triangle is $6.2 + 8.3 + 9.5 = 24$ cm, so each side of the square has length $24/4 = 6$ cm. Therefore, the area of the square is $6^2 = \boxed{36 \text{ cm}^2}$.

12.3.2

(a) | False |. A quadrilateral with four equal sides is a rhombus, but not every rhombus is a square. An example is shown at the right.

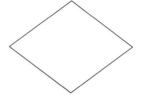

(b) | False |. Just because two sides are equal does not mean that all four angles are equal. The rhombus on the right is an example of a quadrilateral that is not a rectangle, but has at least one pair of equal sides.

(c) | True |. The only restriction on the interior angles in a quadrilateral is that they sum to 360°. So, for example, we can have a quadrilateral with two right angles, a 100° angle, and an 80° angle.

(d) | True |. Each diagonal is the hypotenuse of a right triangle with legs that are the length and the width of the rectangle. Therefore, the Pythagorean Theorem gives the same length for each diagonal.

12.3.3

(a) Since $\overline{EH} \parallel \overline{FG}$, we have $\angle E + \angle F = 180°$, so $\angle F = 180° - \angle E = \boxed{139°}$. Similarly, since $\overline{EF} \parallel \overline{GH}$, we have $\angle E + \angle H = 180°$, so $\angle H = 180° - \angle E = \boxed{139°}$. Finally, since $\overline{EF} \parallel \overline{GH}$, we have $\angle G + \angle F = 180°$, so $\angle G = 180° - \angle F = \boxed{41°}$.

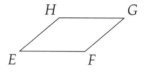

(b) Let $EFGH$ be a parallelogram. Since $\overline{EH} \parallel \overline{FG}$, we have $\angle E + \angle F = 180°$, so $\angle E = 180° - \angle F$. Similarly, since $\overline{EF} \parallel \overline{GH}$, we have $\angle G + \angle F = 180°$, so $\angle G = 180° - \angle F$. Since $\angle E$ and $\angle G$ both equal $180° - \angle F$, we have $\angle E = \angle G$. Similarly, we have $\angle F = \angle H$. Therefore, the opposite angles of a parallelogram are congruent.

12.3.4 The central square has side lengths of 9 feet each, so its area is $9^2 = 81$ square feet. Each pair of opposite corners of the backyard can be pushed together to form a square with side length 9, and each of these squares has area 81 square feet as well. Therefore, the total area with grass is $3(81) = 243$ square feet. Since the whole backyard is a square with side length 27 feet, its area is $27^2 = 729$ square feet. Subtracting the area with grass leaves $729 - 243 = \boxed{486 \text{ square feet}}$ without grass.

12.3.5 As explained in the text, the diagonals of a rhombus are perpendicular and bisect each other, so we draw diagonal $\overline{QS}$. Let M be the intersection point of the diagonals, so we have $PM = PR/2 = 5$. Applying the Pythagorean Theorem to $\triangle PQM$, we have $PM^2 + QM^2 = PQ^2$, so $25 + QM^2 = 49$. Subtracting 25 from both sides gives $QM^2 = 24$, and taking the square root gives $QM = \sqrt{24} = \sqrt{4 \cdot 6} = 2\sqrt{6}$. Therefore, we have $QS = 2(QM) = 4\sqrt{6}$, and

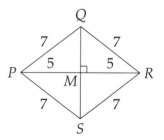

$$[PQRS] = \frac{(QS)(PR)}{2} = \frac{(4\sqrt{6})(10)}{2} = \boxed{20\sqrt{6} \text{ square units}}.$$

12.3.6 The area of the parallelogram is the product of the length of $\overline{AB}$ and the height between $\overline{AB}$ and $\overline{CD}$, so $[ABCD] = (12)(6) = 72$ square units. We can also compute the area of $ABCD$ as the product of the length of $\overline{AD}$ and the height between $\overline{AD}$ and $\overline{BC}$. So, if we let the desired height be h, we have $9h = 72$. Therefore, $h = 72/9 = \boxed{8}$.

12.3.7 Let the other base have length b. Then, the area of the trapezoid is $\frac{b+8}{2} \cdot 4$. Simplifying this expression gives

$$\frac{b+8}{2} \cdot 4 = \frac{4}{2}(b+8) = 2(b+8).$$

We are told that this equals 80, so $2(b+8) = 80$. Dividing both sides by 2 gives $b + 8 = 40$, and subtracting 8 from both sides gives $b = \boxed{32 \text{ inches}}$.

12.3.8 A trapezoid with one $39°$ angle is shown at the right. Since $\overline{WX} \parallel \overline{YZ}$, we have $\angle W + \angle Z = 180°$, so $\angle Z = 180° - \angle W = 141°$. We cannot determine the other two angles of the trapezoid from the given information; all we know about $\angle X$ and $\angle Y$ is that they sum to $180°$. Therefore, the only other angle measure we can determine is $\boxed{\angle Z = 141°}$.

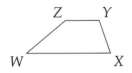

12.3.9 We start with the diagram after we rotated the second trapezoid:

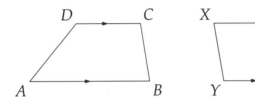

We have $BC = XY$ because the trapezoids are identical. We also have $\angle C + \angle B = 180°$ because $\overline{AB} \parallel \overline{CD}$. Since $\angle B = \angle X$ (identical trapezoids), we therefore know that $\angle C + \angle X = 180°$. So, we push the two trapezoids together so that $\overline{CB}$ and $\overline{XY}$ become the same segment, and sides $\overline{CD}$ and $\overline{XW}$ are on the same line:

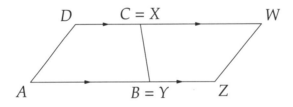

Similarly, sides $\overline{AB}$ and $\overline{YZ}$ are on the same line. Now, quadrilateral $ADWZ$ sure looks like a parallelogram. We know that $\overline{AZ} \parallel \overline{DW}$. We have to check that $\overline{AD} \parallel \overline{WZ}$. Since $\overline{AB} \parallel \overline{CD}$, we have

$$\angle D + \angle A = 180°.$$

Since the trapezoids are identical, we have $\angle A = \angle W$. Therefore, we have

$$\angle D + \angle W = 180°.$$

This tells us that $\overline{AD} \parallel \overline{WZ}$ in the diagram above. Since $\overline{AZ} \parallel \overline{DW}$ and $\overline{AD} \parallel \overline{WZ}$, we know that $ADWZ$ is a parallelogram.

12.3.10 The trapezoid is in bold in the diagram on the left below.

We start by drawing the median dashed. We then cut right triangles off the right and the left sides of the trapezoid with vertical lines through the endpoints of the median. We can slide these triangles into place as shown to complete a rectangle whose dimensions equal the height and the median of the trapezoid.

Review Problems

12.21 The eastward rider travels $9 \cdot 3 = 27$ miles in three hours and the southward rider travels $12 \cdot 3 = 36$ miles. The shortest distance between them then is the hypotenuse of a right triangle with legs of lengths 27 and 36 miles. We can use the Pythagorean Theorem to find the hypotenuse. Or, we could notice that $27 = 3 \cdot 9$ and $36 = 4 \cdot 9$, so, applying the $\{3, 4, 5\}$ Pythagorean triple, we see that the hypotenuse has length $5 \cdot 9 = \boxed{45 \text{ miles}}$.

12.22 The leg has length $4 \cdot 12$ and the hypotenuse has length $4 \cdot 13$. Recalling the Pythagorean triple $\{5, 12, 13\}$, we notice that the other leg has length $4 \cdot 5 = \boxed{20}$.

12.23 In the $\{3, 4, 5\}$ Pythagorean triple, the 3 and the 4 are both leg lengths, but in the given problem, the side with length $4 \cdot 100$ is the hypotenuse, not a leg. We do have $400 = 5 \cdot 80$, but the given leg is not either 3 or 4 times 80, so we cannot use the $\{3, 4, 5\}$ Pythagorean triple to solve the problem. Instead, we just use the Pythagorean Theorem. Letting x be the length of the other leg, we have $x^2 + 300^2 = 400^2$, so $x^2 = 400^2 - 300^2 = 4^2 \cdot 100^2 - 3^2 \cdot 100^2 = 16(100^2) - 9(100^2) = 7(100^2)$. Taking the square root gives $x = \sqrt{7(100^2)} = \boxed{100 \sqrt{7} \text{ cm}}$.

12.24

(a) The area of a right triangle is half the product of the lengths of its legs, so the area of the given triangle is $(7)(24)/2 = \boxed{84 \text{ in}^2}$.

(b) First, we find the hypotenuse length. From the Pythagorean Theorem, or from remembering the $\{7, 24, 25\}$ Pythagorean triple, we find that the hypotenuse of the triangle has length 25 inches. The area of the triangle equals half the product of the hypotenuse and the desired altitude length. Let the desired altitude have length h. Using the area we found in part (a), we have $25h/2 = 84$. Multiplying both sides by 2 gives $25h = 168$, and dividing by 25 gives $h = \boxed{168/25 \text{ inches}}$.

12.25 We find the areas of the two right triangles separately. We start with the triangle on the lower left. Let the missing leg have length a. The Pythagorean Theorem gives $14^2 + a^2 = 28^2$, so $a^2 = 28^2 - 14^2$. We can simplify the computations a bit by noticing that $28 = 2 \cdot 14$. So taking the square root gives us

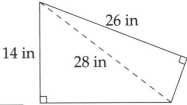

$$a = \sqrt{28^2 - 14^2} = \sqrt{(2 \cdot 14)^2 - 14^2} = \sqrt{2^2 \cdot 14^2 - 14^2}$$
$$= \sqrt{4(14^2) - 14^2} = \sqrt{3 \cdot 14^2} = 14\sqrt{3}.$$

(We might also have noticed that the given leg is half the hypotenuse, so the triangle is a 30-60-90 triangle.) Therefore, the right triangle on the lower left has area $(14)(14\sqrt{3})/2 = 98\sqrt{3}$ in^2.

Turning to the other right triangle, we let b be the length of the missing leg and again apply the Pythagorean Theorem. We find $26^2 + b^2 = 28^2$, so $b^2 = 28^2 - 26^2 = 784 - 676 = 108$. Taking the square root gives $b = \sqrt{108} = \sqrt{36 \cdot 3} = 6\sqrt{3}$. So, the area of the upper right triangle is $(26)(6\sqrt{3})/2 = 78\sqrt{3}$ in^2.

Combining the two triangles gives a total area of $98\sqrt{3} + 78\sqrt{3} = \boxed{176\sqrt{3} \text{ in}^2}$.

12.26 We present two methods for finding the total area.

Method 1: The Hard Way. Because the hypotenuse of an isosceles right triangle is $\sqrt{2}$ times a leg of the triangle, the leg length of each of the triangle sections is $\frac{16}{\sqrt{2}}$. Therefore, the area of each of these sections is

$$\frac{1}{2}\left(\frac{16}{\sqrt{2}}\right)\left(\frac{16}{\sqrt{2}}\right) = \frac{1}{2} \cdot \frac{16^2}{\sqrt{2} \cdot \sqrt{2}} = \frac{1}{2} \cdot \frac{256}{2} = 64 \text{ ft}^2.$$

Combining all four sections gives a total area of $4(64) = \boxed{256 \text{ ft}^2}$.

Method 2: The Easy Way. Pushing together all four right triangles forms a square with side length 16 ft. This square has area $(16)^2 = \boxed{256 \text{ ft}^2}$.

12.27 We have $\angle C = 180° - 30° - 60° = 90°$, so $\triangle ABC$ is a right triangle. Moreover, because the acute angles of the triangle are 30° and 60°, the triangle is a 30-60-90 triangle. Therefore, the short leg of the triangle is opposite the 30° angle and is half the hypotenuse, so $BC = 8/2 = \boxed{4}$. The longer leg is $\sqrt{3}$ times the shorter leg, so $AC = \boxed{4\sqrt{3}}$. The area is half the product of the legs, so $[ABC] = \boxed{8\sqrt{3}}$.

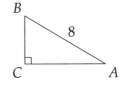

12.28 Either the 54° angle is one of the two congruent angles, or it isn't:

Case 1: The 54° angle is one of the two congruent angles. Then, another angle is 54° and the third angle is $180° - 54° - 54° = 72°$.

Case 2: The 54° angle is not one of the two congruent angles. Then, the other two angles are congruent, and their measures must sum to $180° - 54°$, which equals 126°. Since the two congruent angles sum to 126°, they each measure $126°/2 = 63°$.

(a) Considering the two cases above, the largest possible measure of an angle of the triangle is $\boxed{72°}$.

(b) Considering the two cases above, the smallest possible measure of an angle of the triangle is $\boxed{54°}$.

12.29 Because $\angle A + \angle B = 180°$, we know that $\overline{BC} \parallel \overline{AD}$, so $ABCD$ is a trapezoid. The bases of the trapezoid have lengths 3 and 15, and the height of the trapezoid is 9, so the area of the trapezoid is $\frac{3+15}{2} \cdot 9 = \boxed{81}$.

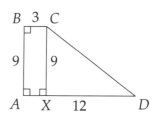

To find the perimeter of the trapezoid, we must find CD. We draw the altitude from C to $\overline{AD}$, forming the right triangle CXD shown in the diagram. Since $BCXA$ is a rectangle, we have $AX = BC = 3$ and $CX = AB = 9$. Therefore, we have $DX = DA - AX = 12$. Applying the Pythagorean Theorem to $\triangle CXD$, or recognizing the $\{9, 12, 15\}$ Pythagorean triple (which is 3 times the $\{3, 4, 5\}$ Pythagorean triple), we find that $CD = 15$, so the perimeter of $ABCD$ is $9 + 3 + 15 + 15 = \boxed{42}$.

12.30 The area of a rhombus is half the product of its diagonals, so the area of the rhombus is $(10)(24)/2 = \boxed{120 \text{ in}^2}$. The diagonals of a rhombus bisect each other, so drawing the diagonals of the rhombus splits the rhombus into four right triangles with legs of length 5 and 12 inches. Recalling the $\{5, 12, 13\}$ Pythagorean triple, or applying the Pythagorean Theorem to these triangles, we find that each side of the rhombus has length 13 inches. So, the perimeter of the rhombus is $4(13) = \boxed{52 \text{ inches}}$.

12.31 Connecting the foot of the ladder to the base of the building completes a right triangle in which part of the building wall is a leg and the ladder is the hypotenuse. From the Pythagorean Theorem, or from the $\{7, 24, 25\}$ Pythagorean triple, we see that the base of the ladder is initially 7 feet from the building. After the top slides down 4 feet, the top is 20 feet from the ground. The ladder is still 25 feet long. Either using the Pythagorean Theorem or recognizing the $\{15, 20, 25\}$ Pythagorean triple (which is 5 times the $\{3, 4, 5\}$ Pythagorean triple), we see that the base of the ladder is 15 feet from the wall after sliding. Therefore, the base of the ladder slid $15 - 7 = \boxed{8 \text{ feet}}$.

12.32 The sum of the angles of a hexagon is $(6 - 2)(180°) = 720°$, so each angle of a regular hexagon is $720°/6 = 120°$. We'll next compute the measure of an obtuse angle of the rhombus. Each obtuse angle of the rhombus shares its vertex with two angles of regular hexagons. The angles around the vertex must sum to $360°$, so the measure of each obtuse angle of the rhombus is $360° - 120° - 120° = 120°$. The rhombus is also a parallelogram, which means its opposite sides are parallel. Therefore, consecutive angles in the rhombus must be supplementary, which means each acute angle has measure $180° - 120° = \boxed{60°}$.

12.33

(a) $\boxed{\text{False}}$. The angles of a quadrilateral add to $360°$. So, if three of them are right angles, the third must measure $360° - 90° - 90° - 90° = 90°$. This tells us that if three of the angles of a quadrilateral are right angles, then the fourth angle must also be a right angle. Therefore, it is impossible for a quadrilateral to have exactly three right angles.

(b) $\boxed{\text{True}}$. As explained in the text, every square is a rectangle and every rectangle is a parallelogram, so every square is a parallelogram.

(c) $\boxed{\text{True}}$. Let the quadrilateral be $ABCD$, so $\angle A = \angle C$ and $\angle B = \angle D$. Therefore, we must have $\angle A + \angle B = \angle C + \angle D$. Since the sum of all four angles must be $360°$, we must have $\angle A + \angle B = \angle C + \angle D = 180°$. Since $\angle A + \angle B = 180°$, we have $\overline{BC} \parallel \overline{AD}$. Similarly we can start with $\angle A = \angle C$ and $\angle D = \angle B$ to find $\angle A + \angle D = \angle B + \angle C$, which gives us $\angle A + \angle D = 180°$, so $\overline{AB} \parallel \overline{CD}$. Therefore, both pairs of opposite sides of $ABCD$ are parallel, so $ABCD$ is a parallelogram.

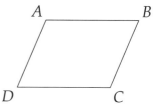

(d) $\boxed{\text{False}}$. For example, the rhombus at the right taken from the textbook has diagonals of different length.

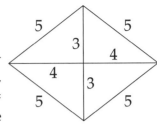

12.34 Let the rhombus be $EFGH$, with $\angle E$ being the given 79° angle. Every rhombus is a parallelogram. Since opposite sides of the rhombus are parallel, each pair of consecutive angles is supplementary. Therefore, we have $\angle E + \angle F = \angle E + \angle H = 180°$, which gives us $\angle F = \angle H = 180° - \angle E = \boxed{101°}$. Finally, the opposite angles of a parallelogram are congruent, so $\angle G = \angle E = \boxed{79°}$.

12.35

(a) Let the quadrilateral formed by connecting the midpoints of the sides of $WXYZ$ be $ABCD$, as shown in the diagram at the right. First, we show that all four sides of $ABCD$ are congruent. Each side of $ABCD$ is the hypotenuse of an isosceles right triangle whose legs are half the side length of the square. Therefore, all four sides of $ABCD$ have length $\sqrt{2}$ times half the side length of the square.

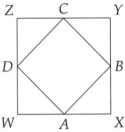

Next, we show that all of the angles of $ABCD$ are right angles. We see that $\angle DAB = 180° - \angle WAD - \angle XAB$. Since $\triangle WAD$ and $\triangle XAB$ are isosceles right triangles, we have $\angle WAD = \angle XAB = 45°$, so $\angle DAB = 180° - 45° - 45° = 90°$. Similarly, all four angles of $ABCD$ are right angles. Since all four angles of $ABCD$ are congruent and all four sides of $ABCD$ are congruent, $ABCD$ is a square.

(b) The side length of $WXYZ$ is $\sqrt{900} = 30$. As described in the previous part, each side of $ABCD$ has length equal to $\sqrt{2}$ times half the side length of $WXYZ$. So, the side length of $ABCD$ is $15\sqrt{2}$, which means

$$[ABCD] = \left(15\sqrt{2}\right)^2 = 15^2\left(\sqrt{2}\right)^2 = 225(2) = \boxed{450}.$$

We also could have figured out that $[ABCD] = [WXYZ]/2$ by drawing the diagonals of $ABCD$. These diagonals divide $WXYZ$ into four smaller squares with area 225 each. Half of each of these squares is inside $ABCD$, while the other half of each is outside $ABCD$. So, the area of $ABCD$ is $\frac{1}{2} \cdot 4 \cdot 225 = 2 \cdot 225 = \boxed{450}$.

12.36 Because opposite sides of a parallelogram are parallel, each pair of consecutive angles adds to 180°. So, the two angles that add to 204° must be opposite each other. The opposite angles of a parallelogram are congruent, so each of these angles must be $204°/2 = 102°$. The other two angles are congruent, and each is supplementary to each of the 102° angles. Therefore, these other two angles have measure $180° - 102° = \boxed{78°}$.

Challenge Problems

12.37 The diagonals of a rectangle are congruent, so $NP = OM$. $\overline{OM}$ is a radius of the circle. Since the area of the circle is 100π, the circle's radius is $\sqrt{100} = 10$. Therefore, $NP = \boxed{10}$.

12.38 Let h be the height of the trapezoid. The height from D of $\triangle ABD$ is also h, as is the height from B of $\triangle BCD$. Since $[ABD] = 2.5[BCD]$, we have $(AB)(h)/2 = 2.5(CD)(h)/2$. Dividing both sides by h and multiplying both sides by 2 gives $AB = 2.5CD$. Substituting this into $AB + CD = 77$, we have $2.5CD + CD = 77$, so $3.5CD = 77$. Multiplying by 2 gives $7CD = 2(77)$, and dividing by 7 gives $CD = 2(77)/7 = 2(11) = 22$. Therefore, $AB = 77 - CD = \boxed{55}$.

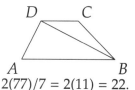

12.39 Let $BM = BN = s$, so the side length of the square is $2s$. Therefore, we have $[MBN] = s^2/2$. We also have $AM = CN = s$, so $[AMD] = [CND] = (s)(2s)/2 = s^2$. Finally, we have

$$[MDN] = [ABCD] - [MBN] - [AMD] - [CND] = 4s^2 - \frac{s^2}{2} - s^2 - s^2 = \frac{3s^2}{2}.$$

Therefore, we have $[MBN]/[MDN] = (s^2/2)/(3s^2/2) = \boxed{1/3}$.

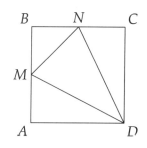

Perhaps a quicker way to see this is to note that $\triangle MBN$ and $\triangle MDN$ share a side, $\overline{MN}$. So, the ratio of the areas of the triangles equals the ratio of their heights to that common side. The altitudes to $\overline{MN}$ of these two triangles together form diagonal $\overline{BD}$. (Note that $BNDM$ is a kite.) See if you can figure out why $\overline{MN}$ divides this diagonal into a $1:3$ ratio.

12.40 Since $\angle CBD = 30°$ and $\angle ABC = 60°$, we have $\angle ABD = 30° + 60° = 90°$. Therefore, we can find AD by first finding BD and AB, and then applying the Pythagorean Theorem to right triangle ABD. From 30-60-90 triangle BCD, we have $BD = CD\sqrt{3} = 6\sqrt{3}$ and $BC = 2CD = 12$. Since $\triangle ABC$ is equilateral, we have $AB = BC = 12$. Applying the Pythagorean Theorem to $\triangle ABD$ gives

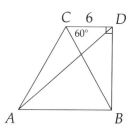

$$AD^2 = AB^2 + BD^2 = 12^2 + \left(6\sqrt{3}\right)^2 = 144 + (6^2)\left(\sqrt{3}\right)^2 = 144 + (36)(3) = 144 + 108 = 252.$$

Taking the square root gives $AD = \sqrt{252} = \sqrt{36 \cdot 7} = \boxed{6\sqrt{7}}$.

12.41 Let h be the height of the trapezoid, so h is also the height from C in $\triangle ACB$. Therefore, we have $[ACB] = (20)(h)/2 = 10h$ and $[ABCD] = h(20 + 12)/2 = 16h$, so $[ACB] : [ABCD] = (10h) : (16h) = 10 : 16 = \boxed{5:8}$.

12.42 Since the radius of the circle is 1 cm, and F is the midpoint of $\overline{AD}$, the side length of the square is 2 cm. The triangle and the square share a side, so the side length of the triangle is also 2 cm. Therefore, the total length of the straight portions of the ant's path is $FA + AB + BC + CE + EG = 1 + 2 + 2 + 2 + 1 = 8$ cm. Next, we tackle the curved portion. Since $ABCD$ is a square, $\angle ADC = 90°$. Since $\triangle CDE$ is equilateral, $\angle CDE = 60°$. Therefore, $\angle ADE = 150°$. So, the portion of the circle that ant doesn't walk on is $\frac{150°}{360°} = \frac{5}{12}$ of a circle, which means that the bold portion of the circle is the other $\frac{7}{12}$ of the circle. The diameter of the circle is 2 cm, so its entire circumference is 2π. Therefore, the bold portion of the circle has length $\frac{7}{12}(2\pi) = \frac{7}{6}\pi$.

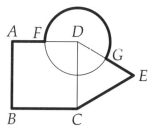

Combining all parts of the bold path, the ant walks a total of $\boxed{8 + \frac{7}{6}\pi \text{ centimeters}}$.

12.43 $\boxed{\text{Yes}}$. In the diagram at the right, the diagonals of $ABCD$ are perpendicular at point X. Therefore, $\overline{BX}$ is the altitude from B to $\overline{AC}$ in $\triangle ABC$, and $\overline{DX}$ is the altitude from D to $\overline{AC}$ in $\triangle ADC$. So, we have

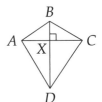

$$[ABCD] = [ABC] + [ADC]$$
$$= \frac{(BX)(AC)}{2} + \frac{(DX)(AC)}{2}$$
$$= \frac{(BX)(AC) + (DX)(AC)}{2} = \frac{(BX + DX)(AC)}{2} = \frac{(BD)(AC)}{2}.$$

So, the area of any quadrilateral with perpendicular diagonals is half the product of its diagonals.

12.44

(a) Not only do the second leg and the hypotenuse differ by 1 in each of the given triples, but in each case the sum of the second leg and the hypotenuse equals the square of the first leg. For example, $4 + 5 = 9 = 3^2$ and $12 + 13 = 25 = 5^2$. So, to find a triple with 9 as a leg length and with the other two side lengths 1 apart, we expect that the sum of these other two lengths equals 9^2, which is 81. If we let the second leg be b, then the hypotenuse is $b + 1$, and we guess that $b + (b + 1) = 81$. Simplifying the left side gives $2b + 1 = 81$. Subtracting 1 from both sides gives $2b = 80$, and dividing by 2 gives $b = 40$. We must still check that the side lengths 9, 40, and 41 satisfy the Pythagorean Theorem. We have $41^2 = 1681$ and $9^2 + 40^2 = 81 + 1600 = 1681$, so indeed, $\boxed{\{9, 40, 41\}}$ is the desired Pythagorean triple.

Even if we hadn't noticed a convenient pattern, we still could have solved the problem with algebra. If we again let the second leg be b, so the hypotenuse is $b + 1$, then the Pythagorean Theorem gives us

$$9^2 + b^2 = (b + 1)^2.$$

We use the distributive property to expand $(b + 1)^2$:

$$(b + 1)^2 = (b + 1)(b + 1) = b(b + 1) + 1(b + 1) = b^2 + b + b + 1 = b^2 + 2b + 1.$$

Our Pythagorean Theorem equation then is

$$9^2 + b^2 = b^2 + 2b + 1.$$

Subtracting b^2 from both sides gives $81 = 2b + 1$, and solving this equation gives $b = 40$, as before.

(b) $\boxed{\text{Yes}}$. Let the first leg have length n, where n is an odd integer greater than 1. As before, suppose the second leg is b, so the hypotenuse is $b + 1$. We guess that $b + b + 1 = n^2$. Subtracting 1 from both sides gives $2b = n^2 - 1$, and dividing by 2 gives $b = \frac{n^2-1}{2}$. Since n is odd, we know that $n^2 - 1$ is even, which means b is an integer. If $b = \frac{n^2-1}{2}$, then the hypotenuse is $b + 1 = \frac{n^2-1}{2} + 1 = \frac{n^2-1}{2} + \frac{2}{2} = \frac{n^2+1}{2}$. We still have to check if the side lengths n, $\frac{n^2-1}{2}$, $\frac{n^2+1}{2}$ satisfy the Pythagorean Theorem. We have

$$\left(\frac{n^2 + 1}{2}\right)^2 = \frac{(n^2 + 1)^2}{2^2} = \frac{(n^2 + 1)(n^2 + 1)}{4} = \frac{n^2(n^2 + 1) + 1(n^2 + 1)}{4}$$
$$= \frac{(n^2)(n^2) + n^2 + n^2 + 1}{4} = \frac{n^4 + 2n^2 + 1}{4},$$

and

$$n^2 + \left(\frac{n^2-1}{2}\right)^2 = n^2 + \frac{(n^2-1)^2}{2^2} = n^2 + \frac{(n^2-1)(n^2-1)}{4} = \frac{4n^2}{4} + \frac{n^2(n^2-1)-1(n^2-1)}{4}$$

$$= \frac{4n^2}{4} + \frac{(n^2)(n^2) - n^2 - n^2 + 1}{4}$$

$$= \frac{4n^2}{4} + \frac{n^4 - 2n^2 + 1}{4}$$

$$= \frac{4n^2 + n^4 - 2n^2 + 1}{4} = \frac{n^4 + 2n^2 + 1}{4}.$$

So, we do indeed have $(n)^2 + \left(\frac{n^2-1}{2}\right)^2 = \left(\frac{n^2+1}{2}\right)^2$. Therefore, for every odd number n greater than 1, there is a Pythagorean triple with n and with two other numbers 1 apart.

(c) $\boxed{\text{No}}$. For example, in the text we saw the triple $\{8, 15, 17\}$. Note that $\{6, 8, 10\}$ is a Pythagorean triple, as are $\{10, 24, 26\}$ and $\{12, 35, 37\}$. Do you see a pattern in these? (Here's a hint: Consider the integer between the two largest numbers in each triple.)

12.45 We saw a convex kite in the text, and we found its area by first showing that the diagonals of the kite were perpendicular. We'll try the same for the concave kite in this problem, but we'll have to extend one of the diagonals in order to make the diagonals intersect. We extend diagonal $\overline{QS}$ past S to meet diagonal $\overline{PR}$ at point T. Since $\triangle PSR$ is isosceles with $PS = SR$, we know that the altitude from S to $\overline{PR}$ meets $\overline{PR}$ at the midpoint of $\overline{PR}$. Similarly, $\triangle PQR$ is isosceles, so we know that the altitude from Q to $\overline{PR}$ also meets $\overline{PR}$ at its midpoint. So, the line through the midpoint of $\overline{PR}$ that is perpendicular to $\overline{PR}$ passes through S and Q. In other words, the diagonals of a concave kite are perpendicular.

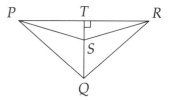

Diagonal $\overline{SQ}$ splits $PQRS$ into two obtuse triangles, $\triangle PSQ$ and $\triangle RSQ$. Since $\overline{PT}$ is the altitude from P in $\triangle PSQ$, we have $[PSQ] = (PT)(SQ)/2$. We also have $PT = TR$, so $[RSQ] = (TR)(SQ)/2 = (PT)(SQ)/2$. So, we have

$$[PQRS] = [PSQ] + [RSQ] = \frac{(PT)(SQ)}{2} + \frac{(PT)(SQ)}{2} = (PT)(SQ).$$

We have $SQ = 7$, so we only need PT. Applying the Pythagorean Theorem to $\triangle PTS$ gives

$$PT^2 + ST^2 = PS^2 = 225.$$

Applying the Pythagorean Theorem to $\triangle PTQ$ gives

$$PT^2 + TQ^2 = PQ^2 = 400.$$

Since $TQ = ST + SQ = ST + 7$, we have $PT^2 + (ST + 7)^2 = 400$. Expanding $(ST + 7)^2$ gives

$$(ST + 7)^2 = (ST + 7)(ST + 7) = ST(ST + 7) + 7(ST + 7) = ST^2 + 7ST + 7ST + 7^2 = ST^2 + 14ST + 49,$$

so $PT^2 + (ST + 7)^2 = 400$ becomes

$$PT^2 + ST^2 + 14ST + 49 = 400.$$

We found earlier that $PT^2 + ST^2 = 225$, so now we have

$$225 + 14ST + 49 = 400.$$

Simplifying the left side gives $274 + 14ST = 400$, so $14ST = 126$ and $ST = 126/14 = 9$. We then have $PT^2 + 9^2 = 225$, so $PT^2 = 225 - 9^2 = 225 - 81 = 144$, which gives $PT = 12$. (You might have used your knowledge of Pythagorean triples to find ST and PT without all that work!)

Finally, the area of $PQRS$ is $(SQ)(PT) = (7)(12) = \boxed{84}$.

12.46 We focus on the post-fold diagram. Let x be the length of $\overline{BF}$. Since $\overline{BF}$ and $\overline{AF}$ together formed a side of the original rectangle, we have $AF = 18 - x$. Therefore, the Pythagorean Theorem applied to $\triangle ABF$ gives us $12^2 + x^2 = (18 - x)^2$. Expanding the square on the right side of this equation gives

$$(18 - x)^2 = (18 - x)(18 - x) = 18(18 - x) - x(18 - x) = 18^2 - 18x - 18x + x^2 = 324 - 36x + x^2.$$

So, the Pythagorean Theorem now gives us

$$12^2 + x^2 = 324 - 36x + x^2.$$

Subtracting 324 and x^2 from both sides gives $-180 = -36x$, so $x = 5$. Therefore, we have $BF = 5$ and $AF = 18 - 5 = 13$. The altitude from E to base $\overline{AF}$ of $\triangle AEF$ has the same length as $\overline{AD}$, so the length of this altitude is 12. Therefore, the area of $\triangle AEF$ is $(12)(13)/2 = \boxed{78 \text{ square inches}}$.

12.47 If we could find the length of any one of the altitudes, then we could find the area of the triangle. Drawing an altitude of the triangle splits the triangle into two right triangles. But which altitude should we draw?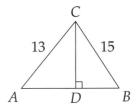

We draw the altitude to the side with length 14, so that the hypotenuses of our two right triangles have lengths 13 and 15, as shown. We do so because we know Pythagorean triples with 13 and with 15 as the hypotenuses. We hope we'll get lucky, and that the corresponding right triangles will fit together to form the desired triangle. The two Pythagorean triples are $\{5, 12, 13\}$ and $\{9, 12, 15\}$. Both have 12 among the leg lengths! If we let $CD = 12$ in the diagram, then $AD = 5$ from right triangle ACD and $BD = 9$ from right triangle BCD. Sure enough, $AD + DB = 14$, as desired! So, the area of the triangle is $(AB)(CD)/2 = (12)(14)/2 = \boxed{84}$. (Of course, this doesn't work out so neatly with most triangles!)

12.48 Let r be the radius of the full circle, so the isosceles right triangle has area $r^2/2$. The hypotenuse of the isosceles right triangle is the diameter of the semicircle. This hypotenuse has length $r\sqrt{2}$, so the radius of the semicircle is $r\sqrt{2}/2$. To find the area of the lune, we must subtract the overlap of the big circle and the semicircle from the semicircle. This overlap region is lightly shaded in the diagram. We have:

$$\text{Lune area} = (\text{Semicircle area}) - (\text{Lightly shaded region}).$$

The semicircle itself has area

$$\frac{1}{2}\left(\frac{r\sqrt{2}}{2}\right)^2 \pi = \frac{1}{2}\left(\frac{r^2\left(\sqrt{2}\right)^2}{2^2}\right)\pi = \frac{1}{2}\left(\frac{(r^2)(2)}{4}\right)\pi = \frac{r^2}{4}\pi,$$

so

$$\text{Lune area} = \frac{r^2}{4}\pi - (\text{Lightly shaded region}).$$

The isosceles right triangle and the lightly shaded region together form a quarter of the large circle. So, we have

$$\text{Isosceles right triangle area} = (\text{Quarter-circle area}) - (\text{Lightly shaded region}).$$

The quarter-circle has area $\frac{1}{4}r^2\pi$, so

$$\text{Isosceles right triangle area} = \frac{r^2}{4}\pi - (\text{Lightly shaded region}).$$

This is the same result as we found for the lune area! Therefore, the area of the lune is the same as the area of the isosceles right triangle. The isosceles right triangle has area $\frac{r^2}{2}$, so the area of the lune is $\frac{r^2}{2}$. Surprisingly, the expression for the area of the lune doesn't have π in it at all!

CHAPTER **13**

Data and Statistics

24® Cards

First Card $(13 \cdot 13 - 1)/7$.

Second Card $7 \cdot 9 - 3 \cdot 13$.

Third Card $5 \cdot 9 - 13 - 8$.

Fourth Card $22/(9 - 7) + 13$.

Exercises for Section 13.1

13.1.1   We start by putting the numbers in increasing order:

$$13, 13, 23, 24, 25, 28, 31, 34, 37, 41.$$

The only repeated number is 13, so the mode is $\boxed{13}$. There is an even number of numbers, so the median of the list is the average of the two middle numbers, which are 25 and 28. Therefore, the median is $\frac{25+28}{2} = \boxed{26.5}$. Finally, there are 10 numbers and they sum to 269, so the average is $\frac{269}{10} = \boxed{26.9}$.

13.1.2   If she had scored higher in her last four games than her average for the first six games, then her average would have gone up. If she had scored lower in her last four games than her average in the first six games, then her average would have gone down. So, she must have scored exactly the same as her original average in order to keep her average the same. This means she bowled $\boxed{143}$ in each of the four games.

13.1.3   Since there are 7 scores, the median is the fourth score when listed from least to greatest. If I score at least 91 on my final three tests, then when I put the tests in order, the lowest four will be $55, 63, 71, 91$, which means 91 will be the median. If I score lower than 91 on any of my tests, then there will be four scores lower than 91, which means that the median will be below 91. Therefore, the highest possible semester grade I can earn is $\boxed{91}$.

13.1.4   Because the sum of the numbers is 350 and their average is 50, there must be $\frac{350}{50} = 7$ numbers. One of the numbers is 100, so the sum of the other 6 numbers is $350 - 100 = 250$. In order to make one of these 6 numbers as large as possible, we make the other 5 numbers as small as possible. The numbers must be different positive integers, so we let the 5 smallest numbers be 1, 2, 3, 4, and 5. The sum of these is 15, which leaves $250 - 15 = \boxed{235}$ for the remaining number.

13.1.5 The sum of these 11 numbers is 66. After removing one number there are 10 numbers left. If the average of the remaining numbers is 6.1, then their sum is $6.1 \cdot 10 = 61$, so the removed number must be $66 - 61 = \boxed{5}$.

13.1.6 *Method 1: Consider sums of weights.* In the original group, the total weight is $72 \cdot 5 = 360$ pounds. In the larger group, the total weight is $73 \cdot 6 = 438$ pounds. So, the sixth child weighs $438 - 360 = \boxed{78}$ pounds.

Method 2: Compare the children's weights to the average. On average, each of the first five children is 1 pound less than the new average, so the new child must be $5 \cdot 1 = 5$ pounds heavier than the new average. Therefore, the new child weighs $73 + 5 = \boxed{78}$ pounds.

13.1.7 In her first 8 games, she scored a total of $7 + 4 + 3 + 6 + 8 + 3 + 1 + 5 = 37$ points. In order for her average after 9 games to be an integer, her total points after nine games must be a multiple of 9. She scored less than 10 points in her ninth game. The only possibility that leaves her with a total that is a multiple of 9 is if she scores 8 points, leaving her with 45 total. Similarly, she must score 5 points in the tenth game in order to have a total (50) that is a multiple of 10. So, the product of her point totals in the ninth and tenth games is $8 \cdot 5 = \boxed{40}$.

13.1.8 Since 8 is the only mode, there are at least two 8s. It is impossible for there to be three 8s, because then 8 would be the median. So, we know that three of the numbers are 5, 8, 8. Since the average of the five numbers is 5, the sum of the numbers is 25. The sum of the three we already found is 21, leaving $25 - 21 = 4$ as the sum of the other two numbers. Since these two numbers must be different (8 is the only mode) and they must be positive, they must be 1 and 3. Therefore, the difference between the largest and the smallest integers is $8 - 1 = \boxed{7}$.

13.1.9

(a) We have $1 + 2 + 3 + 4 + 5 + 6 + 7 = 28$, so the average of the first 7 positive integers is $\frac{28}{7} = \boxed{4}$.

(b) In part (a), we saw that the average of the first 7 positive integers is the middle integer, 4. For each number less than the middle integer, there is another number that is the same amount greater than the middle integer. For example, consider the list

$$1, 2, 3, 4, 5, 6, 7, 8, 9, 10, 11.$$

The middle number is 6. We can pair 5 with 7, 4 with 8, 3 with 9, and so on. The two numbers in each pair are the same distance from 6, with one number less than 6 and one number greater than 6. So, the average of the numbers is 6, since the total distance between 6 and the numbers less than 6 equals the total distance between 6 and the numbers greater than 6. Similarly, when n is odd, the average of the first n positive integers is the middle integer in the list. Since the middle integer is the same distance from the first and last numbers in the list, the middle integer is the average of the first and last numbers in the list. So, the middle integer is $\boxed{(n + 1)/2}$.

(c) The average of the first 4 positive integers is $(1 + 2 + 3 + 4)/4 = 2.5$. The average of the first 6 positive integers is $(1 + 2 + 3 + 4 + 5 + 6)/2 = 3.5$. There's no "middle number" in the list this time, but it looks like the average of each list is the average of the two middle numbers in the list. Let's see why this works.

The average of the two middle numbers is obviously the same distance from the two middle numbers themselves. This average is also the same distance from the next number in each direction. For example, consider the list

$$1, 2, 3, 4, 5, 6, 7, 8, 9, 10.$$

The average of the two middle numbers is 5.5, which is the same distance from 5 as from 6. The number 5.5 is also the same distance above 4 as it is below 7, and it is the same distance above 3 as it is below 8, and so on. So, for each number that is some amount less than 5.5, there is another number that is the same amount greater than 5.5. Therefore, the average of the whole list of numbers is 5.5, since the total distance between 5.5 and the numbers less than 5.5 equals the total distance between 5.5 and the numbers greater than 5.5.

Similarly, the average of the first n integers is the average of the two middle integers in the list. The average of the two middle integers is the same as the average of the first and last integers, which is $\boxed{(n+1)/2}$.

Another way to think about this is to realize that all of the numbers in the list from 1 to n can be paired off into pairs that sum to $n + 1$. The first and last numbers together add to $n + 1$. The second and the next-to-last numbers sum to $n + 1$, and so on. Since there are $\frac{n}{2}$ such pairs, the sum of the whole list is $\frac{n}{2}(n + 1)$. Dividing this by n to get the average, we have $\frac{1}{n} \cdot \frac{n}{2}(n + 1) = \frac{1}{2}(n + 1) = \frac{n+1}{2}$.

(d) The sum of the numbers is

$$21 + 22 + 23 + 24 + 25 + 26 + 27 + 28 + 29 + 30 + 31 = 286,$$

so the average is $286/11 = \boxed{26}$. Notice that this is the middle number in the list!

(e) We can reason exactly as in parts (b) and (c). First, suppose there is an odd number of numbers in the list. For each number greater than the middle one, there is a corresponding number less than the middle one such that these two numbers are the same distance from the average. So, the total distance between the middle number and the numbers less than the middle number equals the total distance between the middle number and the numbers greater than the middle number. Therefore, the average is the middle number in the list. The middle number of the list is the same distance from the first and last numbers of the list. This means the middle number is the average of the first and last numbers of the list, $\frac{a+b}{2}$.

The number line below illustrates the situation. Notice that for each number that is some amount less than $\frac{a+b}{2}$, there is another number the same amount greater than $\frac{a+b}{2}$. So, the middle number "balances" the list.

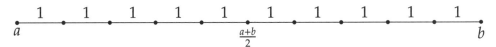

Similar to part (c), if there are an even number of numbers, then we can pair the first number with the last number, the second number with the next-to-last number, and so on. The numbers in each pair have the same average, which is $\frac{a+b}{2}$, and both numbers in each pair are the same distance from this average, one above and one below. So, the average of the whole list of numbers is $\frac{a+b}{2}$, since the total distance between $\frac{a+b}{2}$ and the numbers less than $\frac{a+b}{2}$ equals the total distance between $\frac{a+b}{2}$ and the numbers greater than $\frac{a+b}{2}$.

The number line below illustrates the situation. Again, for each number that is some amount less than $\frac{a+b}{2}$, there is another number the same amount greater than $\frac{a+b}{2}$. That is, $\frac{a+b}{2}$ "balances" the list.

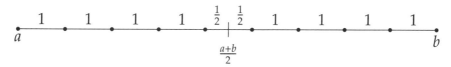

In both cases, the average is the same, $\boxed{\frac{a+b}{2}}$.

Exercises for Section 13.2

13.2.1 The median height tells us the height of the middle player on the team when they are lined up from shortest to tallest. The tallest player must be at least as tall as the middle player, so the tallest player must be at least $\boxed{\text{6 feet, 4 inches}}$ tall. It is indeed possible for the tallest player to have the same height as the median-height player—suppose all the players have the same height.

13.2.2

(a) $\boxed{\text{No}}$. Suppose we form Mediumville by starting with the people in Poorville and then adding in the people from Richville. The people from Poorville have an average wealth of $20,000, so the average before including the people from Richville is $20,000. We expect that if we add people whose average wealth is greater than $20,000, then the average must increase. So, it seems like it is impossible for the average of Mediumville to be $20,000. We'll now make sure this is the case by comparing the total wealth of Poorville and Richville to the total wealth that would be in Mediumville if the average were only $20,000.

Let p the number of people in Poorville and r be the number of people in Richville. So, the total wealth in Poorville is $20,000p$ and the total wealth of Richville is $150,000r$, which means the total wealth in the two cities combined is $20,000p + 150,000r$.

There are $p + r$ people total in Mediumville. So, if the average wealth were $20,000, then the total wealth would be $20,000(p + r)$, which equals $20,000p + 20,000r$. But we know that there must be $20,000p + 150,000r$ total in Mediumville! $20,000r$ is always less than $150,000r$ (since r is positive). So, there has to be more wealth in Mediumville than there would be if the average were just $20,000.

(b) $\boxed{\text{Yes}}$. It's possible that one person in Richville had all of the wealth in Richville, and everyone else had $0. Then, if all the people with $0 merge with Poorville while the rich person stays out, the average wealth of the new city will be lower than that of Poorville. This is because Mediumville will have the same total wealth as the original Poorville, but Mediumville will have more people than Poorville did.

(c) Suppose there are t people in each city. So, the total wealth of Poorville is $20,000t$ and the total wealth of Richville is $150,000t$. Mediumville then has $2t$ people and a total wealth of

$20,000t + \$150,000t = \$170,000t$. Therefore, the average wealth of Mediumville is

$$\frac{\$170,000t}{2t} = \frac{\$170,000}{2} \cdot \frac{t}{t} = \boxed{\$85,000}.$$

Note that $85,000 is the average of Poorville's average wealth ($20,000) and Richville's average wealth ($150,000).

(d) $\boxed{\text{Richville}}$ was larger. In the previous part, we saw that if the two cities had been the same size, then the average wealth of Mediumville would be $85,000. We expect that to get an average wealth that is greater than this, we need to have more rich people.

To make sure that this intuition is correct, we again let there be p people from Poorville and r people from Richville, so that the combined wealth of the two towns is $20,000p + \$150,000r$. Since the population of Mediumville is $p + r$ and the average wealth of Mediumville is $120,000, the total wealth in Mediumville is $120,000(p + r) = \$120,000p + \$120,000r$. The total wealth in Mediumville must equal the combined wealth of Poorville and Richville (since that's where Mediumville came from), so

$$\$20,000p + \$150,000r = \$120,000p + \$120,000r.$$

Subtracting $120,000r from both sides gives

$$\$20,000p + \$30,000r = \$120,000p.$$

Subtracting $20,000p from both sides gives $30,000r = \$100,000p$. Dividing both sides by $30,000 gives

$$r = p \cdot \frac{\$100,000}{\$30,000} = p \cdot \frac{10}{3} = p \cdot \left(3\frac{1}{3}\right).$$

Therefore, r is more than 3 times p, which means that there definitely were more people in Richville!

13.2.3 The median score is the "middle score," which means that Omar scores at least 70 at least half the time. Similarly Nick gets 50 or below at least half the time. Because Omar's average is far below 70, it appears that, although half his scores are at least 70, his bad tests are much more below 70 than his good tests are above 70. So, Omar is the student who usually does OK, but when he does badly, he does very badly. Similarly, because Nick's average is far above 50, even though at least half Nick's scores are 50 or below, we know that his high scores must be quite high, to pull up his average despite all those scores at 50 or below. So, Nick is the student who usually doesn't do very well, but when he does well, he does extremely well.

Exercises for Section 13.3

13.3.1

(a) The most common number of pencils is 2, so the mode is $\boxed{2}$.

(b) Reading the chart, the total number of students is

$$5 + 14 + 17 + 12 + 2 + 4 + 2 + 1 = 57.$$

The total number of pencils is

$$0 \cdot 5 + 1 \cdot 14 + 2 \cdot 17 + 3 \cdot 12 + 4 \cdot 2 + 5 \cdot 4 + 6 \cdot 2 + 7 \cdot 1 = 131.$$

Therefore, the average number of pencils to the nearest hundredth is $131/57 \approx \boxed{2.30}$.

(c) Since there are 57 students, the middle number of pencils is the 29th lowest. Suppose we line the students up in increasing order from left to right based on how many pencils each has. So, the students with no pencils are on the far left, and the student with 7 pencils is on the far right. The 29th student has the median number of pencils. There are $5 + 14 = 19$ students who have fewer than 2 pencils. There are 17 students who have exactly 2 pencils, so the 20th through 36th students have 2 pencils. Most notably, the 29th student has 2 pencils, so the median number of pencils is $\boxed{2}$.

13.3.2

(a) The most common entry is 12, so the mode is $\boxed{12}$. The stem-and-leaf presentation of the data conveniently already has the entries in numerical order from least to greatest. There are 20 entries, so the median is the average of the two middle entries, which are the tenth and the eleventh. The tenth entry is 12 and the eleventh is 14, so the median number of hairs is $(12 + 14)/2 = \boxed{13}$.

To find the average, we find the total number of hairs. To find this sum, we add up the tens and the units separately. The sum of all of the digits to the right of the line is 85. There are no tens contributed to the sum in the first row. In the second row there are 5 entries that have tens digit 1, contributing $5 \cdot 1 = 5$ tens to the sum. The third row contributes $2 \cdot 2 = 4$ tens to the sum, the fourth row contributes $2 \cdot 4 = 8$ tens, the fifth row contributes $2 \cdot 5 = 10$ tens, and the last row contributes $3 \cdot 7 = 21$ tens. This gives a total of 48 tens, or 480. Combining this with the units' sum, we have a total of 565. Therefore, the average is $565/20 = \boxed{28.25}$.

(b) The mode won't change at all, and the median will only go from 13 (the average of the tenth and eleventh entries) to 14 (the eleventh entry). The average will go way up because the sum will increase a great deal, while the number of people in the club only goes up by one. So, the $\boxed{\text{average}}$ will be affected the most.

(c) Using the correct number of hairs, the total number of hairs is $20 \cdot 26.75 = 535$. So, we need to reduce the total number of hairs by 30 by only changing one digit in the table. Since we must change the sum by more than 9, we must change a tens digit, not a units digit. Only changing the tens digit 7 allows us to change the sum by 30, since there are 3 entries on 7's row, but no other rows have exactly 1 or exactly 3 entries. So, the $\boxed{7}$ is incorrect, and we reduce the total number of hairs by 30 by changing the 7 to a $\boxed{6}$.

13.3.3

(a) The largest change in height is between her second and third birthdays, so Spot grew the most in her $\boxed{\text{third}}$ year.

(b) Spot still grew quite a bit between her third and fourth birthdays, but didn't grow much after that. Since she didn't grow much after her fourth birthday, she was probably $\boxed{3}$ when she became an adult. We can't be 100% sure she was 3 when became an adult. It's possible she was still growing an inch a day on her fourth birthday and the day after, and then she stopped growing rapidly. But it is likely that her rapid growth slowed sometime between her 3rd and 4th birthdays.

(c) At the end of her eighth year, Spot is 90 inches tall. So, on average she grew $90/8 = \boxed{11.25}$ inches a year.

13.3.4

(a) Bart beat Bugs by 8%, so he earned $0.08(650) = \boxed{52}$ votes more than Bugs Bunny.

(b) We could figure out how many votes each candidate received, but we don't have to! We know that the percentages received by Bugs Bunny and Mickey Mouse are the same as in the first election, 28% and 8%, respectively. That leaves $100\% - 28\% - 8\% = 64\%$ to be divided among Wile E. Coyote and SpongeBob. Since Wile E. Coyote received three times as many votes as SpongeBob, we know that for every vote SpongeBob earned, Wile earned three. This means than Wile earned $\frac{3}{4}$ of the 64% who voted for a new candidate, or $\frac{3}{4}(64\%) = 48\%$. This leaves SpongeBob with the remaining $64\% - 48\% = 16\%$. The resulting pie chart is shown at the right.

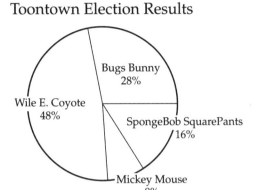

Toontown Election Results

Review Problems

13.16 The total weight of the dogs is $4 \cdot 63 = 252$ pounds. The total weight of all 7 animals is $7 \cdot 41 = 287$ pounds. So, the total weight of all three cats is $287 - 252 = 35$ pounds, which means the average weight of each cat is $\frac{35}{3} = \boxed{11\frac{2}{3}}$ pounds.

13.17 When the 15-year-old joins and the 11-year-old quits, the sum of the ages of the people on the team increases by 4 years. Since there are 10 people on the team, this increases the average age by $\frac{4}{10} = 0.4$ years, to $13.5 + 0.4 = \boxed{13.9}$ years.

13.18 Since the average of the five numbers is 18, their sum is $5 \cdot 18 = 90$. The total increase of the numbers is $1 + 2 + 3 + 4 + 5 = 15$, so the sum of the increased numbers is $90 + 15 = 105$. Therefore, the average of the increased numbers is $\frac{105}{5} = \boxed{21}$.

We also could have noticed that the average increase is the average of 1, 2, 3, 4, and 5, which is 3. Since the average increase is 3, the average of the increased numbers is $18 + 3 = 21$.

13.19 The total score on the first six tests is $6 \cdot 84 = 504$, and the total score on the first eight tests is $8 \cdot 86 = 688$, so the total score on the last two tests is $688 - 504 = 184$. This means the average of the last two scores is $\frac{184}{2} = \boxed{92}$.

We also could have noted that the first six tests are on average 2 points below the final average. So, the first 6 tests are $6 \cdot 2 = 12$ points total below average. This means the final two tests must be a total of 12 points above average. That's $\frac{12}{2} = 6$ points per test on average that these two final scores must exceed the overall average, which means the average on the final two tests must be $86 + 6 = \boxed{92}$.

13.20 $\boxed{\text{Yes}}$. As we saw in Exercise 13.1.9, the average of a group of consecutive numbers equals the middle number if there are an odd number of numbers, and the average of the group equals the average of the two middle terms if there are an even number of numbers. In both cases, the average equals the median.

13.21

(a) *Average:* Suppose there are n numbers in the list. Adding 12 to each number adds $12n$ to the sum of the numbers. Since the sum is increased by $12n$ while the number of numbers stays the same, the average increases by $\frac{12n}{n} = 12$. So, the new average is $14 + 12 = \boxed{26}$.

   Median: Suppose we place the numbers in Larry's list in increasing order. Adding 12 to each number in this list won't change the order; Moe's new list is in increasing order, too. So, the middle number of Moe's list is 12 more than the middle number of Larry's list. This means that Moe's list's median is $21 + 12 = \boxed{33}$. (If there are an even number of numbers, then the median is the average of the middle two numbers. We can use the same argument as in our first paragraph to see that Moe's median is still 12 more than Larry's median.)

   Mode: If two numbers are different in Larry's list, then the corresponding numbers are different in Moe's list. If two numbers are the same in Larry's list, then the corresponding numbers are the same in Moe's list. So, for any number x in Larry's list, the number of times x appears in Larry's list equals the number of times that $12 + x$ appears in Moe's list. Therefore, the number that appears most frequently in Moe's list is 12 more than the number that appears most frequently in Larry's list. This means that the mode of Moe's list is $11 + 12 = \boxed{23}$.

(b) *Average:* We follow essentially the same process as in the previous part. Doubling each number in a sum doubles the sum. So, the sum of the numbers in Curly's list is double the sum of the numbers in Larry's list. The lists have the same number of numbers, so Curly's average is double Larry's average. This means that Curly's list has average $2 \cdot 14 = \boxed{28}$.

   Median: Suppose we place the numbers in Larry's list in increasing order. Multiplying each number in this list by 2 won't change the order; Curly's new list is in increasing order, too. So, the middle number of Curly's list is 2 times the middle number of Larry's list. This means that Curly's list's median is $2 \cdot 21 = \boxed{42}$. (If there are an even number of numbers, then the median is the average of the middle two numbers. We can use the same argument as in our first paragraph to see that Curly's median is still 2 times Larry's median.)

   Mode: The number of times a certain number appears in Larry's list equals the number of times double that number appears in Curly's list. Therefore, the number that appears most frequently in Curly's list is 2 times the number that appears most frequently in Larry's list. This means that the mode of Curly's list is $2 \cdot 11 = \boxed{22}$.

13.22 All 5 of the numbers are very close to 940385980. If we subtract 940385980 from each number, our list becomes

$$8, 14, 23, 1, 11.$$

The average of the numbers in this list is $\frac{8+14+23+1+11}{5} = \frac{57}{5} = 11.4$. Each number in the original list is 940385980 greater than the corresponding number in our list with average 11.4. Therefore, the average of the numbers in the original list is $11.4 + 940385980 = \boxed{940385991.4}$.

CHAPTER 13. DATA AND STATISTICS

13.23

(a) In order to make the average as small as possible, the sum of the chosen numbers must be as small as possible. Because the median of the numbers is 91, we know that five of the nine numbers are at least 91. To make the sum as small as possible, we let all five of these numbers equal 91. The smallest the other four numbers can be is 1, so the smallest the sum of all nine numbers can be is $4 \cdot 1 + 5 \cdot 91 = 459$. This means that the smallest the average can be is $\frac{459}{9} = \boxed{51}$.

(b) There is $\boxed{\text{no maximum}}$. The median only tells us that at least one of the numbers is 91, and four other numbers are at least 91. It doesn't place any limit on how large those four other numbers are, so they can be as large as we like. Therefore, there's no limit on how large the sum of all 9 numbers is, which means there is no limit to how large the average of the numbers can be.

(c) Since the average is 91, the sum of the 9 chosen numbers is $9 \cdot 91 = 819$. We might have one friend choose 811 and each of the other eight friends choose 1. The median of these nine choices is 1 and the sum is $811 + 8 \cdot 1 = 819$. Therefore, it is possible for 1 to be the median. Since the friends must choose positive integers, no one can choose a number smaller than 1. This means that $\boxed{1}$ is the smallest possible median.

(d) As in the previous part, the sum of the chosen numbers is $9 \cdot 91 = 819$. Suppose the median is x. Since the sum of all nine numbers is 819, we must have

$$x = 819 - (\text{sum of the other eight numbers}).$$

Therefore, to make x as large as possible, we must make the sum of the other eight numbers as small as possible.

Since x is the median of the nine numbers, there are four other numbers that are no greater than x and four others that are no less than x. The smallest the numbers in the first group can be is 1, and the smallest the numbers in the second group can be is x. So, the median is as large as possible when four of the numbers are 1 and the other five are the same. If four of the numbers are 1, then the sum of the other five is $819 - 4 = 815$. Since these other five numbers must be the same to make the median as large as possible, the greatest the median can be is $\frac{815}{5} = \boxed{163}$.

As an extra challenge, try to figure out the greatest that the median can be if we have more friends choosing favorite positive integers, but the average of the integers chosen is still 91. What if there are 100 friends? 1000 friends? A million friends?

13.24

(a) $\boxed{\text{Yes, List B}}$. The average of each list is the sum of the list divided by the number of numbers in the list. None of the numbers in List A is greater than 1000, but List B has a number that is greater than fifty million! All the numbers in both lists are positive, so the sum of List B is far greater than the sum of List A. Therefore, the average of List B is far greater than that of List A.

(b) $\boxed{\text{No}}$. We have to put the numbers in order to figure out what the median is in each list. We can't tell at a glance which list is more likely to have a larger middle number. (For those of you keeping score at home, the median of List A is 87 and the median of list B is 65.)

(c) One extreme value, such as the large number in List B, can have a huge effect on the average, but not a very large impact on the median. That is, outliers affect the average more than they affect the median.

13.25 From the chart, we see that 20% of the students do their homework in the dining room. Therefore, there are $0.2 \cdot 880 = \boxed{176}$ students who do their homework in the dining room.

13.26

(a) The number in the upper left corner is 0 because the distance between Adrian's house and Adrian's house is obviously 0. Similarly, each number on the diagonal from upper left to lower right must be 0. This allows us to fill in two of the missing entries with 0 (corresponding to Laurie's house and Walter's house).

Next, we note that the entry in Adrian's row and Dan's column is the same as the entry in Dan's row and Adrian's column, since both of these entries represent the distance between Adrian's house and Dan's house. Similarly, the distance between each pair of houses appears twice in the table. This allows us to fill in all the rest of the missing entries except the ones corresponding to the distance between Walter's house and Jon's house.

Looking at the bottom two rows, we see that the distances from each person to Walter and to Jon differ by 1 mile, so we conclude that Walter's house and Jon's house are 1 mile apart. The completed table is shown below:

	Adrian	Dan	Laurie	Jon	Walter
Adrian	0	3	15	10	11
Dan	3	0	12	7	8
Laurie	15	12	0	5	4
Jon	10	7	5	0	1
Walter	11	8	4	1	0

(b) The largest number in the table is 15, which tells us that $\boxed{\text{Laurie and Adrian}}$ are the farthest apart. The smallest number corresponding to a distance between two different people is 1, so $\boxed{\text{Jon and Walter}}$ are closest.

(c) Since Laurie and Adrian are farthest apart, they are at opposite ends of the road. We can then use the distance between Adrian (or Laurie) and each of the others to place all 5 people along the road. Dan is closest to Adrian, so we place him first, 3 miles away. Next closest is Jon, who is 10 miles from Adrian. Jon is $10 - 3 = 7$ miles farther from Adrian than Dan, so Jon is 7 miles from Dan, as shown. Similarly, Walter is $11 - 10 = 1$ mile from Jon and Laurie is $15 - 11 = 4$ miles from Walter:

(We could also reverse the entire map, placing Adrian at the far right and Laurie at the far left.)

Challenge Problems

13.27 The sum of the first four numbers is $4 \cdot 5 = 20$. The sum of the last four numbers is $4 \cdot 8 = 32$. If we add these two sums, we find that $20 + 32 = 52$ is the sum of all seven numbers plus an extra copy of the middle number. The sum of all seven numbers is $7 \cdot \left(6\frac{4}{7}\right) = 46$, so the middle number must be $52 - 46 = \boxed{6}$.

13.28   In order for there to be a unique mode, one of the numbers must be repeated, but we can't have two numbers be repeated. It doesn't matter which is the repeated value, A or B, so we'll let A be the repeated value. There are four possibilities to consider:

Case 1: 5 is repeated. Our list then is $5, 5, 7, 8, B$. Since the mode is 5, the median and mean must also be 5. Since 7 and 8 together are a total of $2 + 3 = 5$ greater than the mean, we know that B is 5 less than the mean. This tells us that $B = 5 - 5 = 0$, making our list $0, 5, 5, 7, 8$. This list does indeed have mean, median, and mode equal to 5. In this case, $A + B = 5$.

Case 2: 7 is repeated. Our list then is $5, 7, 7, 8, B$. The median and mean must also be 7. Since 8 is 1 greater than the mean and 5 is 2 less than the mean, the remaining number must also be 1 greater than the mean. That makes the list $5, 7, 7, 8, 8$. But the mode of this list is not unique! So, it is impossible for 7 to be the repeated number.

Case 3: 8 is repeated. Our list then is $5, 7, 8, 8, B$. The median and mean must also be 8. Since 5 is 3 less than the mean and 7 is 1 less than the mean, the last number must be $3 + 1 = 4$ greater than the mean. This makes our list $5, 7, 8, 8, 12$. This list has mean, median, and mode equal to 8, and we have $A + B = 20$.

Case 4: A and B are equal. Our list then is $5, 7, 8, A, A$. The mean and median must also be A. The mean of this list is $\frac{5+7+8+A+A}{5}$. Simplifying this and setting it equal to A gives

$$\frac{20 + 2A}{5} = A.$$

Multiplying both sides by 5 gives $20 + 2A = 5A$. Subtracting $2A$ from both sides gives $20 = 3A$, so $A = \frac{20}{3} = 6\frac{2}{3}$. This makes our list $5, 6\frac{2}{3}, 6\frac{2}{3}, 7, 8$, which has mean, median, and mode equal to $6\frac{2}{3}$. In this case, $A + B = 13\frac{1}{3}$.

Combining all of our cases, the possible values of $A + B$ are $\boxed{5, 13\frac{1}{3}, \text{ and } 20}$.

13.29   We start by noting that there are $2 + 4 + 8 + 10 + 7 = 31$ people total.

(a) The number of people who are less than 80 years old is $2 + 4 + 8 = 14$. There are 7 people in their 90s and 10 in their 80s. So, less than half the people are below 80 years old and more than half are below 90 years old. This means that the median age is in the 80s. Therefore, the smallest it can be is $\boxed{80}$, which would indeed occur if all the people in their 80s are exactly 80.

(b) As explained in the first part, the median must be in the 80s. Therefore, the greatest the median age can be is $\boxed{89}$. For example, if all the people in their 80s were 89 years old, the median would be 89.

(c) There are 31 people. To make the average of the ages as small as possible, we must make the sum of the ages as small as possible. The least possible sum occurs when each of the ages is as low as possible, which is the corresponding multiple of 10 for each age. So, the lowest possible sum is

$$2 \cdot 50 + 4 \cdot 60 + 8 \cdot 70 + 10 \cdot 80 + 7 \cdot 90 = 10(2 \cdot 5 + 4 \cdot 6 + 8 \cdot 7 + 10 \cdot 8 + 7 \cdot 9) = 2330.$$

Therefore, the least possible average is $\frac{2330}{31} = \boxed{75\frac{5}{31}} \approx 75.16$.

(d) The average is greatest if all the people are as old as possible. This will occur if all the people in their 50s are 59, all the people in their 60s are 69, and so on. To sum these ages, we add the tens and units separately. We found the sum of these tens in part (c) as 2330. The sum of the units is $31 \cdot 9 = 279$, so the sum of all 31 ages is $2330 + 279 = 2609$, and the desired greatest possible average is $\frac{2609}{31} = \boxed{84\frac{5}{31}} \approx 84.16$. Seeing that this result is exactly 9 greater than the result for part (c), we note that adding 9 to each of the ages in part (c) results in adding 9 to the average, as expected.

13.30

(a) $\boxed{\text{Yes}}$. The total amount by which the above-average numbers in the first list exceeds the average must equal the total amount by which the average exceeds the below-average numbers in the first list. Similarly, the total amount by which the above-average numbers in the second list exceeds the average must equal the total amount by which the average exceeds the below-average numbers in the second list. So, when we combine the two lists, the total amount by which numbers above the common average exceed the common average equals the total amount by which the common average exceeds the below-average numbers. This means that the combined list must have the same average as the common average of the two lists.

(b) $\boxed{\text{Yes}}$. Suppose we start with the first list, in order. The median must be the middle number of this list, or the average of the two middle numbers. When we combine the second list with this first list, there must be just as many new numbers above this median as below, since this median is the middle number (or between the two middle numbers) of the second list, too. So, the common median will be the median of the new list, too.

(c) $\boxed{\text{Yes}}$. The mode is the number that appears most often. If a number appears more than any other in the first list, and it appears more than any other in the second list, then it definitely will have more total appearances across both lists than any other number. So, the mode of the combined list is the same as the mode of each individual list.

(d) $\boxed{\text{Yes}}$. Suppose the lists are 5,5,5,3,3 and 2,2,2,3,3. The mode of the first list is 5 and the mode of the second is 2. The combined list is 2,2,2,3,3,3,3,5,5,5. This list has mode 3.

13.31 We start by noting that the total allowance of my friends is $\$13 + \$17 + \$24 + \$30 = \$84$. We then consider three cases:

Case 1: My weekly allowance is at most $\$17$. The median allowance then is $\$17$. In order for the average also to be $\$17$, the total of all 5 allowances must be $5 \cdot (\$17) = \$85$. So, my allowance must be $\$85 - \$84 = \$1$, which is indeed no more than $\$17$, and I need to have a conversation with my parents.

Case 2: My weekly allowance is at least $\$24$. The median allowance then is $\$24$. In order for the average also to be $\$24$, the total of all 5 allowances must be $5 \cdot (\$24) = \$120$. So, my allowance must be $\$120 - \$84 = \$36$, which is greater than $\$24$, as expected.

Case 3: My weekly allowance is between $\$17$ and $\$24$. The median allowance then is my allowance. Let my allowance be x dollars. The average allowance then is $\frac{84+x}{5}$ dollars. Since this must equal the median, which is my allowance, we must have

$$\frac{84 + x}{5} = x.$$

Multiplying both sides by 5 gives $84 + x = 5x$, so $84 = 4x$ and $x = 21$, which is between 17 and 24 as desired. (My allowance is the average of my four friends' allowances in this case. Is this a coincidence?)

Combining the three cases, the possible values of my weekly allowance are $\boxed{\$1, \$21, \text{ and } \$36}$.

13.32 Since each card receives one vote per person, the four rows must each sum to 124. Similarly, each column must sum to 124. We include a row and a column to our table for these totals, and we have the table shown on the right. Looking at this table, we expect that the erroneous rows are those for Pokemon and Yu-Gi-Oh!, and the erroneous columns are those for 2nd and 3rd. The

	1st	2nd	3rd	4th	Total
Magic	34	56	23	11	124
Pokemon	12	14	38	64	128
SET	57	41	23	3	124
Yu-Gi-Oh!	21	19	44	46	130
Total	124	130	128	124	

Pokemon row total is 4 too high, as is the total for the 3rd place column. So, we can fix both by subtracting 4 from the number of 3rd place votes received by Pokemon. Similarly, we can fix the Yu-Gi-Oh! row and the 2nd place column by subtracting 6 from the number of 2nd place votes Yu-Gi-Oh! received. The resulting table is shown below.

	1st	2nd	3rd	4th	Total
Magic	34	56	23	11	124
Pokemon	12	14	**34**	64	124
SET	57	41	23	3	124
Yu-Gi-Oh!	21	**13**	44	46	124
Total	124	124	124	124	

13.33 The average of the smallest 9 numbers is 1 less than the average of all 10 numbers. So, on average, each of the 9 smallest numbers is 1 less than the average of all 10 numbers. This means that the 9 smallest numbers together are a total of $9 \cdot 1 = 9$ less than the average of all 10 numbers. So, the largest number must be 9 greater than the average of all 10 numbers.

Similarly, the average of the largest 9 numbers is 2 greater than the average of all 10 numbers. So, on average, each of the 9 largest numbers is 2 greater than the average of all 10 numbers. This means that the 9 largest numbers together are a total of $9 \cdot 2 = 18$ greater than the average of all 10 numbers. So, the smallest number must be 18 less than the average of all 10 numbers.

Since the largest number is 9 greater than the average of all 10 numbers, and the smallest number is 18 less than this average number, we know that the smallest and largest are $9 + 18 = \boxed{27}$ apart.

CHAPTER **14**

Counting

24® Cards

First Card $(9 + 10) \cdot 2 - 14.$

Second Card $(21 + 9)/3 + 14.$

Third Card $(20 - 18) \cdot 14 - 4.$

Fourth Card $(10 - 92/14) \cdot 7.$

Exercises for Section 14.1

14.1.1

(a) We subtract 44 from each number in the list to get $1, 2, 3, \ldots, 49$, so there are $\boxed{49}$ numbers. We also could have used the $b - a + 1$ formula we proved in the text for how many numbers are between a and b inclusive: $93 - 45 + 1 = 49$.

(b) Consecutive numbers in the list are 4 apart, and each number is 1 more than a multiple of 4. So, we subtract 1 from each number to allow us to make them all multiples of 4:

$$-28, -24, -20, \ldots, 32, 36.$$

Then, we divide each number by 4 to make the numbers each 1 apart:

$$-7, -6, -5, \ldots, 8, 9.$$

Finally, we add 8 to each number to get $1, 2, 3, \ldots, 17$, which means there are $\boxed{17}$ numbers in the list.

(c) Consecutive numbers in the list are 3 apart, so we divide each number by 3 to get

$$54, 53, 52, \ldots, 23, 22.$$

We then subtract 21 from each number to get $33, 32, 31, \ldots, 3, 2, 1$. So, our list consists of the integers from 1 to 33, but in reverse order. This means there are $\boxed{33}$ numbers in the list.

14.1.2 Since I'm 18th when counting from the front, there are 17 people in front of me. Since I'm 24th when counting from the back, there are 23 people behind me. Counting me, the people in front of me, and the people behind me, there are $1 + 17 + 23 = \boxed{41}$ people in the line.

14.1.3 We start with a Venn diagram, as shown at the right. We place 23 in the overlap region for the students who took both languages. Subtracting these 23 from the 70 students who took French leaves 47 who took only French; we place this in the French-only portion of the diagram. Similarly, $140 - 23 = 117$ took only Spanish. This leaves $220 - 47 - 23 - 117 = \boxed{33}$ who didn't take either language.

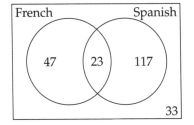

14.1.4 We build a Venn diagram for the percentages of students who own computers or are in the band as shown at the right. We place 10% outside both circles to account for the students who do neither. We place x in the middle for the students who do both. This leaves $80\% - x$ who own computers but are not in the band, and $40\% - x$ who are in band but do not own computers. The total of all the entries in the diagram must be 100%, so we have $(80\% - x) + x + (40\% - x) + 10\% = 100\%$. Simplifying the left side gives $130\% - x = 100\%$, so $x = \boxed{30\%}$.

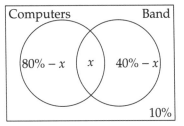

14.1.5 Since $\frac{n}{3}$ is an integer, n must be a multiple of three. However, we need $\frac{n}{3}$ to have three digits. The smallest three-digit number is 100, so the smallest possible value of n is 300. The possible values of n then are $300, 303, 306, \ldots$. But what is the largest possible value of n? Since $3n$ must also have three digits, the largest it can be is 999, which occurs when $n = 333$. Therefore, the possible values of n are $300, 303, 306, \ldots, 333$. Dividing each by 3 gives $100, 101, 102, \ldots, 111$, and subtracting 99 from each gives $1, 2, 3, \ldots, 12$. So, there are $\boxed{12}$ numbers that satisfy the problem.

14.1.6 We don't even need a Venn diagram for this problem! Since 15 students are in the Science Club, and 80% of these students are also in the Math Club, we know that $(0.80)(15) = 12$ students are in both clubs. Let x be the total number of students in the Math Club. Since 30% of the students in the Math Club are in both clubs, we know that $0.3x$ students are in both clubs. But we already know that there are 12 students in both clubs, so $0.3x = 12$. Dividing by 0.3 gives $x = \frac{12}{0.3} = \frac{120}{3} = \boxed{40}$ students in the Math Club.

14.1.7 We draw a Venn diagram with one circle for red cars and one for 4-door cars. The overlap is the red 4-door cars, and the region outside both circles is the white 2-door cars. So, we place 4 outside both circles. We let x be the number of red 4-door cars, so there are $15 - x$ red 2-door cars and $8 - x$ white 4-door cars. There are a total of 24 cars, so $(15 - x) + x + (8 - x) + 4 = 24$. Simplifying the left side gives $27 - x = 24$, so $x = \boxed{3}$ red 4-door cars.

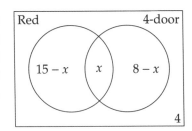

14.1.8   The two-digit multiples of 3 are $12, 15, 18, \ldots, 99$. Dividing each by 3 gives $4, 5, 6, \ldots, 33$, and subtracting 3 from each gives $1, 2, 3, \ldots, 30$. So, there are 30 two-digit multiples of 3. The two-digit multiples of 5 are $10, 15, 20, \ldots, 95$. Dividing each by 5 gives $2, 3, 4, \ldots, 19$, so there are 18 two-digit multiples of 5. It looks like there are $30 + 18 = 48$ two-digit numbers that are multiples of 3 or 5. However, we have to be careful; there are numbers that are multiples of both 3 and 5—these are the multiples of 15. The two-digit multiples of 15 are $15, 30, 45, 60, 75, 90$; there are 6 of them. These are counted among the multiples of 3 and among the multiples of 5, so they are counted twice in our total of $30 + 18$ above. To count them only once, we subtract the count of multiples of 15 once from this total, for a total of $30 + 18 - 6 = \boxed{42}$. We could also use a Venn diagram to do our counting, as shown above.

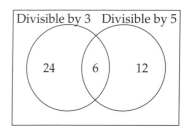

Exercises for Section 14.2

14.2.1   For each of the 7 choices I have for a shirt, I have 4 choices for a tie, giving me $7 \cdot 4 = \boxed{28}$ shirt-and-tie outfits.

14.2.2   Similar to the previous exercise, I have $9 \cdot 9 = 81$ shirt-and-tie outfits, but I refuse to wear the 9 outfits that consist of a shirt and a tie of the same color. This leaves me $81 - 9 = \boxed{72}$ outfits. Another way to think of this is to suppose I choose my shirt first, and then my tie. I have 9 choices for the shirt, but then I only have 8 choices for the tie because I can't pick the one that is the same color as my shirt. So, I have $9 \cdot 8 = 72$ possible outfits, as before.

14.2.3   I have 5 choices for which person stands at the front of the line, then 4 people remaining to choose from for the next spot in line, then 3 people to choose from for the next spot in line, then 2 people for the next-to-last spot, and just 1 person remaining to put at the end of the line. This gives me a total of $5 \cdot 4 \cdot 3 \cdot 2 \cdot 1 = \boxed{120}$ ways to arrange the people in line.

14.2.4   She has 4 choices for each digit, so she can make $4 \cdot 4 \cdot 4 = 4^3 = \boxed{64}$ different three-digit house numbers.

14.2.5   We have 5 choices for the first digit, and 5 choices for the second digit. For the third digit, we can't use either of the two digits we have already used, so there are 8 choices for the third digit. For the fourth digit, we can't use any of the three digits we have already used, so there are 7 choices for the fourth digit. This gives us a total of $5 \cdot 5 \cdot 8 \cdot 7 = \boxed{1400}$ ways to choose the digits.

14.2.6   A three-digit number that reads the same forwards as backwards must have the same hundreds and units digit. We have 9 choices for that digit, since the hundreds digit cannot be 0. We can use any digit as the middle digit, so we have 10 choices for the middle digit. This gives us $9 \cdot 10 = \boxed{90}$ three-digit palindromes.

14.2.7   At first, it might seem that the answer is $5 \cdot 4 \cdot 3 \cdot 2 \cdot 1$, since it looks like I have 5 choices for the first book, 4 for the next, and so on. But we have to have math books on the ends! So, we choose them first. We have 2 choices for which math book to put on the left end, and we have to save the other math book for the right end. That leaves 3 choices for which book to put to the right of the first math book, then 2 choices for the next non-math book, and 1 choice for the last non-math book. Then, we place our other math book on the right end of the shelf. This gives us a total of $2 \cdot 3 \cdot 2 \cdot 1 \cdot 1 = \boxed{12}$ ways to shelve the books.

14.2.8 There are a number of ways to tackle this problem, but perhaps the most clever is to think about making a three-digit number with one 0 as follows:

Step 1: Place the 0. There are 2 places where we can put the 0: in the tens place or the units place.

Step 2: Place digits from 1-9 in the other two places. No matter where we place the 0, there are 9 choices for each of the other two digits in our three-digit number. Therefore, for each possible placement of the 0, there are $9 \cdot 9 = 81$ ways to choose the other two digits to make a three-digit number with one zero.

So, there are 2 choices for where we can put the 0, and each gives us 81 three-digit numbers with one 0. This gives us a total of $2 \cdot 81 = \boxed{162}$ three-digit numbers with one 0.

Exercises for Section 14.3

14.3.1 The 9 one-digit numbers contribute 1 digit each, for a total of 9 digits. The 90 two-digit numbers from 10 to 99 contribute 2 digits each, for a total of $90 \cdot 2 = 180$ digits. There are 51 three-digit numbers from 100 to 150. These contribute a total of $51 \cdot 3 = 153$ digits. Therefore, there are a total of $9 + 180 + 153 = \boxed{342}$ digits.

14.3.2 There are two possibilities for which hat I choose. We tackle these separately:

Case 1: I choose the hat with balls 1 through 15. I then have 15 possibilities for the first ball I choose, and for each of these choices I have 14 possibilities for the second ball. Therefore, there are a total of $15 \cdot 14 = 210$ possible selections in this case.

Case 2: I choose the hat with balls 16 through 25. There are $25 - 16 + 1 = 10$ balls in this hat. So, I have 10 choices for the first ball and 9 for the second, for a total of $10 \cdot 9 = 90$ choices.

Combining the two cases, we have $210 + 90 = \boxed{300}$ different ordered selections.

14.3.3

(a) We go through either B or C to get to D. The number of paths going from A to D through B is $2 \cdot 1$, and the number of paths going from A to D through C is $2 \cdot 3$. Thus the total number of paths from A to D is $(2 \cdot 1) + (2 \cdot 3) = \boxed{8}$.

(b) To get from D to H, we go through one of $E, F,$ or G. The three cases give a total of $(1 \cdot 1) + (2 \cdot 3) + (2 \cdot 1) = \boxed{9}$ paths from D to H.

(c) All paths must go through D. We saw in part (a) that there are 8 paths from A to D. We saw in part (b) that there are 9 paths from D to H. The choice of path from A to D is independent of the choice of path from D to H. Therefore we multiply the number of paths from A to D by the number of paths from D to H to get the number of paths from A to H. The answer is $8 \cdot 9 = \boxed{72}$.

14.3.4 We always have 10 options for the last digit. For each choice of the second digit, we have a different number of options for the first digit. So, we consider each possible second digit as a separate case:

Case 1: The second digit is 0. Then, three times the second digit is 0, so the first digit can be any number from 1 to 9. This gives 9 choices for the first digit, and there are 10 choices for the last digit, so there are $9 \cdot 10 = 90$ three-digit numbers that satisfy this case.

Case 2: The second digit is 1. Then, three times the second digit is 3, so the first digit can be any number from 3 to 9. This gives 7 choices for the first digit, and there are 10 choices for the last digit, so there are $7 \cdot 10 = 70$ three-digit numbers that satisfy this case.

Case 3: The second digit is 2. Then, three times the second digit is 6, so the first digit can be any number from 6 to 9. This gives 4 choices for the first digit, and there are 10 choices for the last digit, so there are $4 \cdot 10 = 40$ three-digit numbers that satisfy this case.

Case 4: The second digit is 3. Then, three times the second digit is 9, so the first digit can only be 9. This gives 1 choice for the first digit, and there are 10 choices for the last digit, so there are $1 \cdot 10 = 10$ three-digit numbers that satisfy this case.

Case 5: The second digit is 4 or greater. Then, three times the second digit is greater than 10, so there are no possibilities for the first digit. Therefore, no three-digit numbers satisfy this case.

Combining our cases, we have a total of $90 + 70 + 40 + 10 = \boxed{210}$ numbers that satisfy the problem.

14.3.5 We list the possible last digits, and the possibilities for the first two digits for each choice of last digit:

Last digit	First two digits
0	–
1	10
2	11, 20
3	12, 21, 30
4	13, 22, 31, 40
5	14, 23, 32, 41, 50
6	15, 24, 33, 42, 51, 60
7	16, 25, 34, 43, 52, 61, 70
8	17, 26, 35, 44, 53, 62, 71, 80
9	18, 27, 36, 45, 54, 63, 72, 81, 90

The third digit can be any of the 10 digits. The answer is $(1 + 2 + 3 + 4 + 5 + 6 + 7 + 8 + 9) \cdot 10 = \boxed{450}$.

There is a quicker solution. Once we choose the last digit, we need the first digit to be between 1 and the chosen last digit, inclusive. (For example, if the last digit is 6, then the first digit must be 1, 2, 3, 4, 5, or 6.) So if the last digit is d (where d is any digit from 0 to 9), then there are d choices for the first digit. The second digit is the last digit minus the first digit, so there is just one choice for the second digit once the first and last digits are chosen. The third digit can be any of the 10 digits. Thus the answer is $(1 + 2 + \cdots + 9) \cdot 10 = \boxed{450}$.

14.3.6 The greatest the sum of the digits of a two-digit integer can be is $9 + 9 = 18$. Therefore, we have four possibilities to consider if the sum of the digits is a perfect square. The sum can be 1, 4, 9, or 16.

Case 1: The sum of the digits is 1. The only two-digit number that satisfies this is 10, so there is 1 possibility in this case.

Case 2: The sum of the digits is 4. For each tens digit from 1 to 4, there is one such number: 13, 22, 31, 40. So, there are 4 possibilities in this case.

Case 3: The sum of the digits is 9. For each tens digit from 1 to 9, there is one such number: 18, 27, 36, 45, 54, 63, 72, 81, 90. So, there are 9 possibilities in this case.

Case 4: The sum of the digits is 16. For each tens digit from 7 to 9, there is one such number: 79, 88, 97. So, there are 3 possibilities in this case.

Combining our cases, we have $1 + 4 + 9 + 3 = \boxed{17}$ numbers that satisfy the problem.

14.3.7 We can make each case be counting squares of a particular size.

Case 1: 1×1 "horizontal" squares
There are sixteen 1×1 squares whose sides are parallel to the sides of the grid, as shown to the right.

Case 2: 2×2 "horizontal" squares
There are nine 2×2 squares whose sides are parallel to the sides of the grid, as shown to the right.

Case 3: 3×3 "horizontal" squares
There are four 3×3 squares whose sides are parallel to the sides of the grid, as shown to the right.

Case 4: 4×4 "horizontal" squares
There is only one 4×4 square whose sides are parallel to the sides of the grid, namely the border of the entire grid.

This gives $16 + 9 + 4 + 1 = 30$ squares whose sides are parallel to the sides of the grid. But we also have squares which are "diagonal," meaning that their sides are not parallel to the sides of the grid.

Case 5: $\sqrt{2} \times \sqrt{2}$ "diagonal" squares
There are 9 $\sqrt{2} \times \sqrt{2}$ squares, as shown to the right.

Case 6: $\sqrt{5} \times \sqrt{5}$ "diagonal" squares
There are 8 $\sqrt{5} \times \sqrt{5}$ squares, as shown to the right. Note that these squares come in two different orientations.

Case 7: $\sqrt{8} \times \sqrt{8}$ "diagonal" squares
There is only 1 $\sqrt{8} \times \sqrt{8}$ square—it is shown at right.

Case 8: $\sqrt{10} \times \sqrt{10}$ "diagonal" squares
There are 2 $\sqrt{10} \times \sqrt{10}$ squares, shown to the right.

This gives a total of $16 + 9 + 4 + 1 + 9 + 8 + 1 + 2 = 50$ squares in the grid.

This *seems* like all of the squares, but how can we be sure that we didn't miss any cases?

We can be pretty certain that we found all the ones whose sides have integer lengths and are parallel to the sides of the grid. We counted all the squares with side lengths 1, 2, 3, and 4, and squares with side lengths 5 and higher obviously don't fit into the grid.

We can also see, because of the Pythagorean Theorem, that all of the diagonal squares must have side length $\sqrt{m^2 + n^2}$, where m and n are positive integers less than 4. We counted the diagonal squares in four of these cases:

m	n	Side Length
1	1	$\sqrt{2}$
1	2	$\sqrt{5}$
1	3	$\sqrt{10}$
2	2	$\sqrt{8}$

Note that the case $m = 1, n = 2$ and the case $m = 2, n = 1$ are identical (although this reminds us that there are two different orientations of squares with side length $\sqrt{5}$). Similarly the cases $m = 1, n = 3$ and $m = 3, n = 1$ are identical. So the only cases that we might have missed are $m = 2, n = 3$ (and its companion $m = 3, n = 2$) and $m = 3, n = 3$. But we can easily see that such squares will not fit into the grid: there is no way to insert into the grid a square with side length $\sqrt{13}$ or with side length $\sqrt{18}$.

Thus we've accounted for all the cases, and counted all the squares; the answer is $\boxed{50}$.

Exercises for Section 14.4

14.4.1 There are 5 ways to choose the first girl for a team and 4 ways to choose the second girl. This gives us $5 \cdot 4$ ways to choose the two girls, but this counts each possible team twice, once for each order in which we choose the girls. Therefore, there are $(5 \cdot 4)/2 = \boxed{10}$ different doubles teams possible.

14.4.2 There are 12 ways to choose the first person to be a co-president, and then 11 ways to choose the second person. This gives us a total of $12 \cdot 11$ ways to choose the two people, but this counts each pair of co-presidents twice, once for each order of choosing them. Therefore, the possible number of pairs of co-presidents is $(12 \cdot 11)/2 = \boxed{66}$.

14.4.3 The greatest number of days for which no pair works together more than once occurs when each student works with each other student exactly once. So, we need to count the number of different pairs of students we can form from a group of 4 students. There are 4 students to choose as the first student and 3 left to choose as the second. This gives us a total of $(4 \cdot 3)/2 = \boxed{6}$ pairs. We divide by 2 because $4 \cdot 3$ counts each pair twice, once for each order in which we can pick the students in a given pair.

14.4.4 Each person plays 5 other people 3 times, so each of the 6 people plays $5 \cdot 3 = 15$ games. In multiplying $6 \cdot 15$, each game is counted twice, once for each player in the game, so we divide by two to get the answer of $(6 \cdot 15)/2 = \boxed{45}$ games.

14.4.5 Each team plays the other 5 teams in its division twice and each of the 6 teams in the other division once. This gives a total of $5 \cdot 2 + 6 \cdot 1 = 16$ games played by each team. There are 12 teams total, and each plays 16 games. In multiplying $12 \cdot 16$, each game is counted twice, once for each team in the game, so we divide by two to get the answer of $(12 \cdot 16)/2 = \boxed{96}$ games.

14.4.6 We'll try the same strategy we used in the book to compute the sum of the first n positive integers. We let S equal the sum of the first n even integers, and then write the sum both forwards and backwards. We then add the two sums:

$$
\begin{array}{rcccccccc}
S & = & 2 & + & 4 & + \cdots + & 2(n-1) & + & 2n \\
+ \quad S & = & 2n & + & 2(n-1) & + \cdots + & 4 & + & 2 \\
\hline
2S & = & 2(n+1) & + & 2(n+1) & + \cdots + & 2(n+1) & + & 2(n+1)
\end{array}
$$

The last line has n copies of $2(n+1)$, and hence $2S = n(2(n+1))$, so $2S = 2n(n+1)$. We divide by 2 to get $S = \boxed{n(n+1)}$.

Notice that our final result is two times our formula for the sum $1 + 2 + 3 + \cdots + n$. Watch what happens when we multiply this sum by 2:

$$2(1 + 2 + 3 + \cdots + n) = 2 + 4 + 6 + \cdots + 2n.$$

So, the sum of the first n positive even integers is indeed twice the sum of the first n positive integers.

14.4.7 There are 12 vertices in the icosahedron, so from each vertex there are potentially 11 other vertices to which we could extend a diagonal. However, 5 of these 11 points are connected to the original point by an edge of the icosahedron, so they are not connected by interior diagonals. So each vertex is connected to 6 other points by interior diagonals. This gives a preliminary count of $12 \cdot 6 = 72$ interior diagonals. However, we have counted each diagonal twice (once for each of its endpoints), so we must divide by 2 to correct for this overcounting, and the answer is $\dfrac{12 \cdot 6}{2} = \boxed{36}$ diagonals.

Exercises for Section 14.5

14.5.1

(a) Rolling a ⚄ is 1 out of 6 possible equally-likely outcomes, so its probability is $\boxed{\dfrac{1}{6}}$.

(b) 3 of the 6 equally-likely possible outcomes are even (namely, 2, 4, and 6), so the probability is $\dfrac{3}{6} = \boxed{\dfrac{1}{2}}$.

(c) A ⚀ or ⚁ can be rolled for success, which is 2 out of 6 possible outcomes, so the probability is $\dfrac{2}{6} = \boxed{\dfrac{1}{3}}$.

14.5.2 There are 5 yellow sides and 8 equally-likely sides in total, so the probability of rolling a yellow side is $\boxed{\dfrac{5}{8}}$.

14.5.3

(a) There are 13 ♠'s and 52 cards total, so the probability that the top card is a ♠ is $\dfrac{13}{52} = \boxed{\dfrac{1}{4}}$.

(b)   There are four 9's and 52 cards total, so the probability that the top card is a 9 is $\frac{4}{52} = \boxed{\frac{1}{13}}$.

(c)   There are $3 \cdot 4 = 12$ face cards and 52 cards total, so the probability that the top card is a face card is $\frac{12}{52} = \boxed{\frac{3}{13}}$.

(d)   There are 26 ways to choose the first card to be black, then 26 ways to choose the second card to be red. There are $52 \cdot 51$ ways to choose any two cards. So the probability is $\frac{26 \cdot 26}{52 \cdot 51} = \boxed{\frac{13}{51}}$.

(e)   There are 4 ways to choose the first card to be a 3, then 4 ways to choose the second card to be an 8. There are $52 \cdot 51$ ways to choose any two cards. So the probability is $\frac{4 \cdot 4}{52 \cdot 51} = \boxed{\frac{4}{663}}$.

(f)   There are 4 ways to choose the first card to be an Ace, then 3 ways to choose the second card to be another Ace. There are $52 \cdot 51$ ways to choose any two cards. So the probability is $\frac{4 \cdot 3}{52 \cdot 51} = \boxed{\frac{1}{221}}$.

14.5.4   In all four parts, there are $2^4 = 16$ possible outcomes, since each of the 4 coins can land 2 different ways (heads or tails).

(a)   There is only 1 way that they can all come up heads, so the probability of this is $\boxed{\frac{1}{16}}$.

(b)   There are 2 possibilities for the dime and 2 for the quarter, so there are $2 \cdot 2 = 4$ successful outcomes, so the probability is $\frac{4}{16} = \boxed{\frac{1}{4}}$.

(c)   If the quarter is heads, there are 8 successful outcomes, since each of the other three coins may come up heads or tails. If the quarter is tails, then the nickel and dime must be heads, so there are 2 successful outcomes, since the penny can be heads or tails. So there are $8 + 2 = 10$ total successful outcomes, and the probability of success is $\frac{10}{16} = \boxed{\frac{5}{8}}$.

14.5.5   There are 4 ways to roll a sum of 5: ⚃ on the first die and ⚀ on the second die, ⚂ on the first die and ⚁ on the second die, ⚁ on the first die and ⚂ on the second die, and ⚀ on the first die and ⚃ on the second die. There are 36 total possibilities, so the probability is $\frac{4}{36} = \boxed{\frac{1}{9}}$.

14.5.6   There are $6 \cdot 6 = 36$ possible rolls. If the product of the rolls is a multiple of 5, then at least one of the two dice shows a 5. We handle separately the case of both dice being 5 and the case of exactly one die being 5. There is one way for both dice to be 5. For each die, there are 5 ways for that die to be a 5 and the other to be not a 5. So, there are $5 + 5 + 1 = 11$ ways for the roll to have at least one 5. This means the desired probability is $\boxed{\frac{11}{36}}$.

   Alternatively, we can compute that there are $5 \cdot 5 = 25$ rolls in which neither die is a 5 (because each die has 5 possibilities to be any number other than 5). Therefore, this leaves $36 - 25 = 11$ rolls that have at least one 5, and again the desired probability is $\frac{11}{36}$.

14.5.7 There are $8 \cdot 8 = 64$ possible rolls. Next, we count the number of rolls in which the product of the numbers showing exceeds 36.

Suppose the first die shows a 4 or smaller. Then, the largest possible product is $4 \cdot 8 = 32$. So, neither die can be 4 or smaller if the product of the dice is greater than 36. We consider each possible result for the first die:

Case 1: The first die shows 5. Only 8 on the second die results in a product greater than 36. There is 1 possibility in this case.

Case 2: The first die shows 6. Only 7 and 8 on the second die result in a product greater than 36. There are 2 possibilities in this case.

Case 3: The first die shows 7. Only 6, 7, and 8 on the second die result in a product greater than 36. There are 3 possibilities in this case.

Case 4: The first die shows 8. Only 5, 6, 7, and 8 on the second die result in a product greater than 36. There are 4 possibilities in this case.

Combining these cases, there are $1 + 2 + 3 + 4 = 10$ rolls that result in a product greater than 36, so the desired probability is $\frac{10}{64} = \boxed{\frac{5}{32}}$.

14.5.8 Because the probability that a randomly-chosen woman is single is $\frac{2}{5}$, we know that $\frac{2}{5}$ of the women are single. Thus, $\frac{3}{5}$ of the women are married. So, for every 2 single women, there are 3 married women. There are just as many men as there are married women, so for every 2 single women, there are 3 married women and 3 married men. As a ratio, we can write this as

$$\text{single women : married women : married men} = 2 : 3 : 3.$$

So, there are 3 married men out of every $2 + 3 + 3 = 8$ people. Therefore, $\boxed{\frac{3}{8}}$ of the people are married men.

Review Problems

14.26

(a) Adding 8 to each number gives $1, 2, 3, \ldots, 32$, so there are $\boxed{32}$ numbers in the list.

(b) Each number is a multiple of 7. Dividing each number by 7 gives $1, 2, 3, \ldots, 98$, so there are $\boxed{98}$ numbers in the list.

(c) The list counts by 3's. We make the list start from 3 by subtracting 0.5 from each number, which gives $3, 6, 9, \ldots, 90$. Dividing each number in the new list by 3 gives $1, 2, 3, \ldots, 30$, so there are $\boxed{30}$ numbers in the list.

(d) The list counts down by 4's. We make each number in the list a multiple of 4 by subtracting 2 from each number, which gives $84, 80, 76, \ldots, 12, 8$. Dividing each number by 4 gives $21, 20, 19, \ldots, 3, 2$. There are 21 numbers from 1 to 21, so our list (which is missing 1) has $\boxed{20}$ numbers.

14.27 Since 6 joined both teams, there are $8 - 6 = 2$ who joined only the track team, and $13 - 6 = 7$ who joined only the math team. This gives us a total of $2 + 6 + 7 = 15$ who are on at least one of the teams, leaving $25 - 15 = \boxed{10}$ who are not on either team. We can represent this information with the Venn diagram at the right.

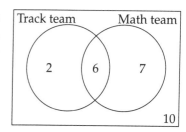

14.28 We use the Venn diagram shown at the right, with one circle for blond hair and one for blue eyes. We place 15 outside both circles for the students who have neither, and we place 3 in the overlap region for those who have both. We let x be the total number of students with blue eyes, so $x - 3$ have blue eyes but not blond hair. There are $3x$ total students with blond hair, so there are $3x - 3$ students with blond hair but not blue eyes. The sum of all four numbers in the diagram equals the total number of students, so $(3x - 3) + 3 + (x - 3) + 15 = 40$. Simplifying the left side gives $4x + 12 = 40$. Subtracting 12 and then dividing by 4 gives $x = 7$, so $\boxed{7}$ students have blue eyes.

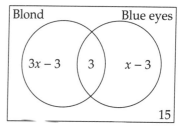

14.29 There are 4 options for shirts, 3 for pants, and 6 for hats, and all of these choices are independent. Thus there are a total of $4 \cdot 3 \cdot 6 = \boxed{72}$ outfits.

14.30 Georgie has 6 choices for which window to enter first. No matter which window he enters, he can't use the same window when leaving, so he has only 5 choices to leave through. Therefore, he has $6 \cdot 5 = \boxed{30}$ ways to enter by one window and leave by another.

14.31 We have 5 choices for the first letter. There are 26 choices for the third letter. The second letter can't be the same as the third. So, once the third letter is chosen, there are only 25 choices left for the second letter. Therefore, there are $5 \cdot 26 \cdot 25 = \boxed{3250}$ such 3-letter combinations.

14.32 We have 9 choices for the first digit (can't start with 0), 10 for the third digit, and 10 for the fourth digit. So, there are $9 \cdot 10 \cdot 10 = 900$ ways to choose these three digits.

Next, we have to count how many ways we can choose the second and fifth digits. We can't choose 3 or greater for the second digit, since then we will have no options for the fifth digit. Thus, since the second digit must be odd and less than 3, it must be 1. This means that the fifth digit must be at least 4, so we have 6 choices for the fifth digit (any digit from 4 to 9 inclusive).

For each of the 6 ways to choose the second and fifth digits, there are 900 ways to choose the other three digits, so there are $6 \cdot 900 = \boxed{5400}$ ways to form the number.

14.33 Pat uses two 2's in the first 19 pages: one for 2 and one for 12. While numbering the pages 20 through 29, he uses 10 2's for the tens digits and one for the units digit of 22. So far, he's up to 13 2's in the first 29 pages. He'll use one 2 for 32, and another 2 every 10 numbers after that. He only has 8 2's left after page 32, and he'll use one every 10 pages, so he'll use his last 2 on page 112. (Note that he won't use any 2's on tens digits between 32 and 112.) However, Pat can still number pages after page 112 until he hits a number with a 2 in it. So, Pat can number all the way through page $\boxed{119}$.

14.34 The units digit must be 2, 4, 6, or 8 (it can't be 0 because then the number would have to be 000, which isn't allowed). If the units digit is a, then the hundreds digits can be any digit from 1 to a, and the middle digit is then necessarily a minus the hundreds digit. Therefore, there are a such numbers with units digit a. (2 ending with 2, 4 ending with 4, and so on.) So there are $2 + 4 + 6 + 8 = \boxed{20}$ such numbers.

14.35   Each team plays each of the other 8 teams 3 times, so each team plays $8 \cdot 3 = 24$ games. There are 9 teams total. In multiplying $9 \cdot 24$, each games is counted twice, once for each team in the game, so we divide by two to get the answer of $(9 \cdot 24)/2 = \boxed{108}$ games.

14.36   Tyler has 3 choices of meat and 4 choices of dessert. There are 4 ways for him to choose his first vegetable and 3 ways to choose his second vegetable, for a preliminary count of $4 \cdot 3$ ways to choose two vegetables. But this counts each possible pair of vegetables twice, once for each order in which they're chosen. So, we have to divide by 2; he has $(4 \cdot 3)/2 = 6$ ways to choose his vegetables.

He has 3 choices of meat, 4 choices of dessert, and 6 choices of pairs of vegetables, so he has $3 \cdot 4 \cdot 6 = \boxed{72}$ different choices for meals.

14.37   We don't need to figure out exactly who beat whom—we just need to count! Each of the 6 players plays 5 games, one against each other player. This produces a total of $(6 \cdot 5)/2 = 15$ games. Since the other 5 players won a total of $4 + 3 + 2 + 2 + 2 = 13$ games, this leaves $15 - 13 = \boxed{2}$ games that Monica won.

14.38   In all of these parts, there are $6 \cdot 6 = 36$ possible rolls of the dice.

(a)   There are 6 different ways to roll doubles ($\boxed{1\,1}$, $\boxed{2\,2}$, ..., $\boxed{6\,6}$), which means the probability of rolling doubles is $\dfrac{6}{36} = \boxed{\dfrac{1}{6}}$.

(b)   We count the number of acceptable outcomes with casework.

Case 1: the first die is a 1. This means the second die can be either 4, 5, or 6. So there are 3 ways.

Case 2: the first die is a 2. This means the second die can be either 3, 4, or 5. So there are 3 ways.

Case 3: the first die is a 3. This means the second die can be either 2, 3, or 4. So there are 3 ways.

Case 4: the first die is a 4. This means the second die can be either 1 or 2. So there are 2 ways.

Case 5: the first die is a 5. This means the second die can only be 1. So there is 1 way.

Case 6: the first die is a 6. No value of the second die will give a total less than 7.

The number of ways to roll greater than 3 but less than 7 is $3 + 3 + 3 + 2 + 1 = 12$, which means the probability of rolling this is $\dfrac{12}{36} = \boxed{\dfrac{1}{3}}$.

(c)   *Solution 1:* There are 6 ways that the first roll can be a 1 ($\boxed{1\,1}$, $\boxed{1\,2}$, ..., $\boxed{1\,6}$), and there are 6 ways that the second roll can be a 1 ($\boxed{1\,1}$, $\boxed{2\,1}$, ..., $\boxed{6\,1}$). However, we counted twice when they are both 1, so the number of ways that at least one of the two dice is 1 is $6 + 6 - 1 = 11$. This means that the probability that at least one of the dice is 1 is $\boxed{\dfrac{11}{36}}$.

Solution 2: There are 5 ways in which the first roll is not 1, and 5 ways in which the second roll is not 1, so there are $5 \cdot 5 = 25$ ways in which neither die shows 1. Therefore there are $36 - 25 = 11$ ways in which one or both dice show 1. So the probability of this is $\boxed{\dfrac{11}{36}}$.

14.39 Since the probability of drawing a red marble is $\frac{2}{3}$, we know that $\frac{2}{3}$ of the marbles in the bag are red. This means that the other $1 - \frac{2}{3} = \frac{1}{3}$ of the marbles are black. Since the probability of drawing a red marble is double the probability of drawing a black marble, there are twice as many red marbles as black marbles. Therefore, there are $2 \cdot 8 = \boxed{16}$ red marbles.

14.40

(a) There are 5 white balls out of 11 total balls, and each of the balls is equally likely to be drawn, so the probability that the ball is white is $\boxed{\frac{5}{11}}$.

(b) First we count the number of ways to draw a pair of balls from the box. There are 11 choices for the first ball and 10 for the second, for a total of $(11 \cdot 10)/2 = 55$ ways to choose the balls. We divide by 2 because $11 \cdot 10$ counts each possible pair of balls twice, once for each order in which we could draw them.

Next, we count the number of ways to choose two white balls. There are 5 choices for the first ball and 4 for the second, for a total of $(5 \cdot 4)/2 = 10$ ways to choose two white balls. Again, we divide by 2 because $5 \cdot 4$ counts each possible pair of white balls twice, once for each order in which we can choose the balls.

Combining these counts, the probability both balls are white is $\frac{10}{55} = \boxed{\frac{2}{11}}$.

14.41 We consider three-digit times and four-digit times separately.

Case 1: Three-digit times. There are 9 choices for the hour digit. For each choice of hour digit, there are 6 choices for the middle digit (only the digits 0–5 may be used). The last digit must be the same as the hour digit, so it only has 1 choice. This gives us $6 \cdot 9 = 54$ three-digit palindromic times.

Case 2: Four-digit times. There are only 3 choices for the hour: 10, 11, or 12. For each choice of hour, the minutes digits must be the reverse of the hour digits: 10:01, 11:11, 12:21. So, there are 3 four-digit palindromic times.

Combining these cases gives us $54 + 3 = \boxed{57}$ palindromic times.

14.42 *Solution 1: Casework.* There are $6 \cdot 6 = 36$ possible pairs of rolls. We can count the outcomes in which Diana gets a higher roll with some casework based on Apollo's roll:

Apollo's roll	Number of higher rolls
⚀	5
⚁	4
⚂	3
⚃	2
⚄	1
⚅	0

Combining these gives $5 + 4 + 3 + 2 + 1 + 0 = 15$ ways Diana can have a higher roll than Apollo. So, the desired probability is $\frac{15}{36} = \boxed{\frac{5}{12}}$.

Solution 2: Some craftiness. Of the 36 possible rolls, there are exactly 6 ways the two can roll the same number. Of the other 30 rolls, half the time Apollo's roll is higher and half the time Diana is higher. So, there are 15 ways in which Diana can roll a higher number than Apollo. As before, this gives us a probability of $\frac{15}{36} = \boxed{\frac{5}{12}}$.

Challenge Problems

14.43 We must have $\frac{1}{2} > \frac{1}{n}$ and $\frac{1}{n} > \frac{3}{100}$. We'll tackle $\frac{1}{2} > \frac{1}{n}$ first. Since n is positive, we can multiply both sides by 2 and by n to get $n > 2$. Next, we tackle $\frac{1}{n} > \frac{3}{100}$. We multiply both sides by 100 and by n, and we have $100 > 3n$. Dividing by 3 gives $\frac{100}{3} > n$. Combining $n > 2$ and $\frac{100}{3} > n$, we have $\frac{100}{3} > n > 2$. Since $\frac{100}{3} = 33\frac{1}{3}$, the integers n that satisfy $\frac{100}{3} > n > 2$ are $\boxed{3, 4, 5, \ldots, 33}$. There are $33 - 3 + 1 = \boxed{31}$ such integers.

14.44

(a) Each row has odd-numbered chairs $1, 3, 5, 7, 9, 11$, for a total of 6 odd-numbered chairs in each row. Since there are 11 rows, there are a total of $11 \cdot 6 = \boxed{66}$ chairs with odd numbers.

(b) If n is odd, each row has odd-numbered chairs $1, 3, 5, \ldots, n - 2, n$. Adding 1 to each number in this list and dividing each result by 2, we get $1, 2, 3, \ldots, \frac{n-1}{2}, \frac{n+1}{2}$. So there are $\frac{n+1}{2}$ odd-numbered chairs in each row and n rows, for a total of $\boxed{\frac{n(n+1)}{2}}$.

If n is even, each row has odd numbered chairs $1, 3, 5, \ldots, n - 3, n - 1$. Adding 1 to each number in this list (to make the numbers even) and then dividing each result by 2 gives $1, 2, 3, \ldots, \frac{n-2}{2}, \frac{n}{2}$.

So there are $\frac{n}{2}$ odd-numbered chairs in each row and n rows for a total of $\boxed{\frac{n^2}{2}}$.

14.45 All 14 people shake hands with 12 other people (everyone except themselves and their spouse). In multiplying $14 \cdot 12$, each handshake is counted twice, so we divide by two to get the answer of $\frac{14 \cdot 12}{2} = \boxed{84}$ handshakes.

14.46 We consider the units digit, the tens digit, and the hundreds digit separately. For simplicity in our computation below, let's assume that we write every number using three digits, including leading zeros if necessary (so that we'd write 22 as "022"). Doing this doesn't change the number of 3's that we write, so it doesn't affect our answer.

Units digit. We count the numbers with 3 as the units digit. There are 5 choices for the hundreds digit (0, 1, 2, 3, 4), and 10 choices for the tens digit, for a total of $5 \cdot 10 = 50$ numbers with 3 as the units digit. (If the hundreds digit is 0, then the resulting number is a one-digit or two-digit number.) Therefore, there are 50 3's in the units place.

Alternatively, we could have noted that we write one 3 in the units place for each block of 10 numbers, for a total of $500/10 = 50$ 3's in the units place.

Tens digit. We count the numbers with 3 as the tens digit. After placing the 3 in the tens place, there are 5 ways to choose the hundreds digit and 10 ways to choose the units digit. So, there are $5 \cdot 10 = 50$ numbers with 3 in the tens place. Therefore, there are 50 3's in the tens place.

Alternatively, we might note that we write the 3 in the tens digit 10 times in each of the 30s, the 130s, the 230s, the 330s, and the 430s.

Hundreds digit. All 100 numbers from 300 to 399 have a 3 in the hundreds place, so there are 100 3's in the hundreds place.

Combining these cases gives a total of $50 + 50 + 100 = \boxed{200}$ 3's.

14.47 Each team plays the four other teams within its division three times, for a total of 12 intra-division games. Each team plays the other 15 teams twice for a total of 30 inter-division games. This gives a total of $12 + 30 = 42$ games per team. However, $20 \cdot 42$ counts every game twice (once for each of the opposing teams) so we must divide by two. The answer is $\dfrac{20 \cdot 42}{2} = \boxed{420}$.

14.48 Each flip has 2 possible outcomes, so there are $2 \cdot 2 \cdot 2 = 8$ different possible outcomes for all three flips. There are two possible cases in which they could have the same number of heads: they could each get 0 heads, or they could each get 1 head.

Case 1: Both get 0 heads. The only way this can happen is if all three flips are tails. There is only 1 way this can happen.

Case 2: Both get 1 head. Keiko's flip must be heads, and Ephraim's flips can be heads then tails, or tails then heads. So, there are 2 possibilities in this case.

Combining both cases, we see that there are 3 ways for the two to get the same number of heads, so the desired probability is $\boxed{\frac{3}{8}}$.

14.49 There are $6 \cdot 6 = 36$ possible rolls of the two dice. It is impossible for the dice to have a product greater than 10 if either die is ⚀. When one die is ⚁, the only way the product is greater than 10 is if the other is ⚅. Either die could be ⚁, so there are 2 ways to do this.

Otherwise, if neither die is ⚀ or ⚁, then they each must show ⚂, ⚃, ⚄, or ⚅, and there are $4 \cdot 4 = 16$ ways for this to occur. Of these 16 possibilities, only ⚂⚂ results in a product less than 10. So, there are 15 ways to produce a product greater than 10 when both dice are greater than ⚁.

This gives us $2 + 15 = 17$ total rolls out of 36 in which the product of the dice is greater than 10, so the probability is $\boxed{\frac{17}{36}}$.

14.50 If Paco's first number is 1 or 2, then any of the 10 numbers can appear on the second spinner, so this gives 10 successful spins for each. If Paco's first number is 3, then any number other than 10 can appear, so there are 9 successful spins. If his first number is 4, then the second spinner must show 1 through 7 (inclusive), so there are 7 more successful spins. Finally, if Paco's first number is 5, then the second spinner must show 1 through 5 (inclusive). So there are a total of $10 + 10 + 9 + 7 + 5 = 41$ successful spins, and $5 \cdot 10 = 50$ possible spins, giving a probability of $\boxed{41/50}$.

14.51 We build a Venn diagram, but this time, we need three circles, one for each sport. We place them so that each pair overlaps, and so that there is a region in which all three overlap. As we often do with 2-circle Venn diagrams, we work from the inside out. Since there are 3 students who like all three sports, we place a 3 in the center.

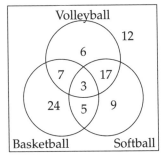

Subtracting these 3 from the 10 who like volleyball and basketball gives us 7 who like volleyball and basketball but not softball. We place a 7 in the region inside the volleyball and basketball circles, but outside the softball circle. Similarly, we place $20 - 3 = 17$ in the region where only volleyball and softball overlap, and we place $8 - 3 = 5$ in the region where only softball and basketball overlap.

Next, we turn to the students who only like basketball. We have 39 total who like basketball. We have 7, 3, and 5, respectively in the three regions basketball shares with other sports, so that leaves $39 - 7 - 3 - 5 = 24$ who only like basketball. We place 24 in the basketball-only region. Similarly, there are $34 - 5 - 3 - 17 = 9$ who only like softball and $33 - 17 - 3 - 7 = 6$ who only like volleyball. Finally, we place 12 outside all 3 circles for the students who don't like any of the sports.

Adding all of the numbers in the diagram, we have $12 + 6 + 24 + 9 + 7 + 5 + 17 + 3 = \boxed{83}$ students.

14.52 We draw a Venn diagram as shown at the right. We have one circle for boys and one for middle school. Let there be x boys in middle school, so we place x in the overlap region as shown.

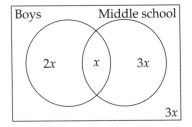

Since there are twice as many boys in high school as in middle school, there are $2x$ boys in high school. We place $2x$ in the region inside the boys circle but not the middle school circle.

There are three times as many middle school girls as middle school boys, so there are $3x$ middle school girls. We place this in the region inside the middle school circle but not the boys circle.

Finally, half the girls are in high school, so there are just as many high school girls as middle school girls. So, there are $3x$ girls in high school, and we place $3x$ outside both circles.

Altogether, we see that there are $x + 2x + 3x + 3x = 9x$ students in the club, so we have $9x = 72$. This gives us $x = \boxed{8}$ middle school boys.

14.53 In any consecutive three seats, at least one of the seats must be occupied. Otherwise, the next person can safely sit in the middle of the three empty seats. We can break the 120 seats into 40 groups of 3 consecutive seats. We need at least one person in each group of 3 seats, so we need at least 40 people. If we seat a person in the middle chair in each of these groups of 3 consecutive seats, then the next person is forced to sit next to someone. Therefore, the fewest number of seats that must be occupied is indeed $\boxed{40}$.

14.54 *Case 1: start at an N in a corner.* If we start at one of the 4 N's in a corner, then we have only 1 choice for the first O, then 3 choices for the second O, then 5 choices for the second N, for a total of $4 \cdot 1 \cdot 3 \cdot 5 = 60$ ways to form NOON, starting from a corner.

Case 2: start at an N on a side. If we start at one of the 8 N's on a side, then we have 2 choices for the first O, then 3 choices for the second O. If we choose the second O adjacent to our original N, then we only have 4 choices for the final N; otherwise (for the other two choices of the second O) we have 5 choices for the final N. Thus we have $8 \cdot 2 \cdot (2 \cdot 5 + 1 \cdot 4) = 224$ ways to form NOON, starting from a side.

Adding our counts from the two cases gives a total of $60 + 224 = \boxed{284}$ ways to form NOON.

14.55 We organize the information with the Venn diagram at the right, with a circle for pit bulls and a circle for males. We place 14 in the non-overlapping portion of the pit bull circle for the 14 female pit bulls. Let there be x male dogs that are not pit bulls, so we place an x in the non-overlapping portion of the male dog circle.

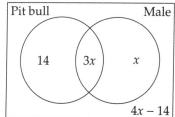

Since $\frac{3}{4}$ of the male dogs are pit bulls, there are 3 male pit bulls for every 1 male dog that is not a pit bull. That is, there are 3 times as many male pit bulls as non-pit bull males. So, the number of male pit bulls is $3x$, which we place in the overlap region.

Since $\frac{1}{2}$ of the dogs at the pound are female, there are just as many females as males. There are $x + 3x = 4x$ males, so there are $4x$ females. We already know that 14 of the females are pit bulls, so the remaining $4x - 14$ are not female, which allows us to complete the Venn diagram shown at the upper right.

The only piece of information left to use is the fact that $\frac{2}{3}$ of the dogs are pit bulls. There are $3x + 14$ pit bulls at the pound, and a total of $8x$ dogs in the pound, so we must have $3x + 14 = \frac{2}{3}(8x)$. Multiplying both sides by 3 gives $3(3x + 14) = 2(8x)$, so $9x + 42 = 16x$. Therefore, we have $7x = 42$ and $x = 6$. Substituting this value of x into our Venn diagram above gives us the Venn diagram at the right, and we have $8x = \boxed{48}$ dogs total.

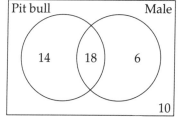

14.56 Since we use 20 vertical toothpicks in each column, there are 21 rows of dots; since we use 10 horizontal toothpicks, there are 11 columns of dots. Then, since there are 20 vertical toothpicks in each column and 11 columns, there are $20 \cdot 11 = 220$ vertical toothpicks. Similarly, since there are 10 horizontal toothpicks in each row and 21 rows, there are $21 \cdot 10 = 210$ vertical toothpicks. This gives a total of $220 + 210 = \boxed{430}$ toothpicks.

14.57 We choose the first block at random, so there are 119 other blocks to choose from for the second block. There are $(4 \cdot 3)/2 = 6$ different ways to choose the two properties that the second block shares with the first block, so there are 6 exclusive cases.

Case 1: Same material and color. The other block must have a different size and different shape. There are 3 choices for the different size and 4 choices for the different shape, for a total of $3 \cdot 4 = 12$ blocks with the same material and color only.

Case 2: Same material and size. The other block must have a different color and different shape. There are 2 choices for the different color and 4 choices for the different shape, for a total of $2 \cdot 4 = 8$ blocks with the same material and size only.

Case 3: Same material and shape. The other block must have a different color and different size. There are 2 choices for the different color and 3 choices for the different size, for a total of $2 \cdot 3 = 6$ blocks with the same material and shape only.

Case 4: Same color and size. The other block must have a different material and different shape. There is 1 choice for the different material and 4 choices for the different shape, for a total of $1 \cdot 4 = 4$ blocks with the same color and size only.

Case 5: Same color and shape. The other block must have a different material and different size. There is 1 choice for the different material and 3 choices for the different size, for a total of $1 \cdot 3 = 3$ blocks with the same color and shape only.

Case 6: Same size and shape. The other block must have a different material and different color. There is 1 choice for the different material and 2 choices for the different color, for a total of $1 \cdot 2 = 2$ blocks with the same size and shape only.

Summing all these cases, there are 35 blocks with exactly two characteristics in common with the original block. There are a total of 119 other blocks, so the probability that the two blocks have exactly two characteristics in common is $\dfrac{35}{119} = \boxed{\dfrac{5}{17}}$.

CHAPTER 15

Problem-Solving Strategies

24® Cards

First Card $6 \cdot 9 - (15 + 15)$.

Second Card $(7 + 11)/2 + 15$.

Third Card $24/12 \cdot 15 - 6$.

Fourth Card $(85 + 3) \cdot 15/55$.

Exercises for Section 15.1

15.1.1   The beads on the necklace repeat the same pattern of 6 bead colors over and over. Since 74 divided by 6 has quotient 12 and remainder 2, the 74th bead is 2 beads into this 6-bead pattern. The second bead of the pattern is $\boxed{\text{orange}}$.

15.1.2   In the first ten terms shown, it looks like there's a pattern of two odd numbers followed by an even number. To see why this pattern will continue, note that the sum of two odd numbers is even. So, two odd numbers must be followed by an even number. Similarly, the sum of an odd and even is odd. So, both "odd then even" and "even then odd" will be followed by an odd number. Putting these observations together, we see that the Fibonacci sequence consists of blocks of two odd numbers and an even number. In the first 60 terms, there are $60/3 = 20$ such blocks, for a total of $20 \cdot 2 = \boxed{40}$ odd numbers.

15.1.3   The last number in each row appears to be a perfect square. To see why this is the case, note that there is 1 number in the first row, 3 numbers in the second row, 5 numbers in the third row, and so on. In each row, we add the next odd number amount of numbers. So, the total number of numbers on or above each row is the sum of consecutive odd numbers starting from 1. As explained in the text, such a sum is always a perfect square. This means the last number in each row is a perfect square.

Since 119 is 2 less than the perfect square 121, we know that 119 is two numbers before the last number of the row that ends with 121. The row above 119's row ends in the previous square, 100, and this final number is directly above the next-to-last number in 119's row, namely 120. So, the number just to 100's left is directly above 119, which means that the number directly above 119 is $\boxed{99}$.

15.1.4 We only care about the units digits of the numbers in the product. The first four numbers are $2 \cdot 4 \cdot 6 \cdot 8$. The product of the units digits of the next four numbers is also $2 \cdot 4 \cdot 6 \cdot 8$. Similarly, the product of the units digits in the rest of the product just repeats this block over and over.

We have $2 \cdot 4 \cdot 6 \cdot 8 = 8 \cdot 6 \cdot 8 = 48 \cdot 8$, which has units digit 4 (since $8 \cdot 8 = 64$). The rest of the product in the problem consists of 28 more blocks of $2 \cdot 4 \cdot 6 \cdot 8$. Each block has a product with units digit 4. The entire product consists of 29 such blocks, each with a product with units digit 4. Therefore, the product has the same units digit as 4^{29}.

4^1 ends in 4, 4^2 ends in 6, and the units digit of 4^3 is back to 4. So, the units digits of the powers of 4 cycle back and forth between 4 and 6, starting with $4^1 = 4$. Since 29 is odd, we see that 4^{29} has 4 as its units digit. Therefore, the original product has units digit $\boxed{4}$.

Exercises for Section 15.2

15.2.1 We tackled the problem by counting backwards from 700 by 31's while looking for a multiple of 15. This is much easier than counting backwards from 700 by 15's while looking for a multiple of 31 for two reasons. First, it is far easier to spot multiples of 15 than it is to spot multiples of 31. Only numbers that end in 0 or 5 can possibly be a multiple of 15. So, we don't have to do anything besides look at the last digit to discard most of the numbers we hit as we subtract 31's. Moreover, we cover a lot more ground as we subtract 31's than when we subtract 15's, so we have fewer numbers to check.

15.2.2 My height is a number of inches that is greater than 24 (2 feet), less than 72 (6 feet), a multiple of 7, and 1 more than a multiple of 6. To find my height, I list the multiples of 7 greater than 24 and less than 72:

$$28, 35, 42, 49, 56, 63, 70.$$

The only number in this list that is 1 more than multiple of 6 is 49. So, I'm $\boxed{4 \text{ feet, 1 inch}}$ tall.

15.2.3 It's easiest to start with the second condition, since there aren't many two-digit numbers that are the product of two consecutive numbers. (It's also much easier to multiply numbers than to find factorizations, which is what we'd have to do if we started by listing numbers whose digits are consecutive numbers.) We build a list of two-digit numbers that satisfy the second condition by multiplying pairs of consecutive numbers. The full list is

$$3 \cdot 4 = 12, \ 4 \cdot 5 = 20, \ 5 \cdot 6 = 30, \ 6 \cdot 7 = 42, \ 7 \cdot 8 = 56, \ 8 \cdot 9 = 72, \ 9 \cdot 10 = 90.$$

The only numbers in this list with consecutive numbers as digits are $\boxed{12 \text{ and } 56}$.

15.2.4 We could start by listing all the two-digit multiples of 3, but that would be a pretty long list. Instead, we start by listing all the two-digit numbers that are 7 more than a multiple of 11. We can do this by starting with 7 and counting by 11's. The two-digit numbers we thus create are

$$18, 29, 40, 51, 62, 73, 84, 95.$$

Next, we note that because the number leaves a remainder of 4 when divided by 5, the number must end in 4 or 9. That leaves only 29 and 84 as possibilities. 29 is not divisible by 3, but $\boxed{84}$ is.

15.2.5 We make an organized list, where we organize the numbers by the hundreds digit. Clearly, there are no numbers starting with 7 or greater that have a digit sum of 6. There is one such number starting with 6, namely 600. Continuing in this manner, we produce the following organized list:

Hundreds digit	Numbers with digit sum 6
6	600
5	510, 501
4	420, 411, 402
3	330, 321, 312, 303
2	240, 231, 222, 213, 204
1	150, 141, 132, 123, 114, 105

Notice how each row is nicely organized. This makes us confident that we didn't miss any, and didn't count any twice. It also reveals a nice pattern. We see that there are $1 + 2 + 3 + 4 + 5 + 6 = \boxed{21}$ three-digit numbers with digit sum 6.

Exercises for Section 15.3

15.3.1 We might think that 5 pieces requires 5 cuts, but a quick sketch makes us realize that we only need 4 cuts. In the diagram at the right, the log is represented by a line segment, and the dots on the log are the cuts.

The lumberjack makes 4 cuts in 20 minutes, so it takes $20/4 = 5$ minutes to make each cut. Just as he needs 4 cuts to make 5 pieces, he needs 6 cuts to make 7 pieces. Each cut takes 5 minutes, so he needs $5 \cdot 6 = \boxed{30 \text{ minutes}}$ to cut the log into 7 pieces.

15.3.2 We start by drawing $\overline{AB}$ with length 11 to represent the 11 cm rod. We can add the 10 cm rod to the end of this either by continuing past B, or by starting at B and going back towards A. Let's try extending past B, as shown at the right. We now have $AC = AB + BC = 21$. So, if we add the 6 cm rod starting at C and going back towards B, the other end of the rod will be $21 - 6 = 15$ cm from A, as desired.

15.3.3 We draw a diagram to figure out how the houses are situated. The resulting diagram is on the right. Each person's house is represented by a point labeled with the first letter of the person's name. Therefore, B is 5 miles to the left of A, and C is 3 miles to the right of A. Since Dolly is 6 miles east of Carnot and 4 miles east of Eli, we know that Eli is $6 - 4 = 2$ miles east of Carnot. Finally, since Frank is 5 miles north of Eli and Greta is 8 miles south of Frank, we know that Greta is 3 miles south of Eli.

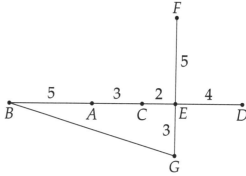

As shown at the right, we find that Greta is 10 miles east and 3 miles south of Belle. Applying the Pythagorean Theorem to $\triangle BEG$, we have $BG^2 = 10^2 + 3^2$, so $BG = \sqrt{109} \approx 10.4$. Therefore, to the nearest tenth of a mile, Belle's house and Greta's house are $\boxed{10.4 \text{ miles}}$ apart.

15.3.4 We start with a quick sketch of the shoes, and we name each person based on the color of their pants. (Yes, those are shoes in the diagram.) We label Green's right shoe "Red," since we know Green's right shoe is red.

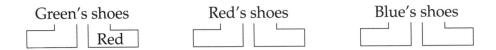

Green's shoes   Red's shoes   Blue's shoes

We know that Green's shoes have different colors, and that neither is green. So, his left shoe must be blue. Also, we know that Blue's right shoe must be red or green, but the red right shoe is already taken by Green. So, Blue's right shoe is green.

Green's shoes   Red's shoes   Blue's shoes

Only the blue shoe remains for Red's right foot. Blue's left shoe can't be green or blue, so it must be Red. Red's left shoe can't be blue (Green has that shoe) and it can't be red, so it must be green. We now have the full arrangement of shoes:

Green's shoes   Red's shoes   Blue's shoes

15.3.5 Below are diagrams for three lines and four lines.

We organize the information we have so far in the table at the right. The number of regions with 2 lines is 2 more than the number of regions with 1 line. The number of regions with 3 lines is 3 more than the number of regions with 2 lines. The number of regions with 4 lines is 4 more than the number of regions with 3 lines. Does this pattern continue?

# of Lines	# of Regions
1	2
2	4
3	7
4	11

To see why this pattern continues, consider what happens when we add a fifth line, as shown in bold at the right. We start the line from the top of the page, dividing the region where we start in two, thereby adding one more new region. Each time we cross one of the four old lines with our new line, we enter another region of the 4-line configuration. Then we cut that region into two pieces, again adding a new region. We therefore create 5 more new regions: one for the region we start in, plus one more for each of the 4 lines we cross. Two lines can only intersect in at most one point, so we can't make any additional new regions.

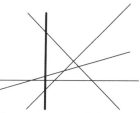

Similarly, we add 6 new regions when we draw the sixth line, 7 new regions when we draw the seventh line, and 8 new regions when we draw the eighth line. This gives a total of

$$2 + 2 + 3 + 4 + 5 + 6 + 7 + 8 = \boxed{37}$$

regions.

Exercises for Section 15.4

15.4.1 Anjou has one-third as much as Granny Smith, so Anjou has $\frac{1}{3}$($63) = $21. Elberta has $2 more than Anjou, so Elberta has $21 + $2 = $\boxed{$23}$.

15.4.2 You ended on floor 7 after going up 2 floors, so you were on floor 5 before going up 2 floors. You were on floor 5 after going down 4 floors, so you were on floor 9 before going down 4 floors. You were on floor 9 after going up 6 floors, so you were on floor $\boxed{3}$ before going up 6 floors.

15.4.3 Working backwards, Joy had 16 coins after taking one-half of the coins out, so she had $2 \cdot 16 = 32$ coins at the start of the last day.

The previous evening, she put 4 coins in the bank, so she had $32 - 4 = 28$ coins in the coin bank before that evening.

That morning, she had these 28 coins after taking one-third of the coins out of the coin bank. Therefore, these 28 coins are $\frac{2}{3}$ of the coins she had at the start of the morning. This means she had $28 \cdot \frac{3}{2} = 42$ coins at the start of the morning. (We also could have reasoned that if 28 is $\frac{2}{3}$ of the coins, then $\frac{1}{3}$ of the coins is 14 coins, so she must have had $28 + 14 = 42$ coins before removing a third of them.)

The previous evening, she had these 42 coins after adding 10 coins. So, she must have had 32 coins before that evening.

Finally, these 32 coins are what she had after removing one-half of her coins. Therefore, she started out with twice this amount, $\boxed{64}$ coins.

15.4.4 Working backwards, the fifth number times 16 is 64, so the fifth number is 4:

$$\underline{?}, \quad \underline{}, \quad \underline{}, \quad \underline{}, \quad \underline{4}, \quad \underline{16}, \quad \underline{64}, \quad \underline{1024}.$$

The fourth number times 4 is 16, so the fourth number is also 4:

$$\underline{?}, \quad \underline{}, \quad \underline{}, \quad \underline{4}, \quad \underline{4}, \quad \underline{16}, \quad \underline{64}, \quad \underline{1024}.$$

Continuing backwards, we find that the third number is 1, the second number is 4, and the first number is $\boxed{\frac{1}{4}}$:

$$\frac{1}{4}, \quad \underline{4}, \quad \underline{1}, \quad \underline{4}, \quad \underline{4}, \quad \underline{16}, \quad \underline{64}, \quad \underline{1024}.$$

15.4.5 Since Serena and Visala together have $180 at the start of the problem, they will always have a total of $180 throughout the problem. At the end, they each have $90. Visala has this $90 after giving $\frac{1}{4}$ of her money to Serena. Therefore, $90 is $\frac{3}{4}$ of the money she had before giving any money to Serena. So, Visala had $\frac{4}{3}(\$90) = \$120$ before giving any money to Serena. Therefore, Visala had $120 after receiving $20 from Serena, which means Visala had $100 before receiving money from Serena. This means that Serena started with the other $180 − $100 = $\boxed{\$80}$.

Review Problems

15.17 Counting backwards by 7s, the 20th, 13th, and 6th are also Mondays. Counting backwards 5 more days, we see that the first day of the month is $\boxed{\text{Wednesday}}$.

15.18 We start by making a grid with one row for each person and one column for each sport. We then include the given information by placing X's in the baseball column for Glen and Harry, and an X in the soccer column for Glen. We see that the only option left for Glen is tennis, and that Kim is the only person who can possibly like baseball.

	Tennis	Baseball	Soccer
Glen		X	X
Harry		X	
Kim			

We place O's in the grid for Glen's and Kim's sports, and X's in the tennis and soccer columns for Kim. That leaves only Harry for soccer, and we have the completed grid at the right. Glen's favorite sport is tennis, Harry's is soccer, and Kim's is baseball.

	Tennis	Baseball	Soccer
Glen	O	X	X
Harry	X	X	O
Kim	X	O	X

15.19 Working backwards, the fifth number is the sum of the second, third, and fourth, so $Q + 86 + 158 = 291$. Simplifying the left side gives $Q + 244 = 291$, so $Q = 47$. The sum of the first three numbers equals the fourth, so $P + Q + 86 = 158$. Since $Q = 47$, we have $P + 133 = 158$, which means $P = \boxed{25}$.

15.20 After writing 4 in the F column, we repeatedly write 6 numbers from right to left, then 6 more from left back to right. This means that, starting with 4, every 12th number is in the F column. So, the following numbers are in the F column:

$$4, 16, 28, 40, 52, 64, 76, 88, 100.$$

We got a little lucky there! 100 is in column $\boxed{F}$.

15.21 1002 is 2 more than a multiple of 5. We start from here and count down by 5's to generate a list of numbers that are 2 more than a multiple of 5. We seek one that is 2 more than a multiple of 3 and is odd. Here's the start of our list:

$$1002, 997, 992, 987, 982, 977, 972, 967.$$

(We could have counted down by 10s from 997, since we know the number must be odd.) 996 is clearly a multiple of 3, so 997 is 1 more than a multiple of 3. Since 987 is 9 less than 996, we know that 987 is a multiple of 3. Finally, 977 is 2 more than 975, which is a multiple of 3. Therefore, the desired integer is $\boxed{977}$.

15.22 Applying the rules in the problem, we have

$$98, 49, 44, 22, 11, 6, 54, 27, 22, 11, 6, 54, \ldots.$$

Once we hit 22 for a second time, we realize that the sequence is going to repeat the 5-term block

$$22, 11, 6, 54, 27$$

over and over. We have to be careful, though, because there are 3 terms in the sequence before we start the repeating block. Therefore, to get to the 98th term of the sequence, we write these first 3 terms and then write the repeating block of 5 terms 19 times. So, the 98th term of the sequence is the last term of the repeating block, which is $\boxed{27}$.

15.23 We work backwards, starting from the final result, 3, and reversing each action.

Adnan's original action	Reverse action	Result
Take square root	Square	9
Add 1	Subtract 1	8
Take square root	Square	64
Subtract 6	Add 6	70
Divide by 2	Multiply by 2	140

Adnan's original number was $\boxed{140}$.

15.24 The first number in column P is the smallest of the first 8 numbers we put in the table. The next number in column P is the smallest of the next 8 numbers, and so on. Therefore, each number in column P after the first is 8 less than the previous number in column P. So, we get the numbers in column P by starting with 93 and counting downward by 8's. Since 93 is 5 greater than a multiple of 8, and we form column P by counting downward from 93 by 8's, column P has all the positive numbers less than 100 that are 5 more than a multiple of 8.

Similarly, each number in column T after the first is 8 less than the previous number in column T. The first number in column T is 97. Since 97 is 1 more than a multiple of 8, and we form column T by counting downward from 97 by 8's, column T has all the positive numbers less than 100 that are 1 more than a multiple of 8. Since 25 is 1 more than a multiple of 8, we know that 25 is in column $\boxed{T}$.

15.25 We work backwards through Luyi's actions in the table below:

Amount after action	Luyi's action	Amount before action
$4	Spent $16 more	$20
$20	Spent one-third of her money	$30
$30	Spent $16 more	$46
$46	Spent half her money	$92

So, she started with $\boxed{\$92}$.

15.26 We find the units digits of the first few powers of 3 and look for a pattern. The first few powers of 3 are $3, 9, 27, 81, 243, 729$. The units digits are $3, 9, 7, 1, 3, 9$. Once we see the second 3, we know that the units digits of the powers of three repeat the 4-term block $3, 9, 7, 1$ over and over. Since 80/4 is 20 with no remainder, we repeat this block exactly 20 times to reach the units digit of 3^{80}. So, the units digit of 3^{80} is the last number in this repeating 4-term block, which is $\boxed{1}$.

15.27 If Toy has $36 at the beginning, then Ami doubles Toy's money in the first step to $72, and Jan doubles it to $144 in the next step. Toy has $36 at the end, so Toy must give away $144 − $36 = $108 in the last step. This amount must double the amount of money Ami and Jan have, so Ami and Jan have a total of 2 · $108 = $216 in the end. Combining this with Toy's $36, the three together must have $\boxed{\$252}$.

15.28 Three-fourths of the chairs are used, so one-fourth of the chairs are not used. Since 6 chairs are not used, there must be 24 chairs total and 18 people seated. Since two-thirds of the people are seated, the other one-third of the people are standing. This means that half as many people are standing as sitting, so 9 people are standing and there are $18 + 9 = \boxed{27}$ people total.

15.29 The number of coins must be 4 more than a multiple of 6, so we make a list of such numbers by starting with 4 and counting by 6's:

$$4, 10, 16, 22, 28, \ldots.$$

The first number in this list that is 3 more than a multiple of 5 is 28. When these 28 coins are divided among 7 people, there are $\boxed{0}$ left over.

15.30 We make a list of the numbers that occur as we press the key over and over.

Press number	Calculation	Resulting Number
	Start	5
1	$\dfrac{1}{1-5} = \dfrac{1}{-4} = -0.25$	-0.25
2	$\dfrac{1}{1-(-0.25)} = \dfrac{1}{1.25} = \dfrac{4}{5} = 0.8$	0.8
3	$\dfrac{1}{1-0.8} = \dfrac{1}{0.2} = 5$	5

Once we see the 5 repeat, we know that the same three numbers will repeat over and over: $5, -0.25, 0.8$. We do have to be careful; the calculator already has the 5 displayed when we press the special key for the first time. So, -0.25 is what results after the first press. Therefore, the sequence of numbers that appears as we press the special key starts $-0.25, 0.8, 5$, rather than starting with 5. Since 100 divided by 3 has a remainder of 1, the 100th number that results is the first in this repeating block, which is $\boxed{-0.25}$.

15.31 When you take the first trip across the river, you will have to leave two of the items on shore. The only two you can leave together without an undesirable (for you) eating incident are the cabbage and the wolf. So, clearly you have to take the goat across first.

Thinking ahead, during your final trip across the river, you will have two items already on the destination shore while you are bringing the third across. Again, these two items can only be the cabbage and the wolf. So, the goat must be the last item brought across, in addition to being the first item brought across. That means you'll have to take the goat across the river in the "wrong" direction at some point.

We organize the information with a picture. We have two sides to our river and a letter for the cabbage (C), goat (G), wolf (W), and you (Y).

Now that you know that the goat will have to take three total trips, you have a pretty good idea of what you'll have to do. After taking the goat across initially, you go back and get the wolf (or the cabbage; either is fine). You bring the wolf to the destination side, and then *take the goat back to the original shore.* You then bring the cabbage over to the wolf's side, and then go back for the goat. As we can see in our table at right, we never have an undesirable eating event, and we get all three items across the river.

CGWY	
CW	GY
CWY	G
C	GWY
CGY	W
G	CWY
YG	CW
	CGWY

15.32 We organize the list of partitions of 7 by the largest number in the partition.

Largest number is 7:
$$7.$$

Largest number is 6:
$$6 + 1.$$

Largest number is 5:
$$5 + 2, 5 + 1 + 1.$$

Largest number is 4:
$$4 + 3, 4 + 2 + 1, 4 + 1 + 1 + 1.$$

Largest number is 3:
$$3 + 3 + 1, 3 + 2 + 2, 3 + 2 + 1 + 1, 3 + 1 + 1 + 1 + 1.$$

Largest number is 2:
$$2 + 2 + 2 + 1, 2 + 2 + 1 + 1 + 1, 2 + 1 + 1 + 1 + 1 + 1.$$

Largest number is 1:
$$1 + 1 + 1 + 1 + 1 + 1 + 1.$$

Counting them all, we find that there are $\boxed{15}$ partitions of 7.

15.33 The key insight is that if the 5-minute person and the 8-minute person (that is, the slow people) travel separately, then those two trips will consume 13 minutes. This leaves only 2 minutes for the other two to get across, and for the torch to get carried back to the original side after the first slow person completes his journey. This is impossible to accomplish, so we look for a process in which the two slowest people cross the bridge together.

The slow people can't be the last to cross, since there would be no way for them to get the torch from the faster people on the other side of the bridge before crossing. So, not only must the slow people cross together, but there must be a fast person on the far side of the bridge to bring the torch back. This inspires our solution: send the fast people across first, and have one return with the torch for the slow people. Then the slow people cross, and the fast person on the far side brings back the torch to fetch the last person. Now that we have a plan, let's see if it works.

We organize the information in the table shown at the right. Initially, all four start on the left side, where we use numbers to show how long each person takes to cross the bridge.

The 5-minute and the 8-minute people must travel across together, but someone else needs to be on the far side to return the torch quickly. Therefore, we send the 1-minute and 2-minute people across first. We show this in our table by placing them in the middle column. We underline the larger number to show that the trip takes 2 minutes. The 1-minute person then brings the torch back, and the 5-minute and 8-minute people then cross.

Left	On Bridge	Right
1 2 5 8		
5 8	1 $\underline{2}$	
5 8		1 2
5 8	1	2
1 5 8		2
1	5 $\underline{8}$	2
1		2 5 8
1	$\underline{2}$	5 8
1 2		5 8
	1 $\underline{2}$	5 8
		1 2 5 8

Finally, the 2-minute person brings the torch back across, and the 1-minute and 2-minute people cross together at the end. (The solo trips of the 1-minute person and 2-minute person could be swapped.) Adding the underlined numbers, we see that the process takes 15 minutes, as required.

15.34 We need a sum of distinct (different) positive squares that equals 50. We search for such a sum by first listing the squares that are less than 50:

$$1, 4, 9, 16, 25, 36, 49.$$

We quickly see one such sum starting with 49, namely $49 + 1$. Therefore, the number 17 satisfies both properties. We then search for a sum starting with 36 and find $36 + 9 + 4 + 1$, so the number 1236 satisfies both properties. Finally, we check 25, and we find $25 + 16 + 9$, which gives us 345. The sum of the first four squares in the list is less than 50, so there are no more possibilities. Therefore, the largest number that satisfies both properties is $\boxed{1236}$.

15.35 In the diagram, C is Cindy's house, J is Jenny's house, H is the halfway point where Cindy turns around, and P is where she finds the phone. Let d be the distance in miles between the two houses, so $CH = \frac{d}{2}$ and $HP = \frac{1}{2}\left(\frac{d}{2}\right) = \frac{d}{4}$. The distance Cindy walks is $CH + HP + PJ = \frac{d}{2} + \frac{d}{4} + \frac{3d}{4} = \frac{3d}{2}$. Since Cindy walks 2 miles per hour for 1 hour, she walks a total of 2 miles. Therefore, we have $\frac{3d}{2} = 2$. Multiplying both sides by $\frac{2}{3}$ gives $d = 2\left(\frac{2}{3}\right) = \boxed{\frac{4}{3} \text{ miles}}$.

Challenge Problems

15.36 For each choice of thousands digit, there are 3 choices remaining for the hundreds digit, then 2 choices remaining for the tens digit, and then finally 1 choice for the units digit. Therefore, there are $3 \cdot 2 \cdot 1 = 6$ numbers with each thousands digit. So, there are 6 that start with 2 and 6 that start with 4. This means that the 17th in the list is the fifth number that starts with 5. We list the six numbers that start with 5 in increasing order:

$$5247, 5274, 5427, 5472, \boxed{5724}, 5742.$$

15.37 After the first pour, there is $\frac{1}{2}$ of the water remaining. On the next pour, $\frac{1}{3}$ is removed, leaving $\frac{2}{3}$ of the $\frac{1}{2}$ container remaining. Therefore, the container is $\frac{1}{2} \cdot \frac{2}{3} = \frac{1}{3}$ full after two pours.

On the next pour, $\frac{1}{4}$ of the amount remaining is poured out, leaving $\frac{3}{4}$ of the $\frac{1}{3}$ container remaining. Therefore, the container is $\frac{1}{3} \cdot \frac{3}{4} = \frac{1}{4}$ full after three pours.

We see a pattern! After the fourth pour, which is $\frac{1}{5}$ of the remaining water, there is $\frac{1}{4} \cdot \frac{4}{5} = \frac{1}{5}$ of the water remaining. Similarly, after the fifth pour, $\frac{1}{6}$ of the water remains. After the sixth pour, $\frac{1}{7}$ of the water remains. After the seventh pour, $\frac{1}{8}$ of the water remains. And after pour $\boxed{8}$, exactly $\frac{1}{9}$ of the water remains.

15.38 Clearly all the numbers from 1 to 9 can be achieved, by pairing 1 with the appropriate number. To count others in an organized manner, we make a table. We cross out all the single-digit numbers as well as any instance of a number after its first appearance:

×	2	3	4	5	6	7	8	9
2	~~4~~	~~6~~	~~8~~	10	12	14	16	18
3		~~9~~	~~12~~	15	~~18~~	21	24	27
4			~~16~~	20	~~24~~	28	32	36
5				25	30	35	40	45
6					~~36~~	42	48	54
7						49	56	63
8							64	72
9								81

We count 27 numbers in the table, and combining these with our 9 one-digit numbers, we have $\boxed{36}$ possible products.

15.39 We make an organized list of pictures. First, we consider the configurations in which the bottom left square is gray, since both of the restrictions on gray squares apply to this square. Neither the square above it nor the square to its right can be white, so those two squares must also be gray. The upper right corner is both above and to the right of a gray square, so it must be gray, too. This gives us the one configuration shown on the right.

Next, we let the bottom left be white and the upper left be gray. The upper right then must be gray, too, but there's no restriction on the bottom right square. This gives us the two configurations shown on the right.

Having covered the cases in which at least one of the left squares is gray, we now consider those in which the two leftmost squares are white. We cannot have the bottom right square gray and the upper right square white, but the other three configurations with both squares on the left white are OK.

This gives us a total of $\boxed{6}$ possible configurations.

15.40 Working forwards from 1, it's not at all clear what a good strategy is to get to 200. But if we work backwards, it becomes more clear. So, instead going from 1 to 200 by adding 1 or doubling at each step, we start from 200 and work our way down to 1 by subtracting 1 or halving at each step.

If we simply halve whenever possible, we get

$$200, 100, 50, 25, 24, 12, 6, 3, 2, 1.$$

This gives us $\boxed{9}$ steps.

To see why we cannot get to 200 in 8 steps, we first note that pressing either [+1] or [×2] on the first step results in the new displayed number being 2. For all other presses, the displayed number is always increased more if [×2] is pressed than if [+1] is pressed. Starting from 2 as the displayed number, if we press [×2] 7 times, we get 256. What if we instead start with 2 displayed, and press [×2] 6 times and [+1] once? If we first press [+1] and then [×2] thereafter, we get $3 \cdot 2^6 = 192$. Pressing [+1] on any other step results and [×2] on the other steps results in even smaller numbers. So, we can't get to 200 in 8 total steps. The highest we can reach in 7 steps is $2^7 = 128$, so we can't reach 200 in 7 or fewer steps, either.

15.41 We make an organized list. We list the numbers in each set from least to greatest. We'll start with the sets that begin with 1:

$$\{1, 1, 40\} \qquad \{1, 2, 20\} \qquad \{1, 4, 10\} \qquad \{1, 5, 8\}.$$

Next, we move on to those that begin with 2 (so the other two numbers multiply to 20, and 1 is not among the numbers):

$$\{2, 2, 10\} \qquad \{2, 4, 5\}.$$

There are no sets in which 4 is the smallest number, since such a set would have a product of at least $4 \cdot 4 \cdot 4 = 64$. Therefore, there are $\boxed{6}$ sets of numbers that satisfy the property in the problem.

15.42 Our path can only be 8 unit steps, and A and B are 8 unit steps apart along the grid. Therefore, each step must be up or to the right. We work backwards to count the number of such paths from A to B. We'll label points in the grid with letters so we can refer to them. To the upper right of a point, we'll place the number of ways to get from that point to point B using only upward or rightward steps. We start by putting 1's along the top and the right, since there's only one path to B from each of these points.

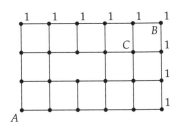

Then, we turn to point C. We can either go up one step from C or right one step from C. We know that after going up one step, there's only one way to finish. Similarly, if we go to the right, there's only one way to finish. So, the number of ways to finish from C is $1 + 1 = 2$.

We place a 2 at C in the grid, and then we continue working backwards. From point D, we can either go up to C, from which we know there are 2 paths, or right, from which point we know there is 1 path. So, there are $2 + 1 = 3$ paths from D. From E, we can either go up to D, from which there are 3 paths, or go right 1, from which there's one way to finish. So, there are $3 + 1 = 4$ paths from E.

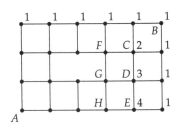

Going down the next column, we see that from F, we can go right to C, from which there are 2 ways to finish, or we can go up 1, from which there is 1 way to finish. This gives $2 + 1 = 3$ paths from F. From G, we can go up to F, from which there are 3 ways to finish, or right to D, from which there are 3 ways to finish. So, there are $3 + 3 = 6$ paths from G. From H, we can go to G or to E, so there are $6 + 4 = 10$ paths from H.

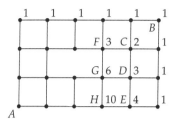

Continuing in this manner, we can work our way backwards throughout the entire grid. Notice that we are careful to remember that we cannot go up from point J in the grid. We can only go rightward from J, so the number of paths from J to B is the same as the number of paths from G to B. We find that there are $\boxed{44}$ paths from A to B.

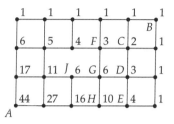

15.43

(a) We make a graph of time (vertical) versus distance from the starting point (horizontal). The bottom of the graph is noon and the top is 8 p.m. On the outbound trip, the graph must go from the bottom left corner (noon at the starting point) to the upper right corner (8 p.m. at the end of his first hike). One possible such graph is shown. His return trip must go from point A in the bottom right corner (noon at the start of the second day) to point B in the upper left (8 p.m. back at the original starting point).

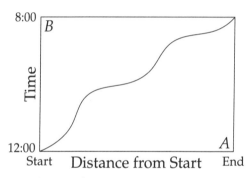

Clearly the graph for the second day must intersect the graph for the first day at some point. At this point, the "distance from starting point" and "time" are the same on both days, so Marco is at the same place on the trail at the same time.

Another way to think about this problem is to imagine that Marco took both trips on the same day. Then, Marco and his alter ego must meet on the path somewhere!

(b) The situation is essentially the same as in part (a). The only difference here is that Marco's return path graph ends at point C shown, mid-way up the left side (4 p.m., back at the original starting point). Points A and C will always be on opposite sides of Marco's first day graph (or C will be on the first day graph if Marco just sits at the start point for four hours on the first day), so the second day graph and the first must intersect.

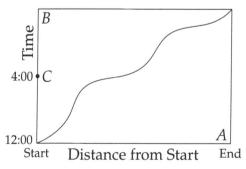

15.44 We start by drawing a picture. Each player is represented by a point, and each team is represented by a triangle connecting the points of the players on that team. Suppose the players on the first team are A, B, and C. We draw triangle ABC to represent this team. Player A must be on another team, but B and C cannot be on this team. So, we need two more players, D and E. We draw team ADE dashed to distinguish it from team ABC.

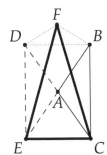

Next, B must be on another team. B cannot be on a team with both D and E, since D and E cannot be together on a second team. So, we need at least one new player, F. We then form team BDF, which we represent with a dotted triangle. Finally, we form team CEF in bold.

We therefore see that it is impossible to have fewer than 6 players, and the diagram at the right above shows that it is possible to have a league with exactly $\boxed{6}$ players.

15.45 Just as the units digit of a product depends only on the units digits of the numbers being multiplied, the last two digits of a product depend only on the last two digits of the numbers being multiplied. So, rather than computing powers of 7, we just start with 7 and only keep track of the final two digits as we multiply by 7 over and over. Here are the first few results:

$$07, 49, 43, 01, 07, 49.$$

We see that the final two digits of powers of 7 repeat in a cycle of 4 terms. Since 2011 divided by 4 has a remainder of 3, the last two digits of 7^{2011} are the third term in this cycle, which is $\boxed{43}$.

15.46 Let A be one of the cities. City A must be connected to at least three of the other cities by the same mode of transportation. To see why, consider what happens if A is not connected to at least three of the other cities by the same mode of transportation. Then, it can be connected to at most 2 by train and at most 2 by airplane. That's a total of 4 connections, but A must be connected directly to all 5 cities. So, A must be connected to at least three of the other cities by the same mode of transportation.

Suppose these three cities are B, C, and D. We illustrate that A is connected to them by the same mode of transportation with solid line segments in the diagram. The dashed lines represent the other mode of transportation. (Actually, A could also be connected to E and/or F by solid lines, but all that matters for the rest of the proof are the connections to B, C, and D.)

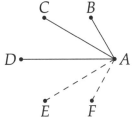

Now, if any two of B, C, and D are also connected by a solid line, then those two cities and A are directly connected to each other by the same mode of transportation. However, if no two of B, C, and D are connected by a solid line segment, then they must all be connected by dashed segments. Then, cities B, C, and D are directly connected to each other by the same mode of transportation.

Therefore, there must be three towns that are directly connected to each other by the same mode of transportation.

15.47 The sum of two four-digit numbers cannot be 20000 or greater. So, if the sum of two four-digit numbers is a five-digit number, then the first digit of the five-digit number is 1. This tells us that M is 1. We place that in the cryptarithm in the right.

	S	E	N	D
	1	O	R	E
1	O	N	E	Y

Next, we note that the sum $S + 1$ in the thousands place, plus a carry from the hundreds place if there is one, must be at least 10. However, it cannot be 11, since we have already used the 1. Therefore, the thousands digit of the sum is 0, so O is 0. Moreover, we know that S is 8 or 9.

We turn to the hundreds digit. If there is a carry from the hundreds place to the thousands in the sum, then $E + 0$ is 9 (and we have a carry from the tens to the hundreds). Then, N would have to be 0, which is already taken. Therefore, there is no carry from the hundreds place to the thousands place. This tells us that S is 9.

$$
\begin{array}{ccccc}
 & S & E & N & D \\
 & 1 & 0 & R & E \\
\hline
1 & 0 & N & E & Y \\
\end{array}
$$

Still looking at the hundreds place, we see that N is one more than E, and that there must be a carry from the tens to the hundreds place. Moreover, since N is one greater than E, the letters in the tens place tell us that either R is 9, or R is 8 and there is a carry in the units digit sum. Since 9 is already taken, we know R is 8.

$$
\begin{array}{ccccc}
 & 9 & E & N & D \\
 & 1 & 0 & R & E \\
\hline
1 & 0 & N & E & Y \\
\end{array}
$$

Now, we know that N is one greater than E, and that there must be a carry in the units place. The greatest D can be is 7, since 8 and 9 are taken. So, E cannot be 2, since this would make a carry in the units place impossible. We continue with guess-and-check, trying values of E to see which works. We only have to check 3, 4, 5, and 6, since $E = 7$ forces N to be 8, which is already taken.

$$
\begin{array}{ccccc}
 & 9 & E & N & D \\
 & 1 & 0 & 8 & E \\
\hline
1 & 0 & N & E & Y \\
\end{array}
$$

Case 1: E is 3. Then D must be 7, which forces Y to be 0. But 0 is already taken, so E cannot be 3.

Case 2: E is 4. Then D must be 6 or 7, which forces Y to be 0 or 1. Again, this is impossible, so E cannot be 4.

Case 3: E is 5. Then D still cannot be 6 (since this forces Y to be 1), but D can be 7. We then can complete the cryptarithm as shown at the right.

$$
\begin{array}{ccccc}
 & 9 & 5 & 6 & 7 \\
 & 1 & 0 & 8 & 5 \\
\hline
1 & 0 & 6 & 5 & 2 \\
\end{array}
$$

Case 4: E is 6. (This is just to make sure there aren't two solutions!) Then, N must be 7, so D can only be 4 or 5. But this forces Y to be 0 or 1. Once again, this is impossible, so E cannot be 6.

We conclude that the cryptarithm stands for $9567 + 1085 = 10652$, so Y is $\boxed{2}$.

15.48 We draw a picture. We represent each person with a point, and we draw a segment between two points if the two people shook hands. Since no one shakes hands with his or her spouse, the person who shook hands with 8 people must have shaken hands with everyone except his or her spouse. Let's call this busy person A, and let A's spouse be B. We then connect A to all the other people as shown. Now, everyone except B has shaken hands with A, so B must be the person who shook hands with 0 people. We note the number of hands shaken by each of A and B with numbers in parentheses.

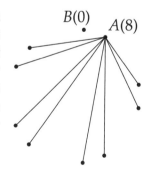

Next, we note that the person who shook 7 hands must have shaken hands with everyone except B and his or her own spouse. We call this person C, and connect this person to everyone except the spouse and B. Now, everyone but B and C's spouse has shaken hands at least twice, which means C's spouse must be the person who shook hands with just 1 person. We'll call this spouse D.

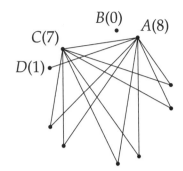

Continuing in the same manner, the person who shook hands with 6 others must have shaken hands with everyone not labeled so far (besides his or her spouse), and the spouse of this person must be the person who shook hands with 2 others. We label these people $E(6)$ and $F(2)$.

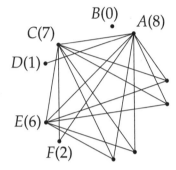

Next, the person who shook hands with 5 others must have shaken hands with the two unlabeled people who are not his or her spouse. The spouse of this person must have shaken 3 hands. We label these $G(5)$ and $H(3)$. We now have all the handshakes accounted for, and we see that both people in the remaining couple shook 4 hands. The numbers 0 through 8 were each spoken once when Kyle asked everyone except himself how many hands they shook. So, Kyle must have the repeated number of handshakes, which is $\boxed{4}$.

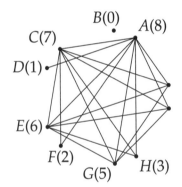